Mathematical Modelling and Characterization of Cylindrical Structures

Online at: https://doi.org/10.1088/978-0-7503-5638-1

Mathematical Modelling and Characterization of Cylindrical Structures

Farzad Ebrahimi
Mechanical Engineering Department, Imam Khomeini International University, Qazvin, Iran

Rajendran Selvamani
Department of Mathematics, Karunya Institute of Technology and Sciences, Coimbatore, Tamil Nadu, India

IOP Publishing, Bristol, UK

ISBN 978-0-7503-5638-1 (ebook)
ISBN 978-0-7503-5636-7 (print)
ISBN 978-0-7503-5639-8 (myPrint)
ISBN 978-0-7503-5637-4 (mobi)

DOI 10.1088/978-0-7503-5638-1

Version: 20251001

IOP ebooks

British Library Cataloguing-in-Publication Data: A catalogue record for this book is available from the British Library.

Published by IOP Publishing, wholly owned by The Institute of Physics, London

IOP Publishing, No.2 The Distillery, Glassfields, Avon Street, Bristol, BS2 0GR, UK

US Office: IOP Publishing, Inc., 190 North Independence Mall West, Suite 601, Philadelphia, PA 19106, USA

Contents

Preface

In the current scenario, the technology promotion of modelling and characterization of cylindrical structures and their diverse applications are well proposed among several scientists and researchers. This book provides an initial exhaustive text to cover linear elasticity theory for static and dynamic stability of thermoelastic and magneto-electro-elastic cylinders, panels and plates. The authors have drawn on their own scientific and research experience to write this book. The text is written from writers' literature and reaserch experience in waves of cylindrical structure. The aim of this book is to present analytical models of thermoelastic and magneto-electro-elastic cylindrical structures in terms of buckling and vibration, and it also depicts side-effects like thermal and fluid environment etc, alongside other side-effects. Mathematical and computational aspects are given prime importance in this book. A mathematical framework for the wave propagation problem of cylindrical panel, hollow cylinder, polygonal and arbitrary cross-sectional plate manufactured from various materials, including thermoelastic composites, piezoelectric materials, magneto-electro-elastic materials, magnetostrictive materials, and structrues pasted with LEMV/CFRP is exposed in this book. Both classical and refined elastic linear deformation cylinder and plate hypotheses are employed to formulate the wave propagation problem using the well-known Bessel fuctions and Fourier collocation principle. Additionally, the influences of fluid environment on the mechanical behaviours of cylindrical and plate structures are covered using both two-dimensional and three-dimensional elasticity in linear form. Impacts of various terms, such as stress, strain, displacement, wave number, frequency, aspect ratio, different types of temperature change, piezoelectric and magnetic potential on the dispersion curves of cylindrical and plate structures, are included via numerous examples. There are few books in the literature that explain the theory and advancement of linear elasticity continuum mechanics controlling on cylindrical and plate structures. The material in this book is the research work done by the authors in the field of cylindrical and plate structure during the past years. This analytical study is supported with extensive numerical examples of different materials like zinc, copper and cobalt iron oxide. This book consists of ten chapters.

Chapter 1 begins with the introduction and application of wave propagation in cylindrical panels. A literature survey relating to isotropic, transversely isotropic and magneto-thermoelastic cylindrical panels is presented and is given with its scope, objectives and background of the problem. In the background, firstly it discusses the history of wave propagation and cylindrical panels. Secondly, it emphasizes the development of elastic constants with different material properties and introduces the two conventional theories of thermoelasticity and magneto-thermo-elasticity.

Chapter 2 deals with the mathematical formulation of the problem using the linear three-dimensional theory of elasticity. The stress–strain relations and strain–displacement relations for the isotropic and transversely isotropic materials are discussed. The equations of motion are decoupled using the three wave potential functions and the solution of the simplified partial differential equation is obtained by Bessel function with complex argument. Analytical formulations of thermo-elastic, generalized thermoelastic and magneto-thermoelastic cylindrical panels are derived. By introducing the rotational parameter in the equation of motions the mathematical formulation is obtained for a rotating cylindrical panel. The damping of a generalized thermo-elastic plate is studied by introducing proper reduction in the formulation. The heat conduction equations are reduced for two non-conventional equations such as the Lord–Shulman (LS) and Green–Lindsay (GL) theories of thermo-elasticity.

Chapter 3 investigates wave propagation in homogenous isotropic thermoelastic cylindrical panels. The frequency equations are obtained by using the boundary conditions and the numerical computations are carried out for non-dimensional length to mean radius ratio versus dimensionless frequency for the material zinc using MATLAB programming. The boundary conditions and frequency equations for the generalized thermoelastic cylindrical panels are discussed and numerical computations have been carried out for non-dimensional frequency. Finally, the numerical calculations are carried out for damping of generalized thermoelastic waves in a homogeneous isotropic plate and the computed non-dimensional thermoelastic damping factor is plotted as the dispersion curves for the plate with thermally insulated and isothermal boundaries.

Chapter 4 focusses on the mathematical modelling of waves in a homogeneous isotropic rotating cylindrical panel. The frequency equations are calculated using the traction-free boundary conditions. The dispersion curves are drawn for the non-dimensional frequency with respect to circumferential wave number for the material copper. This study is extended to the computation of a rotating thermoelastic cylindrical panel by discussing the phase velocity for the material zinc.

Wave propagation of a homogeneous transversely isotropic magneto-thermo-elastic cylindrical panel is discussed in chapter 5. The computation is carried out for frequency, phase velocity and attenuation coefficient with respect to various panel parameters. The numerical calculation for a generalized magneto-therm-elastic cylindrical panel is studied for the thermomechanical coupling factor and specific loss.

Chapter 6 exposes the modelling of elastic waves in a hollow fibre of circular cross-section immersed in fluid composed of piezoelectric material. The secular equations of the model are obtained from the no-slip boundary conditions at the fluid–solid interfaces. The numerically computed non-dimensional frequencies, non-dimensional wave number, phase velocity and attenuation are drawn in the form of dispersion curves for longitudinal and flexural modes of vibrations for the material lead zirconate titanate-4.

Wave propagation in a transversely isotropic piezothermoelastic bar of circular cross-section is studied in chapter 7. The displacement functions to represent three displacement components on the basis of three-dimensional generalized piezothermoelasticity for transversely isotropic media are considered. The frequency equations are obtained for LS and GL theories of thermoelasticity. The computed thermomechanical coupling, electro-mechanical coupling, frequency shift, specific loss and frequency are presented in the form of dispersion curves.

In chapter 8, the dispersion analysis of magneto-electro elastic plate of arbitrary cross-sections immersed in fluid is studied by imposing the Fourier expansion collocation method (FECM) on the irregular boundaries. Displacement potential functions along radial, axial and circumferential directions, electric field vector and magnetic fields are used to uncouple the equations of motion. The frequency equations are obtained from the arbitrary cross-sectional boundary conditions. The computed physical quantities such as radial stress, hoop strain, non-dimensional frequency, magnetic potential, and electric potential are plotted in the form of dispersion curves and their characteristics are discussed.

The propagation of non-homogeneous waves in a visco-thermoelastic piezo-electric resonator plate is studied with the application of linear elasticity theory in chapter 9. The considered plate is a resonator by nature. The stress–strain equations are formulated using non-homogeneous form of three-dimensional linear elasticity theories. The solution of the problem is derived via the first and second types of Bessel function. The irregular boundary conditions are evaluated using the FECM. The numerical results are carried out for various shapes of resonator plates such as triangle, square, pentagon and hexagon. The dispersion curves are presented for the physical variables.

In chapter 10, we create a mathematical model to investigate the effect of initial stress on wave propagation in a hollow infinite multilayered composite cylinder. The inner and outer thermopiezoelectric layers are connected together by a linear elastic material with voids (LEMV) layer in the elastic cylinder. The equations of elasticity describe the model, the influence of the initial stress, and the framework of a linearized, three-dimensional thermoelasticity theory. The components of displacement are acquired by finding the analytical solutions of the motion equations. The perfect-slip boundary conditions and Bessel function solutions are used to generate the frequency equations that include the interaction between the composite hollow cylinders. By comparing LEMV with carbon fibre reinforced polymer, the numerical calculations for the material PZT-5A and the computed non-dimensional frequency versus various parameters are plotted as a dispersion curve (CFRP). It is obvious from the graph that compression and tension are examined in the presence of hydrostatic stress. In chapter 10, the axisymmetric waves of a piezoelectric rod coated with a thin film and submerged in an inviscid fluid are investigated using the constitutive equation of linear theory of elasticity and piezoelectricity. To decouple the equations of motion, potential functions are introduced. A flawlessly conductive material is applied to the rod's surface area. The frequency equations for longitudinal and flexural vibration modes are developed and quantitatively investigated for PZT-4 ceramic. Dispersion curves are used to represent non-dimensional frequency, phase velocity, and electric displacement variations.

Author biographies

Farzad Ebrahimi

Farzad Ebrahimi is an Associate Professor in the Department of Mechanical Engineering, IKIU, Qazvin, Iran. His research interests include mechanical behaviours of nano-engineered systems, mechanics of composites and nano composites, functionally graded materials, viscoelasticity, and smart materials and structures. Dr Ebrahimi has authored more than 400 high-quality peer-reviewed research articles in his fields of interest.

Rajendran Selvamani

Rajendran Selvamani is an Associate Professor in the Department of Mathematics, Karunya University, Coimbatore, India. His research interests include solid mechanics, numerical methods, partial differential equations, and modeling of macro and nanostructures. He has 25 years of teaching experience and published more than 106 international research papers, and is the author of two books in engineering mathematics.

IOP Publishing

Mathematical Modelling and Characterization of Cylindrical Structures

Farzad Ebrahimi and Rajendran Selvamani

Chapter 1

Introduction

1.1 Scope and objective

The propagation of mechanical disturbances in solids has piqued the interest of many fields of physical sciences and engineering in recent years. Using the three-dimensional linear theory of elasticity, the wave propagation in cylindrical structures is explored for homogeneous isotropic and transversely isotropic materials. The coupled equations are separated into motion equations via potential functions. The stressless, isothermal, thermally insulated, and magnetic boundary conditions are taken to achieve the frequency equations. The frequency equations are directly analysed using a Bessel function with a complex argument. With appropriate inclusions, the analysis is extended to thermoelastic, generalized thermoelastic, and magneto-thermoelastic cylindrical structures. Analytical solutions are discussed in distinct chapters to understand the effect of various stimulating forces of cylindrical structures. Finally, numerical examples are used to demonstrate the theory developed in each chapter. The majority of numerical examples are chosen to demonstrate the theory's practical characteristics in various chapters. The objective of this investigation is to develop a procedure for analyzing the wave propagation of cylindrical structures, which are becoming increasingly common in a range of complex structural designs. It is primarily intended for the investigation of the impacts of various mechanical parameters on the natural frequency, phase velocity, attenuation coefficient, and damping analysis of the cylindrical structures.

1.2 Background

1.2.1 History of elastic wave propagation

The study of wave propagation in elastic solids has a long and illustrious history. In terms of engineering applications, a strong interest in wave propagation effects arose in the early forties, when the special technical needs of the period necessitated

doi:10.1088/978-0-7503-5638-1ch1

knowledge on the performance of structures under high rates of loading. In the 18th century, famous mathematicians like Cauchy and Poisson advocated for this viewpoint. Because of applications in the field of geophysics, interest in the study of waves in elastic substances resurfaced in the nineteenth century. Love (1952) developed the theory of waves in a thin layer overlaying a solid and showed that such waves accounted for certain anomalies in seismogram records.

1.2.2 History of cylindrical panel

Cylindrical structures have an important structural role in civil, architectural, aeronautical, and marine engineering. The dynamic behaviour, stability, and strength are vividly decided by such structures. Circular cylindrical structures are frequently used as protective tank walls, thick cylindrical covers, and roof structures of large-span and open-space structures, among other applications. In the design of intelligent buildings, aircraft modelling, submarine structures, pressure vessels, chemical pipes, and as a component in automotive suspensions, the knowledge of wave propagation is prevalent. So, engineers must therefore comprehend the vibration behaviour of such cylindrical structures in order to create more dependable and cost-effective designs.

The rotating cylindrical panel has a problem in its applications in high-speed steam and gas turbines, planetary landing, and many other domains. At the moment, applied mathematicians are showing a lot of interest in dynamical elasticity approaches, because the traditional quasi-static approach ignores a lot of key aspects of the problems. This approach is based on the premise that the inertia terms in the equations of motion can be discarded. This assumption holds true only when stresses and displacements vary. However, there are a variety of situations in engineering and technology when this assumption may not hold true, and the inertia terms in the equations of motion may have resulted in cases of significant mathematical complexity. Furthermore, it is well acknowledged that studying the thermal, magnetic, and rotation effects on elastic wave propagation has implications for a variety of seismological applications.

1.2.3 Elastic constants and basic definitions

The velocity of propagation of waves in an elastic body is determined by the elastic constants of the material that makes up the body. When the material is isotropic or transversely isotropic, mathematical methods can be used to answer for the link between the velocity of various types of wave motion and the elastic constants. The vibration of isotropic and transversely isotropic materials is becoming increasingly important in the growth of various engineering sectors, such as ultrasonic non-destructive inspection for concrete structures and health monitoring of ailing infrastructure.

1.2.3.1 Stress

A body is said be in a state of 'stress' whenever one part of the body exerts forces on adjacent parts of the body and vice versa. The strength of the forces acting in a body can be characterized in terms of a stress, which is the force per unit area acting on the surface of an infinite decimal element in the body due to adjacent elements; and it is short range. There can be also a body force, which is the force per unit volume acting throughout the body on all its elements and it is a long-range force. To accurately describe the state of stress in a real three-dimensional stressed body, we need to define a 'stress tensor' by making the following assumptions:

1. The body is in a homogeneous state of stress.
2. The material is in equilibrium position.
3. There are no body forces or torques.

The nine stress components can also be expressed in matrix form as follows:

$$\begin{bmatrix} \sigma_{xx} & \tau_{xy} & \tau_{xz} \\ \tau_{yx} & \sigma_{yy} & \tau_{yz} \\ \tau_{zx} & \tau_{zy} & \sigma_{zz} \end{bmatrix} \tag{1.1}$$

where the forces σ_{xx}, σ_{yy}, σ_{zz} acting normal to the surface are called normal stress and the forces τ_{xy}, τ_{yz}, τ_{zx} acting parallel (tangential) to the surface are called shear stress. Breakdown of the symmetry matrix will occur only in the presence of body torques.

1.2.3.2 Strain

Whenever forces are exerted on a real elastic body, it undergoes deformation in the form of a change of shape or volume. In discussing the strain in a real body we make the following assumptions:

1. Rigid body motions do not constitute deformation of the body.
2. Only small deformations and displacements are considered; such as those permitted by linear theory of elasticity.

The dimensionless quantity strain is expressed as the elongation of an element with respect to its original length. Material expressing normal strain changes in dimensions but not shape; it is the change in length, area, or volume divided by the original length, area, and volume. Change in shape (and possibly dimensions) is usually measured in terms of change in angle between sizes of the rectangle as it is distorted.

Mathematically the components are expressed in terms of a matrix as follows:

$$\begin{bmatrix} \varepsilon_{xx} & \varepsilon_{xy} & \varepsilon_{xz} \\ \varepsilon_{yx} & \varepsilon_{yy} & \varepsilon_{yz} \\ \varepsilon_{zx} & \varepsilon_{zy} & \varepsilon_{zz} \end{bmatrix} \tag{1.2}$$

where ε_{xx}, ε_{yy}, ε_{zz} are called a normal strain and ε_{xy}, ε_{yz}, ε_{zx} are called a shear strain and it is worth noting that the strain matrix defined in equation (1.2) can vary from point to point in a body (homogeneous or non-homogeneous) and also depends on time.

1.2.3.3 Hooke's law

A solid body becomes strained when subjected to stress. Provided the stress is below a certain limiting value, the elastic limit, the strains disappear when the stresses are removed and the solid is back to its original shape. Moreover, when the stresses and strains are small enough, we often find that the two are linearly related and the constant of proportionality is called an elastic modulus. This linear relation of stress and strain is called Hooke's law. The most general linear relationship, which connects stress to strain, can be expressed as

$$\varepsilon_{ij} = C_{ijkl}\tau_{kl}.$$

for an elastic anisotropic media, where C_{ijkl} are the coefficients of deformation and is a fourth-order tensor. Solving for stresses, we get

$$\tau_{kl} = \left[C_{ijkl}\right]^{-1}\varepsilon_{ij} \tag{1.3}$$

where C_{ijkl} = moduli of elastic constants or elastic constants. There are 81 elastic constants for the most general case. If the body is in a state of strain, the only strain components different from zero are $\varepsilon_{12},\ \varepsilon_{21}$, then $\tau_{ij} = \bar{C}_{ij12}\varepsilon_{12} + \bar{C}_{ij21}\varepsilon_{21}$. But $\varepsilon_{12} = \varepsilon_{21}$

$$\tau_{ij} = \left(\bar{C}_{ij12} + \bar{C}_{ij21}\right)\varepsilon_{12}.$$

Now define $C_{ij12} = (\bar{C}_{ij12} + \bar{C}_{ij21})$ which is symmetric

$$C_{ijkl} = C_{ijlk} \tag{1.4}$$

Thus, the tensor C_{ijkl} is unchanged when the last two indices are interchanged. This reduces the number of independent elastic constants by 27 to a total of 54. If the state of strain in a body is such that the only non-zero strain components is ε_{11}, then $\tau_{ij} = C_{ij11}\varepsilon_{11}$ and $\tau_{12} = C_{1211}\varepsilon_{11}, \tau_{21} = C_{2111}\varepsilon_{21}$ but $\tau_{12} = \tau_{21}, C_{1211} = C_{2111}$ and in general

$$C_{ijkl} = C_{jikl}. \tag{1.5}$$

Thus the tensor C_{ijkl} is also unchanged if the indices are interchanged. Because of this symmetry the number of independent elastic constants reduces by 18 to a total of 36.

Combining equations (1.4) and (1.5), we get

$$C_{ijkl} = C_{ijlk}. \tag{1.6}$$

The generalized Hooke's law to a three-dimensional elastic body states that the six components of stress are linearly related to the six components of strain. Thus stress and strain relationship can be written in matrix form, where the six components of stress and strain are organized into column vectors and is given by

$$\begin{bmatrix} \tau_{11} \\ \tau_{22} \\ \tau_{33} \\ \tau_{12} \\ \tau_{23} \\ \tau_{13} \end{bmatrix} = \begin{bmatrix} C_{11} & C_{12} & C_{13} & C_{14} & C_{15} & C_{16} \\ C_{21} & C_{22} & C_{23} & C_{24} & C_{25} & C_{26} \\ C_{31} & C_{32} & C_{33} & C_{34} & C_{35} & C_{36} \\ C_{41} & C_{42} & C_{43} & C_{44} & C_{45} & C_{46} \\ C_{51} & C_{52} & C_{53} & C_{54} & C_{55} & C_{56} \\ C_{61} & C_{62} & C_{63} & C_{64} & C_{65} & C_{66} \end{bmatrix} \begin{bmatrix} \varepsilon_{11} \\ \varepsilon_{22} \\ \varepsilon_{33} \\ \varepsilon_{12} \\ \varepsilon_{23} \\ \varepsilon_{13} \end{bmatrix} \tag{1.7}$$

i.e. $\{\tau\} = [C]\{\varepsilon\}$

The elastic constants are different from one material to another material, if the symmetry of a crystal is increased, the symmetry of the tensor of elastic moduli is raised and consequently, the number of different components is decreased. Let the z-axis be parallel to the fibre elements and let the x and y-axes lie in a plane perpendicular to the element orientation. If the fibres are uniformly distributed in the matrix, then it is obvious that the elastic properties at any point in the x–y-plane, which is the plane of symmetry, are independent of the direction in that plane. A body of this nature is said to be transversely isotropic. The independent elastic constants reduce to five for a transversely isotropic material, then equation (1.7) becomes,

$$\begin{bmatrix} \tau_{11} \\ \tau_{22} \\ \tau_{33} \\ \tau_{12} \\ \tau_{23} \\ \tau_{13} \end{bmatrix} = \begin{bmatrix} C_{11} & C_{12} & C_{13} & 0 & 0 & 0 \\ C_{12} & C_{11} & C_{13} & 0 & 0 & 0 \\ C_{13} & C_{13} & C_{33} & 0 & 0 & 0 \\ 0 & 0 & 0 & C_{44} & 0 & 0 \\ 0 & 0 & 0 & 0 & C_{44} & 0 \\ 0 & 0 & 0 & 0 & 0 & C_{66} \end{bmatrix} \begin{bmatrix} \varepsilon_{11} \\ \varepsilon_{22} \\ \varepsilon_{33} \\ \varepsilon_{12} \\ \varepsilon_{23} \\ \varepsilon_{13} \end{bmatrix} \tag{1.8}$$

where $c_{66} = (c_{11} - c_{12})/2$.

If all the three mutually perpendicular planes are isotropic then the elastic constants are reduced to two. For an isotropic material the independent elastic constants will be of the form

$$C_{12} = C_{13} = \frac{\nu E}{(1+\nu)(1-2\nu)} = \lambda$$

$$C_{44} = C_{66} = \frac{E}{2(1+\nu)} = \mu$$

$$C_{11} = C_{33} = \frac{(1-\nu)E}{(1+\nu)(1-\nu)} = \lambda + 2\mu \tag{1.9}$$

where λ and μ are Lamé's constants, E and ν are, respectively, Young's modulus and Poisson's ratio.

1.2.3.4 Frequency equation

The frequency equation is an expression relating to ω to p in waveguides. The frequency equations are generally derived from the equations of motion and the boundary conditions of the system.

1.2.3.5 Frequency spectrum

A collection of branches is the frequency spectrum of the system. A complete frequency spectrum includes branches extending into the imaginary and complex wave number domains.

1.2.3.6 Phase velocity

A travelling, mechanical wave in one dimension is defined by an expression of the type $f = f(\omega t - pz)$, where f, is a function of the spatial coordinate z, time t, wave number p, and radian frequency ω. The argument $\omega t - pz$ is the phase of the wave function. Points of constant phase are propagated with the phase velocity

$$c = \omega/p. \tag{1.10}$$

The phase velocity is the real part of c and the attenuation is the imaginary part of c.

1.2.3.7 Attenuation coefficient

The attenuation coefficient is the quantity that characterizes how easily a material or medium can be penetrated by a beam of light, sound, particles or matter. It is defined as

$$1/c = 1/s + iq/\omega \tag{1.11}$$

where c is the phase velocity, ω is angular frequency and q is the attenuation coefficient.

1.2.3.8 Dispersion

If the phase velocity depends on the wave number, the system is said to be dispersive. Dispersion occurs in elastic materials and in elastic waveguides, where a waveguide is any extended body with a cross-section of finite dimensions.

1.2.3.9 Specific loss

Specific loss is the most direct method to defining internal friction for a material. The specific loss is the ratio of the amount of energy (ΔE) dissipated in a specimen through a stress cycle to the elastic energy (E) stored in that specimen of maximum strain. In the case of a sinusoidal plane wave of small amplitude, the specific loss equals 4π times the absolute value of the imaginary part of the wave number p to the real part of p

$$\text{Specific loss} = \left|\frac{\Delta E}{E}\right| = 4\pi \left|\frac{\mathrm{Im}(p)}{\mathrm{Re}(p)}\right|. \tag{1.12}$$

1.2.3.10 Relative frequency shift

The frequency shift of the wave due to rotation is defined as $\Delta\omega = \omega(\Omega) - \omega(0)$, Ω being the angular rotation, the relative frequency shift is given by

$$\text{Relative frequency shift} = \left|\frac{\Delta\omega}{\omega}\right| = \left|\frac{\omega(\Omega) - \omega(0)}{\omega(0)}\right| \tag{1.13}$$

where $\omega(\Omega)$ represents frequency with rotation and $\omega(0)$ indicates frequency without rotation.

1.2.3.11 Thermomechanical coupling

In the design of sensors and surface acoustic damping wave filters, the coupling effect between thermal and elasticity in a generalized thermoelastic material provides a mechanism for sensing thermomechanical disturbance from measurements of induced magnetic potentials, and for altering structural responses through applied magnetic fields. Thermomechanical coupling is defined as

$$\kappa^2 = \left| \frac{V_1 - V_2}{V_2} \right| \tag{1.14}$$

where V_1 is the phase velocity of the wave under thermally insulated boundary and V_2 the phase velocity of the wave under an isothermal boundary.

1.2.3.12 Thermoelastic damping factor

The thermoelastic damping factor (Q^{-1}) is defined with respect to the anguar frequency ω as

$$Q^{-1} = 2\left| \frac{\mathrm{Im}(\omega)}{\mathrm{Re}(\omega)} \right| \tag{1.15}$$

where $\mathrm{Re}(\omega)$ is real part of ω and $\mathrm{Im}(\omega)$ is the imaginary part of ω, respectively.

1.2.4 Thermoelasticity

The theory of thermoelasticity is concerned with the effect of temperature on the distribution of stress and strain in an elastic solid, as well as the inverse effect of deformation on the temperature distribution. Many modern designs, such as gas and steam turbines, jets and rockets, high-speed aircraft, and nuclear reactors, rely heavily on the thermomechanical effects caused by the interaction of temperature and deformation forces. Hyperbolic heat transport is gaining popularity as a theoretical and practical solution to various problems involving the rapid supply of thermal energy. In contrast, when relaxation effects are factored into the constitutive equation representing heat flux, the heat conduction equation becomes a hyperbolic equation, implying that heat transport has a finite speed. The generalized theory of thermoelasticity, which takes into consideration the coupling between the temperature and strain fields, is a modified theory of thermoelasticity.

1.2.5 Magneto-thermoelasticity

The theory of magneto-thermoelasticity, which deals with the interaction among elastic, temperature, and magnetic fields, is crucial in geophysics for understanding the effect of the Earth's magnetic field on seismic waves, acoustic wave damping in a magnetic field, and nuclear device emission of electromagnetic radiation. The coupling effects between the elastic, magnetic, and temperature for their application in sensing and actuation have sparked a lot of interest with the development of active material systems. Pumping and stirring of liquid metal coolants in nuclear reactors and molten metal in industrial metallurgy can be caused by thermoelectric currents in the presence of magnetic fields. In the nuclear field, the extraordinarily high temperatures and temperature gradients generated inside nuclear reactors have an impact on their design and operation. The magneto-thermoelasticity theory has a wide range of applications in industry, particularly in nuclear devices where a primary magnetic field occurs. The goal of this study is to create a method for analysing wave propagation in magneto-thermoelastic cylindrical panels, which are becoming more common in a range of complex structural designs.

1.3 Literature review

This section was reproduced from Ponnusamy (2013).

1.3.1 Wave propagation in cylinder/shell

Love (1952) explained the mathematical theory of elasticity which is occupied with an attempt to reduce calculating the state of strain, or relative displacement, within a solid body which is subject to the action of an equilibrating system of forces, or is in a state of slight internal relative motion, and with endeavours to obtain results which woukd be practically important in application to architecture. Wave propagation phenomena, such as radiation, reflection, refraction and diffraction with pertinent mathematical techniques were discussed by Achenbach (1973). Mirsky (1965) analyzed the wave propagation in a transversely isotropic circular cylinder of infinite length and presented the numerical results. Vibration of a circular cylinder of transversely isotropic material was studied by Chakravorthy (1956). Gazis (1959,) has studied the most general form of harmonic waves in a hollow cylinder of infinite length.

Acoustic scattering by transversely isotropic cylinders was studied by Ahmad and Rahman (2000). In their study, they explained the scattering of angular pattern with critical angles using normal mode method. Wave propagation in a transversely isotropic cylinder was discussed by Farhang *et al* (2007). The static analysis cannot predict the behaviour of the material due to the thermal stresses changes which are rapid. Therefore, in the case of a suddenly applied load, thermal deformations and the role of inertia becomes more important. This thermoelastic stress response being significant leads to the propagation of thermoelastic stress waves in solids. The theory of thermoelasticity is well established by Nowacki (1975). Lord and Shulman (1967) and Green and Lindsay (1972) modified the Fourier law and constitutive relations, so as to get the hyperbolic equation for heat conduction by taking into account the time needed for acceleration of heat flow and relaxation of stresses. A special feature of the Green–Lindsay model is that it does not violate the classical Fourier's heat conduction law. Dhaliwal and Sherief (1980) extended generalized thermoelasticity to anisotropic elastic bodies.

Experimentally and conceptually Hallam and Ollerton (1973) investigated the thermal stresses and deflections that occurred in a composite cylinder owing to a uniform rise in temperature, and contrasted the findings using a particular application of the frozen stress technique of photoelasticity. Erbay and Suhubi

(1986) have studied the longitudinal wave propagation of a thermoelastic cylinder. Chen (2005) analyzed the point temperature solution for a penny-shaped crack in an infinite transversely isotropic thermo piezo elastic medium subjected to a concentrated thermal load applied arbitrarily at the crack surface using the generalized potential theory.

The dynamic coupled thermoelasticity problem for a half-space was explored by taking into consideration the finiteness of the heat propagation velocity (Popov 1967). By taking into consideration the thermal relaxation period, Banerjee and Pao (1974) studied the propagation of planar harmonic waves in infinitely extended anisotropic materials. Florence and Goodier (1963) investigated the thermoelastic problem of a penny-shaped insulated crack disrupting uniform heat flow. The propagation of plane harmonic waves in homogeneous anisotropic heat conducting solids was investigated by Chadwick (2002).

In the framework of generalized thermoelasticity, Sharma and Sidhu (1986) investigated the propagation of plane harmonic thermoelastic waves in homogeneous transversely isotropic, cubic crystals, and anisotropic materials. The interplay of magnetic, thermal, and strain forces has an impact on a variety of applications in biomedical engineering and geometric studies. Gao and Noda (2004) have studied the thermal-induced interfacial cracking of magneto-electro-elastic materials under uniform heat flow. Ponnusamy (2007, 2011) has studied, respectively, the wave propagation in a generalized thermoelastic solid cylinder of arbitrary cross-section and dispersion analysis of generalized thermoelastic plate of polygonal cross-sections using the Fourier expansion collocation method. Later, in (2012, 2013) he extended his study, respectively, to the stress wave propagation in electro-magneto-elastic plate of arbitrary cross-sections and wave propagation in a piezoelectric solid bar of circular cross-section immersed in fluid using the Fourier expansion collocation method. Dai and Wang (2006) investigated the magneto-thermo-electro elastic transient response in a piezoelectric hollow cylinder subjected to complex loadings.

Haskins and Walsh (1957) studied the vibrations of transversely isotropic ferro-electric cylindrical shells. Hou and Leung (2004) studied the transient responses of magneto-electro-elastic hollow cylinders. Free vibration of an infinite magneto-electro-elastic cylinder was discussed by Buchanan (2003). The magneto-thermoelastic interaction in an infinite isotropic elastic cylinder exposed to periodic loading was studied by Mukhopodhyay and Roychoudhury (1997). Suhubi (1964) investigated the magneto-thermo-visco-elastic interaction in a cylindrical body. The generalized magneto-thermo waves in an infinite elastic solid with a cylindrical cavity were studied by Dhaliwal and Saxena (1991).

Ezzat (1997) studied the generalized magneto-thermo waves by a thermal shock in a perfectly conducting half-space. The Laplace transform technique was used by Sherief and Ezzat Sherief and Ezzat (1996) to determine the distribution of thermal stresses and temperature in a generally thermoelastic electrically conducting half-space subjected to sudden thermal shock and permeated by a primary uniform magnetic field. In an orthotropic thermoelastic cylinder, Wang and Dai (2004) reported magneto-thermo dynamic stresses and perturbation of the magnetic field vector. Novik (1996) estimates the disturbance of the electromagnetic and thermal fields of a model low-resistivity geo-block during deformation using magneto-thermoelastic theory.

Ashida and Tauchert (2001) presented the temperature and stress analysis of an elastic circular cylinder in contact with heated rigid stamps. Later, Ashida (2003) analyzed the thermally induced wave propagation in a piezoelectric plate. Tso and Hansen (1995) studied the wave propagation through cylinder/plate junctions. An excellent collection of works on vibration of shells were published by Leissa (1997). Ip *et al* (1996) studied the vibration analysis of orthotropic cylindrical thin shells with free ends by the Rayleigh–Ritz method. Free vibrations of thin cylindrical shells having finite lengths with freely supported and clamped edges were discussed by Yu and Syracuse (1995). Loy and Lam (1997) analyzed the cylindrical shells using generalized differential quadrature method. Zhang and Zhang (1990) investigated the vibrational power flow in a cylindrical shell.

For an isotropic cylindrical shell buried at a depth below the free surface of the ground, Wong *et al* (1986) gave its dynamic response from the point of view of three-dimensional elastic theory. Upadhyay and Mishra (1988) dealt with the non-axisymmetric dynamic behaviour of buried orthotropic cylindrical shells excited by a combination of P-, SV- and SH-waves. Bhimraddi (1984) developed a higher order theory for the free vibration analysis of a circular cylindrical shell. Free vibration study of layered cylindrical shells by collocation with splines was studied by Viswanathan and Navaneethakrishnan (2003). In their study they analyzed the effect of neglecting the coupling between the flexural and extensional displacements and the influences of the relative layer thickness, a length parameter and a total thickness parameter on the frequencies. The exact elasticity solution for laminated anisotropic cylindrical shells was studied by Bhaskar and Varadan (1993). Later, Bhaskar and Ganapathysaran (2003) discussed the elasticity solution for laminated orthotropic cylindrical shells subjected to localized longitudinal and circumferential moments. Vibration analysis of thin cylindrical shells using wave propagation approach was discussed by Zhang *et al* (2001a). Later, Zhang (2002) investigated the parametric analysis of frequency of rotating a laminated composite cylindrical shell using the wave propagation approach. Body wave propagation in rotating thermoelastic media was investigated by Sharma and Grover (2009). Sinha *et al* (1992) discussed the axisymmetric wave propagation in a circular cylindrical shell immersed in fluid in two parts.

Noor and Burton (1990) assessed the computational models for multilayered composite shells. In Part I, the theoretical analysis of the propagating modes are discussed and in Part II, the axisymmetric modes excluding torsional modes are obtained theoretically and experimentally and are compared. The exact solutions for laminated cylindrical shells in cylindrical bending were obtained by Ren (1987). Later, in 1989 (Ren and Owen 1989) the simply supported laminated circular cylindrical shell roof structures were analyzed by him. The effect of rotation, magneto field, thermal relaxation time and pressure on the wave propagation in a generalized visco-elastic medium under the influence of time harmonic source was discussed by Abd-Alla and Bayones (2011).

The propagation of waves in conducting piezoelectric solid was studied for the case when the entire medium rotates with a uniform angular velocity by Wauer (1999). A vibration of longitudinally polarized ferroelectric cylindrical tubes was studied by Martin (1963). Roychoudhury and Mukhopadhyay (2000) studied the effect of rotation and relaxation times on plane waves in generalized thermo-visco-elasticity. Hua and Lam (1998) studied the frequency characteristics of a thin rotating cylindrical shell using the general differential quadrature method.

The transient thermoelastic waves in an anisotropic hollow cylinder owing to localized heating were recently investigated Ravi *et al* (2011). They utilized the integral transform method in the frequency and wave number domain to get the modal wave forms, using the generalized thermoelastic theory provided by Lord and Shulman (2007). Later, based on the Green–Naghdi model of coupled thermo-elasticity, Hosseini and Hosseini (2012) established an analytical solution for thermoelastic waves in a thick hollow cylinder. Using the fast Laplace inverse transform method, they investigated thermomechanical behaviour in the time domain.

1.3.2 Wave propagation in a cylindrical plate/panel

Heyliger and Ramirez (2000) analyzed the free vibration characteristics of laminated circular piezoelectric plates and discs by using a discrete-layer model of the weak form of the equations of periodic motion. Vibration of completely free composite plates and cylindrical panels by a higher order theory were discussed by Arcangelo and Soldatos (1999). They analyzed the application of the Ritz approach on the energy functional of the Love-type version of a unified shear deformable shell theory. Sharma and Pathania (2005) investigated the generalized wave propagation in circumferential curved plates. Modelling of circumferential waves in a cylindrical thermoelastic plate with voids was discussed by Sharma and Kaur (2010). The thermal deflection of an inverse thermoelastic problem in a thin isotropic circular plate was presented by Gaikwad and Deshmukh (2005). Yu *et al* (2010) analyzed the generalized thermoelastic waves in functionally graded plates without energy dissipation. Sharma *et al* (2004) developed the Rayleigh–Lamb waves in a magneto-thermoelastic homogeneous isotropic plate. Three-dimensional solutions for shape control of a simply supported rectangular hybrid plate were obtained by Kapuria *et al* (1999).

The vibration of a circular plate laterally supported by an elastic foundation was investigated by Leissa (1981), which indicates that the effect of the Winkler foundation merely increases the square of the natural frequency of the plate by a constant. Kamal and Duruvasula (1983) discussed a circular plate embedded on an elastic medium, in which the governing differential equation was formulated using the Chebyshev–Lanczos techniques. Benhard (1999) studied the buckling frequency for a clamped plate embedded in an elastic medium. Wang (2005) studied the fundamental frequency of a circular plate supported by a partial elastic foundation using the finite element method.

An iterative approach predicting the frequency of isotropic cylindrical shell and panel was studied by Soldatos and Hadjigeorgiou (1990). Free vibration of composite cylindrical panels with random material properties was developed by Singh *et al* (2002). In this work, the effect of variations in the mechanical properties of laminated composite cylindrical panels on its natural frequency was obtained by modelling these as random variables. Zhang *et al* (2001b) employed a wave propagation method to analysis the frequency of cylindrical panels. Liew *et al* (2007) discussed the dynamic stability analysis of composite laminated cylindrical panels via the mesh-free kp-Ritz method. Free vibration analysis of curved and twisted cylindrical thin panels was investigated by Hu and Tsuij (1999). In their study, they discussed the non-linear strain displacement relations of the model which are derived based on the general thin shell theory, and a numerical method for analyzing the free vibrations of curved and twisted cylindrical thin panels. These are presented by means of the principle of virtual work for the free vibration using the Rayleigh–Ritz method, assuming two-dimensional polynomial functions as displacement functions.

Jianqiao and Soldatos (1994) studied the three-dimensional vibration of laminated cylindrical panels with symmetric or anti-symmetric cross-ply lay-up. Free vibration analysis of a simply supported cross-ply cylindrical panel, with arbitrarily located lateral surface point supports and using a recursive approach, was presented by Jianqiao and Soldatos (1996). Vibration of functionally graded multilayered orthotropic cylindrical panel under thermomechanical load was analyzed by Wang (2008). Three-dimensional vibration of a homogenous transversely isotropic thermoelastic cylindrical panel was investigated by Sharma (2001). Free vibrations of transversely isotropic piezoelectric circular cylindrical panels were studied by Ding *et al* (2002).

Vibration analysis of bimodulus laminated cylindrical panels was studied by Khan *et al* (2009). It is also becoming increasingly necessary to apply powerful numerical tools such as finite element or boundary element techniques to these problems. Moreover, it is well recognized that the investigation of the thermal effects on elastic wave propagation has a bearing on many seismological applications. Effects of temperature changes on damping properties of sandwich cylindrical panels were discussed by Xia and Lukasiewich (1996).

Prevost and Tao (1983) performed a genuine finite element analysis of problems with relaxation effects. The Galerkin finite element was used for the coupled thermoelasticity problem in beams by Eslami and Vahedi (1989). Using the extended power series technique, Huang and Tauchert (1991) developed the analytical solution for cross-ply laminated cylindrical panels of finite length exposed to mechanical and thermal loads. Later, Sharma *et al* (2004) investigated the three-dimensional vibration analysis of a piezothermoelastic cylindrical panel and discussed the numerical result for PZT 5 A.

Study of a cylindrical panel embedded on an elastic medium is important for design of structures such as atomic reactors, steam turbines, submarine structures with wave loads, or for the impact effects due to superfast trains, or for jets and other devices operating at elevated temperatures. Natural frequency of a cylindrical panel on a Kerr foundation was studied by Cai *et al* (2004) and they used Bessel functions with complex argument directly for the complex eigenvalue case. Selvadurai (1979) presented the most general form of a soil model used in practical applications. Paliwal (1996) presented an investigation on the coupled free vibrations of an isotropic circular cylindrical shell on Winkler and Pasternak foundations by employing a membrane theory.

Radial vibration of a row of cylindrical panel of finite length using the concept of wave propagation in a periodic structure was studied by Chitaranjan *et al* (2002). In their study they discussed the natural frequencies of cylindrically curved panels of a given length and curvature but with different subtended angles with respect to the same phase frequency curve. A modified mixed variational principle in a cylindrical coordinate system for an elastic cylindrical panel with damping was established, and the state vector formulation with damping parameters was derived from the variational principle by Guanghui *et al* (2008). Abouhamze and Shakeri (2007) discussed a multi-objective optimization strategy for optimal stacking sequence of laminated cylindrical panels. This is presented with respect to the first natural frequency and critical buckling load using the weighted summation method. They used a trained neural network to evaluate the fitness function in the optimization process and in this way increased the procedure speed.

Detailed experimental results and analytical results are presented on chaotic vibration of a shallow cylindrical panel subjected to gravity and periodic excitation was obtained by Nagai *et al* (2007). Awrejcewicz *et al* (2007) have studied the chaotic vibrations of flexible infinite length cylindrical panels using the Kirchhoff–Love model. Palazotto and Linnemann (1991) investigated the vibration and buckling characteristics of composite cylindrical panels incorporating the effects of a higher order shear theory. Elasticity solution for the free vibration analysis of laminated cylindrical panels using the differential quadrature method was studied by Alibeigloo and Shakeri (2007). They applied the differential quadrature method to the state space formulations along the axial or circumferential directions of a panel in order to get new state equations about state variables at discrete points. Recently, the three-dimensional exact solutions for free vibrations of simply supported magneto-electro-elastic cylindrical panels were investigated by Yun *et al* (2010).

1.4 Research needs

From the above literature review, it is obvious that the free-wave propagation in isotropic, transversely isotropic, and magneto-thermoelastic cylindrical panels have not been analysed by any of the researchers. In addition, due to its application in engineering disciplines, the study of wave propagation in isotropic and anisotropic materials is gaining the attention of notable academics. The properties of wave propagation in isotropic, transversely isotropic, and magneto-thermoelastic cylindrical panels are discussed in this study.

References

Abd-Alla A M and Bayones F S 2011 Effect of rotation in a generalized magneto thermo visco elastic media *Adv. Theor. Appl. Mech.* **4** 15–42

Abouhamze M and Shakeri M 2007 Multi objective stacking sequence optimization of laminated cylindrical panels using the genetic algorithm and neural network *Compos. Struct.* **81** 253–63

Achenbach J D 1973 *Wave Propagation in Elastic Solids* (North Holland: Amsterdam)

Ahmad F and Rahman A 2000 Acoustic scattering by transversely isotropic cylinders *Int. J. Eng. Sci.* **38** 325–35

Alibeigloo A and Shakeri M 2007 Elasticity solution for the free vibration analysis of laminated cylindrical panels using the differential quadrature method *Compos. Struct.* **81** 105–13

Arcangelo M and Soldatos K P 1999 Vibration of completely free composite plates and cylindrical shell panels by a higher-order theory *Int. J. Mech. Sci.* **41** 891–918

Ashida F 2003 Thermally-induced wave propagation in piezoelectric plate *Acta Mech.* **161** 1–16

Ashida F and Tauchert T R 2001 A general plane-stress solution in cylindrical coordinates for a piezoelectric plate *J. Solids Struct.* **30** 4969–85

Awrejcewicz J, Krysko V A and Nazar'iantz V 2007 Chaotic vibrations of flexible infinite length cylindrical panels using the Kirchhoff–Love model *Commun. Nonlinear Sci. Numer. Simul.* **12** 519–42

Banerjee D K and Pao Y H 1974 Thermo-elastic waves in anisotropic solids *J. Acoust. Soc. Am.* **56** 1444–54

Benhard 1999 Buckling eigen values for a clamped plate embedded in an elastic medium and related questions *SIAM J. Math. Anal.* **24** 327–40

Bhaskar K and Ganapathysaran N 2003 Elasticity solution for laminated orthotropic cylindrical shells subjected to localized longitudinal and circumferential moments *J. Press. Vessel Technol.* **125** 26–35

Bhaskar K and Varadan T K 1993 Exact elasticity solution for laminated anisotropic cylindrical shells *J. Appl. Mech.* **60** 41–7

Bhimraddi A A 1984 A higher order theory for free vibration analysis of circular cylindrical shell *Int. J. Solids Struct.* **20** 623–30

Buchanan G R 2003 Free vibration of an infinite magneto-electro-elastic cylinder *J. Sound Vib.* **268** 413–26

Cai J B, Chen W Q, Ye G R and Ding H J 2004 On natural frequencies of a transversely isotropic cylindrical panel on a Kkerr foundation *J. Sound Vib.* **232** 997–1004

Chadwick P 2002 Basic prosperities of plane harmonic waves in a pre-stressed heat conducting elastic material *J. Therm. Stresses* **2** 193–214

Chakravorthy 1956 Vibration of a circular cylinder of transversely isotropic material *J. Appl. Mech.* **22** 220–7

Chen W Q 2005 Point temperature solution for a penny-shaped crack in an infinite transversely isotropic thermo-piezo-elastic medium *Eng. Anal. Bound. Elem.* **29** 524–32

Chitaranjan P, Parthan S and Mukherjee S 2002 Vibration of multi supported curved panel using the periodic structure approach *Int. J. Mech. Sci.* **44** 269–85

Dai H L and Wang X 2006 Magneto thermo electro elastic transient response in a piezoelectric hollow cylinder subjected to complex loadings *Int. J. Solids Struct.* **43** 5628–564

Dhaliwal R S and Saxena H X 1991 The generalized magneto thermo waves in an infinite elastic solid with a cylindrical cavity *J. Therm. Stresses* **14** 353–69

Dhaliwal R S and Sherief H H 1980 Generalized thermo elasticity for anisotropic media *Q. J. Appl. Math. Mech.* **38** 1–8

Ding H J, Xu R Q and Chen W Q 2002 Free vibration of transversely isotropic piezoelectric circular cylindrical panels *Int. J. Mech. Sci.* **44** 191–206

Erbay E S and Suhubi E S 1986 Longitudinal wave propagation in a thermo elastic cylinder *J. Therm. Stresses* **9** 279–95

Eslami M R and Vahedi H 1989 Coupled thermo elasticity beam problems *Am. Inst. Aeronaut. Astronaut. J.* **27** 662–5

Ezzat M A 1997 Generation of generalized magneto thermo waves by a thermal shock in a perfectly conducting half space *J. Therm. Stresses* **20** 613–33

Farhang H, Enjilela E, Sinclair A N and Abbas Mirnezami S 2007 Wave propagation in transversely isotropic cylinders *Int. J. Solid Struct.* **44** 5236–46

Florence A L and Goodier J N 1963 The thermo elastic problem of uniform heat flow disturbed by a penny shaped insulated crack *Int. J. Eng. Sci.* **1** 533–40

Gaikwad M K and Deshmukh K C 2005 Thermal deflection of an inverse thermo elastic problem in a thin isotropic circular plate *Appl. Math. Model.* **29** 797–804

Gao C F and Noda N 2004 Thermal-induced interfacial cracking of magneto electro elastic materials *Int. J. Eng. Sci.* **42** 1347–60

Gazis D C 1959 Three dimensional investigation of the propagation of waves in hollow circular cylinders *J. Acoust. Soc. Am.* **31** 568–78

Green A E and Lindsay K A 1972 Thermo-elasticity *J. Elast.* **2** 1–7

Guanghui Q, Liu Y, Guo Q and Zhang D 2008 Dynamic analysis of three dimensional laminated plates and panels with damping *Int. J. Mech. Sci.* **50** 83–91

Hallam C B and Ollerton E 1973 Thermal stresses in axially connected circular cylinders *J. Strain Anal.* **8** 160–7

Haskins J F and Walsh J L 1957 Vibrations of ferroelectric cylindrical shells with transversely isotropy-I radially polarized case *J. Acoust. Soc. Am.* **29** 729–34

Heyliger P R and Ramirez G 2000 Free vibration of laminated circular piezoelectric plates and discs *J. Sound Vib.* **229** 935–56

Hosseini S M and Hosseini M 2012 Analytical solution for thermo elastic wave propagation analysis in thick hollow cylinder based on Green-Naghdi model of coupled thermo elasticity *J. Therm. Stresses* **35** 363–76

Hou P F and Leung Y T 2004 The transient responses of magneto-electro-elastic hollow cylinders *Smart Mater. Struct.* **13** 762–76

Hu X X and Tsuij T 1999 Free vibration analysis of curved and twisted cylindrical thin panels *J. Sound Vib.* **219** 63–88

Hua L I and Lam K Y 1998 Frequency characteristics of a thin rotating cylindrical shell using general differential quadrature method *Int. J. Mech. Sci.* **40** 443–59

Huang N N and Tauchert T R 1991 Thermo elastic solution for cross-ply cylindrical panels *J. Therm. Stresses* **14** 227–37

Ip K H, Chan W K, Tse P C and Lai T C 1996 Vibration analysis of orthotropic cylindrical thin shells with free ends by the Rayleigh–Ritz method *J. Sound Vib.* **195** 117–35

Jianqiao Y E and Soldatos K P 1994 Three dimensional vibration of laminated cylindrical panels with symmetric or anti-symmetric cross-ply lay-up *Compos. Eng.* **4** 429–44

Jianqiao Y E and Soldatos K P 1996 Three-dimensional vibration of laminated composite plates and cylindrical panels with arbitrarily located lateral surface point supports *Int. J. Mech. Sci.* **8** 271–81

Kamal K and Duruvasula S 1983 Bending of circular plate on elastic foundation *J. Eng. Mech.* **109** 1293–8

Kapuria S, Dumir P C and Sengupta S 1999 Three-dimensional solution for shape control of a simply supported rectangular hybrid plate *J. Therm. Stresses* **22** 159–76

Khan K, Patel B P and Nath Y 2009 Vibration analysis of bimodulus laminated cylindrical panels *J. Sound Vib.* **321** 166–83

Leissa A W 1981 Plate vibration research: complicating effects *Shocks Vib. Dig.* **13** 19–36

Leissa A W 1997 Theory of elasticity *ASME J. Vib. Acoust.* **119** 89–95

Liew K M, Lee Y Y, Ng T Y and Zhao X 2007 Dynamic stability analysis of composite laminated cylindrical panels via the mesh-free kp-Ritz method *Int. J. Mech. Sci.* **49** 1156–67

Lord H W and Shulman Y 1967 A generalized dynamical theory of thermo-elasticity *J. Mech. Phys. Solids* **5** 299–309

Love A E H 1952 *A treatise on the mathematical theory of elasticity* 4th edn (Cambridge: Cambridge University Press)

Loy C T and Lam K Y 1997 Analysis of cylindrical shells using generalized differential quadrature *Shock Vib.* **6** 193–8

Martin G E 1963 Vibrations of longitudinally polarized ferroelectric cylindrical tubes *J. Acoust. Soc. Am.* **35** 510–20

Mirsky I 1965 Wave propagation in transversely isotropic circular cylinders, Part I: Theory, Part II: Numerical results *J. Acoust. Soc. Am.* **37** 1016–26

Mukhopodhyay S B and Roychoudhury S B 1997 Magneto-thermo-elastic interactions in an infinite isotropic elastic cylinder *Int. J. Eng. Sci.* **35** 437–44

Nagai K, Maruyama S, Muruta T and Yamaguchi T 2007 Experiment and analysis on chaotic vibrations of a shallow cylindrical shell-panel *J. Sound Vib.* **305** 492–520

Noor A K and Burton W S 1990 Assessment of computational models for multi-layered composite shells *Appl. Mech. Rev. Trans.* **43** 67–96

Novik O B 1996 Seismically-driven magneto-thermo-elastic effect. Earth science sections *MAIK Nauka Interperiodica Publ.* **5** 1083–3552

Nowacki W 1975 *Dynamical Problems of Thermo Elasticity* (Leyden: Noordhoff)

Palazotto A N and Linnemann P E 1991 Vibration and buckling characteristics of composite cylindrical panels incorporating the effects of a higher order shear theory *Int. J. Solids Struct.* **28** 341–61

Paliwal D N 1996 Free vibrations of circular cylindrical shell on Winkler and Pasternak foundations *Int. J. Press. Vessel Pip.* **69** 79–89

Ponnusamy P 2007 Wave propagation in a generalized thermo elastic solid cylinder of arbitrary cross-section *Int. J. Solids Struct.* **44** 5336–48

Ponnusamy P 2011 Wave propagation in thermo-elastic plate of arbitrary cross-sections *Multidiscip. Model. Ma. Struct.* **7** 1573–605

Ponnusamy P 2012 Dispersion analysis of generalized thermo elastic plate of polygonal cross-section *Appl. Math. Model.* **36** 3343–58

Ponnusamy P and Selvamani R 2013 Wave propagation in a transversely isotropic magneto thermo elastic cylindrical panel *Eur. J. Mech.* A*39 76–85*

Popov E B 1967 Dynamic coupled problem of thermo elasticity for a half-space taking into account of the finiteness of the heat propagation velocity *J. Appl. Mech.* **31** 349–56

Prevost J H and Tao D 1983 Finite element analysis of dynamic coupled thermo elasticity problems with relaxation times *J. Appl. Mech. -T. ASME* **50** 817–22

Ravi C, Subhendu K, Datta A, Shah H and Hao B 2011 Transient thermo elastic waves in an anisotropic hollow cylinder due to localized heating *Int. J. Solids Struct.* **48** 3063–74

Ren J G 1987 Exact solutions for laminated cylindrical shells in cylindrical bending *J. Compos. Sci. Technol.* **29** 69–187

Ren J G and Owen D R J 1989 Vibration and buckling of laminated plates *Int. J. Solids Struct.* **25** 95–106

Roychoudhury S K and Mukhopadhyay S 2000 Effect of rotation and relaxation times on plane waves in generalized thermo visco elasticity *Int. J. Math. Math. Sci.* **23** 497–505

Selvadurai A P S 1979 The interaction between a uniformly loaded circular plate and an isotropic elastic halfspace: a variational approach *J. Struct. Mech.* **7** 231–46

Sharma J N 2001 Three dimensional vibration of a homogenous transversely isotropic thermo elastic cylindrical panel *J. Acoust. Soc. Am.* **110** 648–53

Sharma J N and Grover G 2009 Body wave propagation in rotating thermo elastic media *Mech. Res. Commun.* **36** 715–21

Sharma J N and Kaur D 2010 Modeling of circumferential waves in cylindrical thermo elastic plates with voids *Appl. Math. Model.* **34** 254–65

Sharma J N, Pal M and Chand D 2004 Three dimensional vibration analysis of a piezo-thermo elastic cylindrical panel *Int. J. Eng. Sci.* **42** 1655–73

Sharma J N and Pathania V 2005 Generalized thermo elastic wave propagation in circumferential direction of transversely isotropic cylindrical curved plate *J. Sound Vib.* **281** 1117–31

Sharma J N and Sidhu R S 1986 On the propagation of plane harmonic waves in anisotropic generalized thermo elasticity *Int. J. Eng. Sci.* **24** 1511–6

Sherief H H and Ezzat M A 1996 Thermal-shock problem in magneto-thermo-elasticity with thermal relaxation *Int. J. Solids Struct.* **33** 4449–59

Singh B N, Yadav D and Iyengar N G R 2002 Free vibration of composite cylindrical panels with random material properties *Compos. Struct.* **58** 435–42

Sinha B K, Plona T J, Kostek S and Chang S 1992 Axisymmetric wave propagation in a fluid loaded cylindrical shells *J.Acoust. Soc. Am.* **92** 1132–55

Soldatos K P and Hadjigeorgiou V P 1990 Three dimensional solution of the free vibration problem of homogeneous isotropic cylindrical shells and panels *J. Sound Vib.* **137** 369–84

Suhubi E S 1964 Longidutional vibration of a circular cylinder coupled with a thermal field *J. Mech. Phys. Solids* **12** 69–75

Tso Y K and Hansen C H 1995 Wave propagation through cylinder/plate junctions *J. Sound Vib.* **186** 447–61

Upadhyay P C and Mishra B K 1988 Non-axisymmetric dynamic response of buried orthotropic cylindrical shells *J. Sound Vib.* **121** 149–60

Viswanathan K K and Navaneethakrishnan P V 2003 Free vibration study of layered cylindrical shells by collocation with splines *J. Sound Vib.* **260** 807–27

Wang C M 2005 Fundamental frequencies of a circular plate supported by a partial elastic foundation *J. Sound Vib.* **285** 1203–9

Wang X 2008 Vibration of functionally graded multilayered orthotropic cylindrical panel under thermo mechanical load *Mech. Mater.* **40** 235–54

Wang X and Dai H L 2004 Magneto-thermo-dynamic stress and perturbation of magnetic field vector in a hollow cylinder *J. Therm. Stresses* **3** 269–88

Wauer J 1999 Waves in rotating and conducting piezoelectric media *J. Acoust. Soc. Am.* **106** 626–36

Wong K C, Datta S K and Shah A H 1986 Three-dimensional motion of buried Pipeline. I: analysis *ASCE J. Eng. Mech.* **112** 1319–37

Xia Z Q and Lukasiewich S 1996 Effects of temperature changes on damping properties of sandwich cylindrical panels *Int. J.Solids Struct.* **33** 835–49

Yu Y Y and Syracuse N Y 1995 Free vibrations of thin cylindrical shells having finite lengths with freely supported and clamped edges *J. Appl. Mech.* **22** 547–52

Yu J, Zhang X and Xue T 2010 Generalized thermo elastic waves in functionally graded plates without energy dissipation *Compos. Struct.* **93** 32–9

Yun W, Xu R, Ding H and Chen J 2010 Three-dimensional exact solutions for free vibrations of simply supported magneto-electro-elastic cylindrical panels *Int. J. Eng. Sci.* **48** 1778–96

Zhang X M 2002 The parametric analysis of frequency of rotating laminated composite cylindrical shell using wave propagation approach *Comput. Meth. Appl. Mech. Eng.* **191** 2027–43

Zhang X M, Liu G R and Lam K Y 2001a Vibration analysis of thin cylindrical shells using wave propagation approach *J. Sound Vib.* **239** 397–403

Zhang X M, Liu G R and Lam K Y 2001b Frequency analysis of cylindrical panels using a wave propagation approach *J. Appl. Acoust.* **62** 527–43

Zhang X M and Zhang W H 1990 Vibrational power flow in a cylindrical shell *Proc. of ASME PVP (New Orleans, LA)*

IOP Publishing

Farzad Ebrahimi and Rajendran Selvamani

Chapter 2

Mathematical formulation

2.1 Introduction

In this chapter, the basic governing equations for the mathematical analysis of cylindrical panels are addressed, including stress–strain and strain–displacement relationships. Navier's equations of motion, which are the fundamental relations for the material under consideration, are used to discuss the equations of motion. Three potential functions are added to uncouple the displacement equations of motion for a homogeneous isotropic cylindrical panel. Bessel functions are used to obtain the solutions to the displacement equations.

We consider a homogeneous elastic cylindrical panel of length L having inner and outer radii a and b respectively, with centre angle α. R is the mean radius of the cylindrical panel with origin O and the geometry of the problem is shown in figure 2.1.

2.2 Stress–strain relations

The stress–strain relation for an isotropic material by generalized Hooke's law is given by

$$\sigma_{rr} = \lambda(e_{rr} + e_{\theta\theta} + e_{zz}) + 2\mu e_{rr}$$

$$\sigma_{\theta\theta} = \lambda(e_{rr} + e_{\theta\theta} + e_{zz}) + 2\mu e_{\theta\theta}$$

$$\sigma_{zz} = \lambda(e_{rr} + e_{\theta\theta} + e_{zz}) + 2\mu e_{zz}$$

$$\sigma_{r\theta} = \mu e_{r\theta}, \quad \sigma_{rz} = \mu e_{rz}, \quad \sigma_{\theta z} = \mu e_{\theta z}. \tag{2.1}$$

doi:10.1088/978-0-7503-5638-1ch2

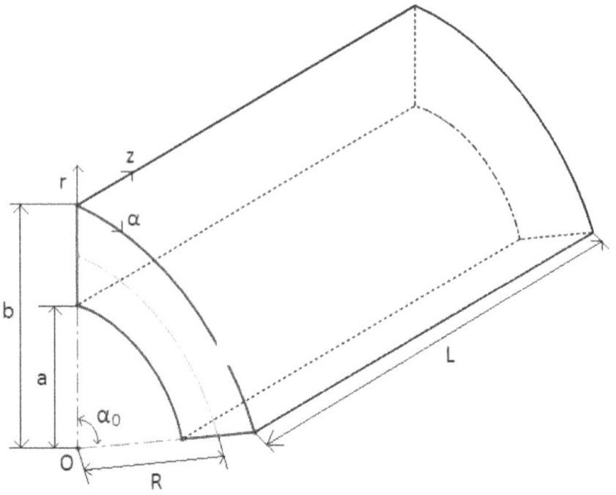

Figure 2.1. Geometry of a homogeneous elastic cylindrical panel.

2.3 Strain–displacement relations

The strain–displacement relations are given as

$$e_{rr} = \frac{\partial u}{\partial r}, \quad e_{\theta\theta} = \frac{u}{r} + \frac{1}{r}\frac{\partial v}{\partial \theta}, \quad e_{zz} = \frac{\partial w}{\partial z}$$

$$e_{r\theta} = \frac{1}{2}\left(\frac{\partial v}{\partial r} - \frac{v}{r} + \frac{1}{r}\frac{\partial u}{\partial \theta}\right), \quad e_{rz} = \frac{1}{2}\left(\frac{\partial w}{\partial r} + \frac{\partial u}{\partial z}\right), \quad e_{z\theta} = \frac{1}{2}\left(\frac{\partial v}{\partial z} + \frac{1}{r}\frac{\partial w}{\partial \theta}\right). \tag{2.2}$$

2.4 Equations of motion and solution

In cylindrical coordinates the three-dimensional stress equations of motion from Newton's second law in the absence of body force for a linearly elastic medium are given by Berliner and Solecki (1996) as

$$\sigma_{rr,\,r} + r^{-1}\sigma_{r\theta,\,\theta} + \sigma_{rz,\,z} + r^{-1}(\sigma_{rr} - \sigma_{\theta\theta}) = \rho u_{,tt}$$

$$\sigma_{r\theta,\,r} + r^{-1}\sigma_{\theta\theta,\,\theta} + \sigma_{,rzz} + \sigma_{\theta z,\,z} + 2r^{-1}\sigma_{r\theta} = \rho v_{,tt}$$

$$\sigma_{rz,r} + r^{-1}\sigma_{\theta z,\theta} + \sigma_{zz,z} + r^{-1}\sigma_{r\theta} = \rho w_{,tt}. \tag{2.3}$$

Substitution of the equations (2.1) and (2.2) in equation (2.3), gives the following three displacement equations of motions

$$(\lambda + 2\mu)(u_{,rr} + r^{-1}u_{,r} - r^{-2}u) + \mu r^{-2}u_{,\theta\theta} + \mu u_{,zz} + r^{-1}(\lambda + \mu)v_{,r\theta} - r^{-2}(\lambda + 3\mu)v_{,\theta} + (\lambda + \mu)w_{,rz} = \rho u_{,tt}$$

$$\mu(v_{,rr} + r^{-1}v_{,r} - r^{-2}v) + r^{-2}(\lambda + 2\mu)v_{,\theta\theta} + \mu v_{,zz} + r^{-2}(\lambda + 3\mu)u_{,\theta} + r^{-1}(\lambda + \mu)u_{,r\theta} + r^{-1}(\lambda + \mu)w_{,\theta z} = \rho v_{,tt}$$

$$(\lambda + 2\mu)w_{,zz} + \mu(w_{,rr} + r^{-1}w_{,r} + r^{-2}w_{,\theta\theta}) + (\lambda + \mu)u_{,rz} + r^{-1}(\lambda + \mu)v_{,\theta z} + r^{-1}(\lambda + \mu)u_{,z} = \rho w_{,tt}. \tag{2.4}$$

The partial differentiation with respect to the variables is indicated by the comma in the subscripts. The aforementioned equations represent the system's governing equations of motion. Because we are dealing with a guided wave problem, these equations must be augmented by physically permissible boundary conditions in order to properly pose the problem. In the three displacement components u, v, w., the system of equation (2.4) is coupled. By describing the components of the displacement vector in terms of derivatives of scalar and vector potentials in the form of Sharma (2001) and assuming the solution of equation (2.4) in the following form, these equations may be uncoupled.

$$u = r^{-1}\psi_{,\theta} - \phi_{,r} \quad v = -r^{-1}\phi_{,\theta} - \psi_{,r} \quad w = -W_{,z}\,. \tag{2.5}$$

Substituting equation (2.5) in equation (2.4) yields the following second order partial differential equation with constant coefficients

$$\left((\lambda + 2\mu)\,\nabla_1^2 + \mu\frac{\partial^2}{\partial z^2} - \rho\frac{\partial^2}{\partial t^2}\right)\phi - (\lambda + \mu)\frac{\partial^2 W}{\partial z^2} = 0$$

$$\left(\mu\,\nabla_1^2 + (\lambda + 2\mu)\frac{\partial^2}{\partial z^2} - \rho\frac{\partial^2}{\partial t^2}\right)W - (\lambda + \mu)\nabla_1{}^2\phi = 0 \tag{2.6}$$

$$\left(\nabla_1{}^2 + \frac{\partial^2}{\partial z^2} - \frac{\rho}{\mu}\frac{\partial^2}{\partial t^2}\right)\psi = 0. \tag{2.7}$$

Equation (2.7) in ψ gives a purely transverse wave. A shear horizontal wave is one that is polarized in planes perpendicular to the z-axis. The solution of the equation of motion in both space and time is required in general wave propagation analysis, however, in many circumstances this is unnecessary or even prohibitive as a time-stepping algorithm can be computationally expensive. Solving in the frequency domain with a small excitation or response frequency ω helps simplify linear system analysis. So we can assume that the disturbance is time harmonic through the factor $e^{j\omega t}$ and hence, the system of equations (2.6) and (2.7) becomes

$$\left((\lambda + 2\mu)\,\nabla_1^2 + \mu\frac{\partial^2}{\partial z^2} + \rho\omega^2\right)\phi - (\lambda + \mu)\frac{\partial^2 W}{\partial z^2} = 0$$

$$\left(\mu\,\nabla_1^2 + (\lambda + 2\mu)\frac{\partial^2}{\partial z^2} - \rho\omega^2\right)W - (\lambda + \mu)\nabla_1{}^2\phi = 0 \tag{2.8}$$

$$\left(\nabla_1{}^2 + \frac{\partial^2}{\partial z^2} - \frac{\rho}{\mu}\omega^2\right)\psi = 0. \tag{2.9}$$

Equation (2.8) is coupled partial differential equations of the three displacement components. To uncouple equations (2.8) and (2.9), we can write three displacement potential functions which satisfy the simply supported boundary conditions followed by Sharma (2001) as

$$\psi(r, \theta, z, t) = \bar{\psi}(r)\sin(m\pi z)\cos(n\pi\theta/\alpha)e^{j\omega t}$$

$$\phi(r, \theta, z, t) = \bar{\phi}(r)\sin(m\pi z)\sin(n\pi\theta/\alpha)e^{j\omega t}$$

$$W(r, \theta, z, t) = \bar{W}(r)\sin(m\pi z)\sin(n\pi\theta/\alpha)e^{j\omega t}. \tag{2.10}$$

Introducing the dimensionless quantities such as

$$r' = \frac{r}{R},\ z' = \frac{z}{L},\ \delta = \frac{n\pi}{\alpha},\ t_L = \frac{m\pi R}{L},\ \bar{\lambda} = \frac{\lambda}{\mu},\ \in_0 = \frac{1}{2 + \bar{\lambda}},$$

$$\Omega^2 = \rho\omega^2 R^2/\mu \tag{2.11}$$

and substituting equations (2.10) and (2.11) in equations (2.8) and (2.9), we obtain the following second order partial differential equations

$$(\nabla_2^2 + h_3)\bar{\phi} - h_2\bar{W} = 0$$

$$(\nabla_2^2 - h_1)\bar{W} - (1 + \bar{\lambda})\nabla_2^2\bar{\phi} = 0 \tag{2.12}$$

$$(\nabla_2^2 + k_1{}^2)\bar{\psi} = 0 \tag{2.13}$$

where $h_1 = (2 + \bar{\lambda})(t_L^2 - \Omega^2)h_2 = \in_0(1 + \bar{\lambda})t_L^2\ h_3 = (\Omega^2 - \in_0 t_L^2)$.

The system of equations given in equation (2.12) has a non-trivial solution only if the determinant of the coefficients is equal to zero, that is

$$\begin{vmatrix} (\nabla_2^2 + h_3) & -h_2 \\ (1 + \bar{\lambda})\nabla_2^2 & (\nabla_2^2 - h_1) \end{vmatrix}(\bar{\phi}, \bar{W}) = 0. \tag{2.14}$$

Equation (2.14), on simplification, reduces to the following differential equation. In addition, α_1 and α_2 ($\mathrm{Re}(\alpha_1 \geqslant 0)$ and $\mathrm{Re}(\alpha_2 \geqslant 0)$) are the two roots of the equation

$$(\nabla_2^4 + E_1\nabla_2^2 + E_2)(\bar{\phi}, \bar{W}) = 0 \tag{2.15}$$

where $E_1 = -h_1 + h_2(1 + \bar{\lambda}) + h_3\ E_2 = -h_1 h_3$.

The solutions of equation (2.15) are

$$\bar{\phi}(r) = \sum_{i=1}^{2}[A_i J_\delta(\alpha_i r) + B_i Y_\delta(\alpha_i r)]$$

$$\bar{W}(r) = \sum_{i=1}^{2} a_i[A_i J_\delta(\alpha_i r) + B_i Y_\delta(\alpha_i r)] \tag{2.16}$$

here, $(\alpha_i r)^2$ $(i = 1, 2)$ are the non-zero roots of the algebraic equation

$$(\alpha_i r)^4 + E_1(\alpha_i r)^2 - E_2 = 0. \tag{2.17}$$

The arbitrary constant a_i $(i = 1, 2)$ is obtained as

$$a_i = \frac{(1 + \bar{\lambda})(\alpha_i r)^2}{(\alpha_i r)^2 + h_3}. \tag{2.18}$$

Equation (2.13) is a Bessel equation with its possible solutions as

$$\bar{\psi} = \begin{cases} A_3 J_\delta(k_1 r) + B_3 Y_\delta(k_1 r), & k_1^2 > 0 \\ A_3 r^\delta + B_3 r^{-\delta}, & k_1^2 = 0 \\ A_3 I_\delta(k_1' r) + B_3 K_\delta(k_1' r), & k_1^2 < 0 \end{cases} \tag{2.19}$$

where $k_1'^2 = -k_1^2$ and J_δ and Y_δ are Bessel functions of the first and second kind of order δ, respectively, when k_1 is real or complex, and I_δ and K_δ are modified Bessel functions of first and second kind of order δ,respectively, when k_1 is imaginary. A_3 and B_3 are two arbitrary constants. Generally, $k_1^2 \neq 0$, so the case $k_1^2 \neq 0$ has not been discussed in this part. For convenience, we consider the case of $k_1^2 > 0$ and the derivation for the case of $k_1^2 < 0$ are similar.

The solution of equation (2.13) where $k_1^2 > 0$ is

$$\bar{\psi}(r) = A_3 J_\delta(k_1 r) + B_3 Y_\delta(k_1 r) \tag{2.20}$$

where $k_1{}^2 = (2 + \bar{\lambda})\Omega^2 - t_L{}^2$.

2.4.1 Analysis of thermoelastic cylindrical panel

The analytical formulation for a thermoelastic cylindrical panel is discussed using the three-dimensional stress equations of motion (2.3) together with the heat conduction equation as in Ponnusamy (2007)

$$K(T_{,rr} + r^{-1}T_{,r} + r^{-2}T_{,\theta\theta} + T_{,zz}) = \rho c_v T_{,t} + T_0 \frac{\partial}{\partial t}(\beta(e_{rr} + e_{\theta\theta} + e_{zz})). \tag{2.21}$$

Then the stress–strain relation in equation (2.1) is modified as follows

$$\sigma_{rr} = \lambda(e_{rr} + e_{\theta\theta} + e_{zz}) + 2\mu e_{rr} - \beta T$$

$$\sigma_{\theta\theta} = \lambda(e_{rr} + e_{\theta\theta} + e_{zz}) + 2\mu e_{\theta\theta} - \beta T$$

$$\sigma_{zz} = \lambda(e_{rr} + e_{\theta\theta} + e_{zz}) + 2\mu e_{zz} - \beta T \tag{2.22}$$

where $\beta = (3\lambda + 2\mu)\gamma_0$, upon using the equations (2.22) and (2.2) in equations (2.3) and (2.21), gives the following equations

$$(\lambda + 2\mu)(u_{,rr} + r^{-1}u_{,r} - r^{-2}u) + \mu r^{-2}u_{,\theta\theta} + \mu u_{,zz}$$
$$+ r^{-1}(\lambda + \mu)v_{,r\theta} - r^{-2}(\lambda + 3\mu)v_{,\theta} + (\lambda + \mu)w_{,rz} - \beta T_{,r} = \rho u_{,tt}$$

$$\mu(v_{,rr} + r^{-1}v_{,r} - r^{-2}v) + r^{-2}(\lambda + 2\mu)v_{,\theta\theta} + \mu v_{,zz} + r^{-2}(\lambda + 3\mu)u_{,\theta} + r^{-1}(\lambda + \mu)u_{,r\theta} + r^{-1}(\lambda + \mu)w_{,\theta z} - \beta T_{,\theta} = \rho v_{,tt}$$

$$(\lambda + 2\mu)w_{,zz} + \mu(w_{,rr} + r^{-1}w_{,r} + r^{-2}w_{,\theta\theta}) + (\lambda + \mu)u_{,rz} + r^{-1}(\lambda + \mu)v_{,\theta z} + r^{-1}(\lambda + \mu)u_{,z} - \beta T_{,z} = \rho w_{,tt}$$

$$K(T_{,rr} + r^{-1}T_{,r} + r^{-2}T_{,\theta\theta} + T_{,zz}) - \rho c_v T_{,t} - T_0\frac{\partial}{\partial t}(\beta(u_{,r} + r^{-1}u + r^{-1}v_{,\theta} + w_{,z})) = 0. \tag{2.23}$$

The second order partial differential equation (2.23) can be written as

$$\left((\lambda + 2\mu)\,\nabla_1^2 + \mu\frac{\partial^2}{\partial z^2} - \rho\frac{\partial^2}{\partial t^2}\right)\phi - (\lambda + \mu)\frac{\partial^2 W}{\partial z^2} = \beta T$$

$$\left(\mu\,\nabla_1^2 + (\lambda + 2\mu)\frac{\partial^2}{\partial z^2} - \rho\frac{\partial^2}{\partial t^2}\right)W - (\lambda + \mu)\nabla_1{}^2\phi = \beta T$$

$$\nabla_1{}^2 T + \frac{\partial^2 T}{\partial z^2} - \frac{1}{K}\frac{\partial T}{\partial t} + \frac{\beta T_0(j\omega)}{\rho c_v K}\left(\nabla_1{}^2\phi + \frac{\partial^2 W}{\partial z^2}\right) = 0 \tag{2.24}$$

$$\left(\nabla_1{}^2 + \frac{\partial^2}{\partial z^2} - \frac{\rho}{\mu}\frac{\partial^2}{\partial t^2}\right)\psi = 0. \tag{2.25}$$

Equation (2.25) in ψ gives a purely transverse wave, which is not affected by temperature. This wave is polarized in planes perpendicular to the z-axis. We assume that the disturbance is time harmonic through the factor $e^{j\omega t}$. To uncouple equation (2.24), we can take equation (2.10) together with the temperature displacement as

$$T(r,\,\theta,\,z,\,t) = \bar{T}(r)\sin(m\pi z)\sin(n\pi\theta/\alpha)e^{j\omega t}. \tag{2.26}$$

By introducing the dimensionless quantity $\bar{\mathrm{T}} = T/T_0$ together with equation (2.11), we obtain the following system of equations

$$(\nabla_2^2 + h_1)\bar{\phi} + h_2\bar{W} - h_4\bar{T} = 0$$

$$(\nabla_2^2 + h_3)\bar{W} + (1 + \bar{\lambda})\,\nabla_2^2\,\bar{\phi} + (2 + \bar{\lambda})h_4\bar{T} = 0$$

$$(\nabla_2{}^2 - t_L{}^2 + \epsilon_2\Omega^2 - j\,\epsilon_3)\bar{T} + j\,\epsilon_1\,\Omega\,\nabla_2{}^2\bar{\phi} - j\,\epsilon_1\,\Omega t_L{}^2\bar{W} = 0 \tag{2.27}$$

$$(\nabla_2^2 + k_1{}^2)\bar{\psi} = 0 \tag{2.28}$$

where $\epsilon_1 = \frac{T_0 R\beta^2}{\rho^2 c_v CK}\epsilon_2 = \frac{c^2}{c_v K}\epsilon_3 = \frac{C_1 R}{K}$

$$h_1 = (2 + \bar{\lambda})(t_L^2 - \Omega^2)\;h_2 = \epsilon_4(1 + \bar{\lambda})t_L{}^2$$

$$h_3 = (\Omega^2 - \in_4 t_L{}^2)\, h_4 = \frac{\beta T_0 R^2}{\lambda + 2\mu}\, h_5 = \in_1 \Omega. \tag{2.29}$$

A non-trivial solution of the algebraic equation (2.27) exists only when the determinant of coefficients of equation (2.27) is equal to zero

$$\begin{vmatrix} (\nabla_2^2 + h_1) & -h_2 & h_4 \\ (1+\bar{\lambda})\nabla_2^2 & (\nabla_2^2 + h_3) & (2+\bar{\lambda})h_4 \\ jh_5\nabla_2^2 & -jh_5 t_L{}^2 & (\nabla_2^2 - t_L{}^2 + \in_2\Omega^2 - j\Omega \in_3) \end{vmatrix} (\bar{\phi}, \bar{W}, \bar{T}) = 0. \tag{2.30}$$

Equation (2.30), on simplification reduces to the following differential equation

$$(\nabla_2^6 + E_3\, \nabla_2^4 + E_4\, \nabla_2^2 + E_5)(\bar{\phi}, \bar{W}, \bar{T}) = 0 \tag{2.31}$$

where $E_3 = -h_1 + h_2(1 + \bar{\lambda}) + h_3 - h_4 h_5 j t_L{}^2 + \in_2\Omega^2 - j \in_3 \Omega$

$$\begin{aligned} E_4 = &- h_1 h_3 - h_1 h_4 h_5 j - h_2 h_4 h_5 j(2 + \bar{\lambda}) + t_L^2(h_1 - h_2 - h_3) + h_4 h_5 j t_L^2 \\ &+ h_2\Omega^2 \in_2 (1 + \bar{\lambda}) - h_2 j \in_3 \Omega\, (1 + \bar{\lambda}) - h_2 t_L^2 \bar{\lambda} + h_3\Omega\, (\Omega \in_2 - j \in_3) \\ &+ h_1\Omega\, (j \in_3 - \Omega \in_2) \end{aligned}$$

$$E_5 = h_1 h_3(t_L^2 + j \in_3 \Omega - \in_2\Omega^2) + jh_3 h_4 h_5 t_L^2(2 + \bar{\lambda}). \tag{2.32}$$

The solution of equation (2.31) is

$$\bar{\phi}(r) = \sum_{i=1}^{3} A_i J_\delta(\alpha_i r)\phi(r)\sin(m\pi z)\sin(n\pi\theta/\alpha)e^{j\omega t}$$

$$\bar{W}(r) = \sum_{i=1}^{3} A_i b_i J_\delta(\alpha_i r)\chi(r)\sin(m\pi z)\sin(n\pi\theta/\alpha)e^{j\omega t}$$

$$\bar{T}(r) = \sum_{i=1}^{3} A_i c_i J_\delta(\alpha_i r)T(r)\sin(m\pi z)\sin(n\pi\theta/\alpha)e^{j\omega t}. \tag{2.33}$$

Here, $(\alpha_i r)^2 (i = 1, 2, 3)$ are the non-zero roots of the algebraic equation

$$(\alpha_i r)^6 - E_3(\alpha_i r)^4 + E_4(\alpha_i r)^2 - E_5 = 0. \tag{2.34}$$

The arbitrary constant b_i and $c_i (i = 1, 2, 3)$ are obtained from

$$b_i = \left[\frac{(1+\bar{\lambda})\alpha_i{}^2 - (2+\bar{\lambda})\alpha_i{}^2 - h_1}{h_2(2+\bar{\lambda}) - \alpha_i{}^2 - h_3}\right]$$

$$c_i = \left(\frac{\bar{\lambda} + 2}{\beta T_0 R^2}\right) \left[\frac{\in_0\alpha_i{}^2 + (\in_0(h_1 + h_3) + \in_0(1+\bar{\lambda})h_2)\alpha_i{}^2 + \in_0 h_1 h_3 + \alpha_i{}^2) - h_1 h_3}{\in_0 h_3 + \in_0\alpha_i{}^2 - h_2}\right]. \tag{2.35}$$

Equation (2.28) is a Bessel equation and its solution is obtained based on the concept discussed in equation (2.19) as

$$\bar{\psi}(r) = A_4 J_\delta(k_1 r)\psi(r)\sin(m\pi z)\cos(n\pi\theta/\alpha)e^{j\omega t} \tag{2.36}$$

where $k_1{}^2 = (2 + \bar{\lambda})\Omega^2 - t_L{}^2$.

2.4.2 Analysis of generalized thermoelastic cylindrical panel

By incorporating the relaxation periods τ_0 and τ_1 into the heat conduction equation (2.21), the fundamental formulation of three-dimensional wave propagation in a homogeneous isotropic generalized thermoelastic cylindrical panel is explored.

The three-dimensional stress equation of motion for a generalized thermoelastic cylindrical panel is addressed in cylindrical coordinates by taking equation (2.3) with the heat conduction equation for a linearly elastic medium in the absence of body force as follows.

$$\begin{aligned} &K(T_{,rr} + r^{-1}T_{,r} + r^{-2}T_{,\theta\theta} + T_{,zz}) = \rho c_v[T + \tau_0 T_{,tt}] \\ &+ T_0\left[\frac{\partial}{\partial t} + \delta_{1k}\tau_0\frac{\partial^2}{\partial t^2}\right](\beta(e_{rr} + e_{\theta\theta} + e_{zz})). \end{aligned} \tag{2.37}$$

The stress–strain relation of isotropic generalized thermoelastic material are given as

$$\begin{aligned} \sigma_{rr} &= \lambda(e_{rr} + e_{\theta\theta} + e_{zz}) + 2\mu e_{rr} - \beta(T + \delta_{2k}\tau_1 T_{,t}) \\ \sigma_{\theta\theta} &= \lambda(e_{rr} + e_{\theta\theta} + e_{zz}) + 2\mu e_{\theta\theta} - \beta(T + \delta_{2k}\tau_1 T_{,t}) \\ \sigma_{zz} &= \lambda(e_{rr} + e_{\theta\theta} + e_{zz}) + 2\mu e_{zz} - \beta(T + \delta_{2k}\tau_1 T_{,t}). \end{aligned} \tag{2.38}$$

Substituting equations (2.38) and (2.2) in equations (2.3) and (2.37) gives the following three-dimensional displacement equations of motion and heat conduction equation

$$\begin{aligned} &(\lambda + 2\mu)(u_{,rr} + r^{-1}u_{,r} - r^{-2}u) + \mu r^{-2}u_{,\theta\theta} + \mu u_{,zz} + r^{-1}(\lambda + \mu)v_{,r\theta} \\ &- r^{-2}(\lambda + \mu)v_{,\theta} + (\lambda + \mu)w_{,rz} - \beta(T_{,r} + \tau_1 T_{,rt}) = \rho u_{,tt} \end{aligned}$$

$$\begin{aligned} &\mu(v_{,rr} + r^{-1}v_{,r} - r^{-2}v) + r^{-2}(\lambda + 2\mu)v_{,\theta\theta} + \mu v_{,zz} + r^{-2}(\lambda + 3\mu)u_{,\theta} \\ &+ r^{-1}(\lambda + \mu)u_{,r\theta} + r^{-1}(\lambda + \mu)w_{,\theta z} - \beta(T_{,\theta} + \tau_1 T_{,\theta t}) = \rho v_{,tt} \end{aligned}$$

$$\begin{aligned} &(\lambda + 2\mu)w_{,zz} + \mu(w_{,rr} + r^{-1}w_{,r} + r^{-2}w_{,\theta\theta}) + (\lambda + 2\mu)u_{,rz} \\ &+ r^{-1}(\lambda + \mu)v_{,\theta z} + r^{-1}(\lambda + \mu)u_{,z} - \beta(T_{,z} + \tau_1 T_{,zt}) = \rho w_{,tt} \end{aligned}$$

$$\begin{aligned} &K(T_{,rr} + r^{-1}T_{,r} + r^{-2}T_{,\theta\theta} + T_{,zz}) - \rho c_v(T_{,t} + \tau_0 T_{tt}) - T_0\left(\frac{\partial}{\partial t} + \delta_{1k}\tau_0\frac{\partial^2}{\partial t^2}\right) \\ &(\beta(u_{,r} + r^{-1}u + r^{-1}v_{,\theta} + w_{,z})) = 0. \end{aligned} \tag{2.39}$$

Using equation (2.5) in equation (2.39), we find that ϕ, W, T satisfies the following equations.

$$\left((\lambda + 2\mu)\,\nabla_1^2 + \mu\frac{\partial^2}{\partial z^2} - \rho\frac{\partial^2}{\partial t^2}\right)\phi - (\lambda + \mu)\frac{\partial^2 W}{\partial z^2} = \beta(T + \tau_1 T_{,t})$$

$$\left(\mu\,\nabla_1^2 + (\lambda + 2\mu)\frac{\partial^2}{\partial z^2} - \rho\frac{\partial^2}{\partial t^2}\right)W - (\lambda + \mu)\,\nabla_1^2\,\phi = \beta(T + \tau_1 T_{,t})$$

$$\nabla_1^2\, T + \frac{\partial^2 T}{\partial z^2} - \frac{\tau}{c_v K}\frac{\partial^2 T}{\partial t^2} - \frac{1}{K}\frac{\partial T}{\partial t} + \frac{\beta T_0(j\omega)}{\rho c_v K}\left(\nabla_1^2\,\phi + \frac{\partial^2 W}{\partial z^2}\right) = 0 \quad (2.40)$$

$$\left(\nabla_1^2 + \frac{\partial^2}{\partial z^2} - \frac{\rho}{\mu}\frac{\partial^2}{\partial t^2}\right)\psi = 0. \quad (2.41)$$

Equation (2.41) in ψ gives a purely transverse wave, which is not affected by temperature. This wave is polarized in planes perpendicular to the z-axis. We assume that the disturbance is time harmonic through the factor $e^{j\omega t}$. We obtain the modified system of equations (2.40) and (2.41) as given below

$$(\nabla_2^2 + h_3)\bar{\phi} - h_2\bar{W} + h_6\bar{T} = 0\,(\nabla_2^2 - h_1)\bar{W} + (1 + \bar{\lambda})\,\nabla_2^2\,\bar{\phi} + (2 + \bar{\lambda})h_6\bar{T} = 0$$

$$(\nabla_2^2 - t_L{}^2 + \epsilon_4\Omega^2 - j\,\epsilon_3\,)\bar{T} + j\,\epsilon_1\,\Omega\,\nabla_2{}^2\bar{\phi} - j\,\epsilon_1\,\Omega t_L{}^2\bar{W} = 0 \quad (2.42)$$

$$(\nabla_2^2 + k_1{}^2)\bar{\psi} = 0. \quad (2.43)$$

Rewriting equation (2.42), results in the following determinant form of equations

$$\begin{vmatrix} (\nabla_2^2 + h_3) & -h_2 & h_6 \\ (1 + \bar{\lambda})\nabla_2^2 & (\nabla_2^2 - h_1) & (2 + \bar{\lambda})h_6 \\ jh_5\nabla_2^2 & -jh_5 t_L{}^2 & (\nabla_2^2 - t_L{}^2 + \epsilon_4\Omega^2 - j\Omega\,\epsilon_3\,) \end{vmatrix}(\bar{\phi}, \bar{W}, \bar{T}) = 0 \quad (2.44)$$

where $\epsilon_1 = \frac{T_0 R\beta^2}{\rho^2 c_V CK}\epsilon_3 = \frac{CR}{K}\epsilon_4 = \frac{\tau C^2}{c_V K}h_1 = (2 + \bar{\lambda})(t_L^2 - \Omega^2)$

$$h_2 = \epsilon_4(1 + \bar{\lambda})t_L{}^2\, h_3 = (\Omega^2 - \epsilon_4 t_L{}^2)\, h_5 = \epsilon_1\Omega\, h_6 = \frac{\beta T_0 R^2\tau}{\lambda + 2\mu}.$$

Equation (2.44), on simplification reduces to the following differential equation

$$(\nabla_2^6 + E_6\,\nabla_2^4 + E_7\,\nabla_2^2 + E_8)(\bar{\phi}, \bar{W}, \bar{T}) = 0 \quad (2.45)$$

where

$$E_6 = -h_1 + h_2(1 + \bar{\lambda}) + h_3 - h_6 h_5 j t_L^2 + \epsilon_4\Omega^2 - j\,\epsilon_3\,\Omega$$

$$
\begin{aligned}
E_7 = & - h_1h_3 - h_1h_4h_5j - h_2h_6h_5j(2 + \bar{\lambda}) + t_L{}^2(h_1 - h_2 - h_3) \\
& + h_6h_5jt_L{}^2 + h_2\Omega^2 \in_4 (1 + \bar{\lambda}) \\
& - h_2j \in_3 \Omega(1 + \bar{\lambda}) - h_2t_L{}^2\bar{\lambda} + h_3\Omega(\Omega \in_2 - j \in_3) + h_1\Omega(j \in_3 - \Omega \in_2)
\end{aligned}
$$

$$
E_8 = h_1h_3(t_L{}^2 + j \in_3 \Omega - \in_4\Omega^2) + jh_3h_6h_5t_L^2(2 + \bar{\lambda}). \tag{2.46}
$$

The solutions of equation (2.45) are

$$
\bar{\phi}(r) = \sum_{i=1}^{3}(A_iJ_\delta(\alpha_ir) + B_i\,Y_\delta(\alpha_ir))
$$

$$
\begin{aligned}
\bar{W}(r) &= \sum_{i=1}^{3}d_i(A_iJ_\delta(\alpha_ir) + B_i\,Y_\delta(\alpha_ir)) \\
\bar{T}(r) &= \sum_{i=1}^{3}e_i(A_iJ_\delta(\alpha_ir) + B_i\,Y_\delta(\alpha_ir))
\end{aligned} \tag{2.47}
$$

where d_i and $e_i (i = 1, 2, 3)$ are computed from

$$
d_i = \left[\frac{(1 + \bar{\lambda})\delta_i{}^2 - (2 + \bar{\lambda})\alpha_i{}^2 - h_1}{h_2(2 + \bar{\lambda}) - \alpha_i{}^2 - h_3}\right]
$$

$$
e_i = \left(\frac{\bar{\lambda} + 2}{\beta T_0 R^2}\right)\left[\frac{\in_0\alpha_i{}^2 + (\in_0(h_1 + h_3) + \in_0(1 + \bar{\lambda})h_2)\alpha_i{}^2 + \in_0 h_1h_3 + \alpha_i{}^2) - h_1h_3}{\in_0 h_3 + \in_0\alpha_i{}^2 - h_2}\right]. \tag{2.48}
$$

Equation (2.43) is a Bessel equation and its solution is obtained from equation (2.19) as

$$
\bar{\psi}(r) = A_4J_\delta(k_1r) + B_4\,Y_\delta(k_1r), \tag{2.49}
$$

where $k_1{}^2 = (2 + \bar{\lambda})\Omega^2 - t_L{}^2$.

2.4.3 Analysis of a generalized thermoelastic plate

This section has been reprinted from Selvamani and Ponnusamy (2012) with permission from Materials Physics and Mechanics.

Consider a thin homogeneous, isotropic, thermally conducting elastic plate of radius R with uniform thickness d and temperature T_0 in the undisturbed state initially. The system displacements and stresses are defined in the polar coordinates r and θ for an arbitrary point inside the plate, with u denoting the displacement in the radial direction of r and v the displacement in the tangential direction of θ. The in-plane vibration and displacements of the plate embedded on an elastic medium are obtained by assuming that there is no vibration and a displacement along the z-axis in the cylindrical coordinate system (r, θ, z).

The two-dimensional stress equations of motion and heat conduction equation in the absence of body force for a linearly elastic plate is reduced from equations (2.3) and (2.38) as follows

$$\sigma_{rr,\,r} + r^{-1}\sigma_{r\theta,\,\theta} + r^{-1}(\sigma_{rr} - \sigma_{\theta\theta}) = \rho u_{,tt}$$

$$\sigma_{r\theta,\,r} + r^{-1}\sigma_{\theta\theta,\,\theta} + 2r^{-1}\sigma_{r\theta} = \rho v_{,tt}$$

$$K(T_{,rr} + r^{-1}T_{,r} + r^{-2}T_{,\theta\theta}) - \rho c_{\nu}(T + \tau_0 T_{,tt}) = \beta T_0\left(\frac{\partial}{\partial t} + \delta_{1k}\tau_0\frac{\partial^2}{\partial t^2}\right)[e_{rr} + e_{\theta\theta}]. \tag{2.50}$$

The strain–displacement relations for the isotropic plate are given by

$$\sigma_{rr} = \lambda(e_{rr} + e_{\theta\theta}) + 2\mu e_{rr} - \beta(T + \delta_{2k}\tau_1 T_{,t})$$

$$\sigma_{\theta\theta} = \lambda(e_{rr} + e_{\theta\theta}) + 2\mu e_{\theta\theta} - \beta(T + \delta_{2k}\tau_1 T_{,\,t})$$

$$\sigma_{r\theta} = 2\mu e_{r\theta} \tag{2.51}$$

where δ_{ij} is the Kronecker delta function. In addition, we can replace $k = 1$ for the L–S theory and $k = 2$ for the G–L theory. The thermal relaxation times τ_0 and τ_1 satisfies the inequalities $\tau_0 \geqslant \tau_1 \geqslant 0$ for the G–L theory only.

2.4.3.1 Lord–Shulman (L–S) theory for a plate

Based on the Lord–Shulman theory of thermo-elasticity, the three-dimensional rate-dependent temperature with one relaxation time is obtained by replacing $k = 1$ in the heat conduction equation (2.50)

$$K(T_{,rr} + r^{-1}T_{,r} + r^{-2}T_{,\theta\theta}) = \rho c_v[T + \tau_0 T_{,tt}] + \beta T_0\left[\frac{\partial}{\partial t} + \tau_0\frac{\partial^2}{\partial t^2}\right](e_{rr} + e_{\theta\theta}). \tag{2.52}$$

The stress–strain relation is replaced by

$$\begin{aligned}\sigma_{rr} &= \lambda(e_{rr} + e_{\theta\theta}) + 2\mu e_{\theta\theta} - \beta T\\ \sigma_{\theta\theta} &= \lambda(e_{rr} + e_{\theta\theta}) + 2\mu e_{\theta\theta} - \beta T\\ \sigma_{r\theta} &= 2\mu e_{r\theta}.\end{aligned} \tag{2.53}$$

By substituting equation (2.53) into equations (2.50) and (2.52),we can get the following displacement equations

$$(\lambda + 2\mu)(u_{,rr} + r^{-1}u_{,r} - r^{-2}u) + r^{-2}\mu u_{,\theta\theta} + r^{-1}(\lambda + \mu)v_{,r\theta} + r^{-2}(\lambda + 3\mu)v_{,\theta} - \beta T_{,r} = \rho u_{,tt}$$

$$\mu(v_{,rr} + r^{-1}v_{,r} - r^{-2}v) + r^{-1}(\lambda + \mu)u_{,r\theta} + r^{-2}(\lambda + 3\mu)u_{,\theta} + r^{-2}(\lambda + 3\mu)v_{,\theta} + r^{-2}(\lambda + 2\mu)v_{,\theta\theta} - \beta T_{\theta} = \rho v_{,tt}. \tag{2.54}$$

The symbols and notations involved have the same meanings as defined in earlier sections. Since the heat conduction equation of this theory is of the hyperbolic wave type, it can automatically ensure the finite speed of propagation for heat and elastic waves.

2.4.3.2 Green–Lindsay (G–L) theory for a plate

The second generalization to the coupled thermoelasticity with two relaxation times called the Green–Lindsay theory of thermoelasticity is obtained by setting $k = 2$ in equation (2.50)

$$K(T_{,rr} + r^{-1}T_{,r} + r^{-2}T_{,\theta\theta}) = \rho c_v[T + \tau_0 T_{,tt}] + \beta T_0\frac{\partial}{\partial t}(e_{rr} + e_{\theta\theta}). \tag{2.55}$$

The stress–strain relation is replaced by

$$\begin{aligned}\sigma_{rr} &= \lambda(e_{rr} + e_{\theta\theta}) + 2\mu e_{\theta\theta} - \beta(T + \tau_1 T_{,t})\\ \sigma_{\theta\theta} &= \lambda(e_{rr} + e_{\theta\theta}) + 2\mu e_{\theta\theta} - \beta(T + \tau_1 T_{,t})\\ \sigma_{r\theta} &= 2\mu e_{r\theta}.\end{aligned} \tag{2.56}$$

By substituting equation (2.56) into equations (2.50) and (2.55), the displacement equation can be reduced as

$$\begin{aligned}&(\lambda + 2\mu)(u_{,rr} + r^{-1}u_{,r} - r^{-2}u) + r^{-2}\mu u_{,\theta\theta} + r^{-1}(\lambda + \mu)v_{,r\theta} + r^{-2}(\lambda + 3\mu)v_{,\theta}\\ &- \beta(T_{,r} + \tau_1 T_{,rt}) = \rho u_{,tt}\\[2ex] &\mu(v_{,rr} + r^{-1}v_{,r} - r^{-2}v) + r^{-1}(\lambda + \mu)u_{,r\theta} + r^{-2}(\lambda + 3\mu)u_{,\theta} + r^{-2}(\lambda + 3\mu)v_{,\theta}\\ &+ r^{-2}(\lambda + 2\mu)v_{,\theta\theta} - \beta(T_{,\theta} + \tau_1 T_{,\theta t}) = \rho v_{,tt}.\end{aligned} \tag{2.57}$$

In the previous sections, the symbols and notations were specified. The generalized thermoelasticity theories are considered to be more realistic than the conventional theory in dealing with practical problems involving very large heat fluxes and/or short time intervals, such as those occurring in laser units and energy channels, due to available experimental evidence in favour of the finiteness of heat propagation speeds.

The strain–displacement relations will be reduced as given by

$$e_{rr} = u_{,r}\ e_{\theta\theta} = r^{-1}(u + v_{,\theta})\ e_{r\theta} = v_{,r} - r^{-1}(v - u_{,\theta}). \tag{2.58}$$

By substituting equations (2.51) and (2.58) into equation (2.50), the following displacement equations of motions are obtained

$$\begin{aligned}&(\lambda + 2\mu)(u_{,rr} + r^{-1}u_{,r} - r^{-2}u) + \mu r^{-2}u_{,\theta\theta} + r^{-1}(\lambda + \mu)v_{,r\theta}\\ &+ r^{-2}(\lambda + 3\mu)v_{,\theta} - \beta(T_{,r} + T\delta_{2k}\tau_1 T_{,rt}) = \rho u_{,tt}\\[2ex] &\mu(v_{,rr} + r^{-1}v_{,r} - r^{-2}v) + r^{-2}(\lambda + 2\mu)v_{,\theta\theta} + r^{-2}(\lambda + 3\mu)u_{,\theta}\\ &+ r^{-1}(\lambda + \mu)u_{,r\theta} - \beta(T_{,\theta} + T\delta_{2k}\tau_1 T_{,\theta t}) = \rho v_{,tt}\\[2ex] &K(T_{,rr} + r^{-1}T_{,r} + r^{-2}T_{,\theta\theta}) - \rho c_v(T + \tau_0 T_{,tt})\\ &= \beta T_0\left(\frac{\partial}{\partial t} + \tau_0\delta_{1k}\frac{\partial^2}{\partial t^2}\right)[u_{,r} + r^{-1}(u + v_{,\theta})].\end{aligned} \tag{2.59}$$

To uncouple equation (2.59), the mechanical displacement u, v along the radial and circumferential directions are reduced from equation (2.5)

$$u = \phi_{,r} + r^{-1}\psi_{,\theta}\ v = r^{-1}\phi_{,\theta} - \psi_{,r}\ . \tag{2.60}$$

Substituting equation (2.57) into equation (2.56) yields the following second order partial differential equation with constant coefficients

$$\{(\lambda + 2\mu)\ \nabla^2 + \rho\omega^2\}\phi - \beta(T + \delta_{2k}\tau_1 T_{,t}) = 0$$

$$\{k\ \nabla^2 - \rho c_v j\omega\eta_0\}T + \beta T_0(j\omega\eta_1)\ \nabla^2\ \phi = 0 \tag{2.61}$$

$$\left(\nabla^2 + \frac{\rho}{\mu}\omega^2\right)\psi = 0. \tag{2.62}$$

Equations (2.61) are coupled partial differential equations with two displacements and heat conduction components. We assume that the vibration and displacements along the axial direction z are zero to decouple these equations. Hence, the solutions of equation (2.61) can be presented in the following form

$$u(r,\theta,t) = \bar{\phi}(r)\exp(j(p\theta - \omega\, t))\ v(r,\theta,t) = \bar{\psi}(r)\exp(j(p\theta - \omega t))$$

$$T(r,\theta,t) = (\lambda + 2\mu/\beta a^2)\bar{T}(r)\exp(j(p\theta - \omega t)) \tag{2.63}$$

where p is the wave number, substituting equation (2.63) into equations (2.61) and introducing the dimensionless quantities such as $x = r/a$, $\Omega^2 = \rho\omega^2 a^2/\mu$, $\bar{\lambda} = \lambda/\mu\bar{d} = \rho c_v\mu/\beta\ T_0$, we can get the following partial differential equation with constant coefficients

$$\{(2 + \bar{\lambda})\ \nabla_1^2 + \Omega^2\}\phi - (2 + \bar{\lambda})\eta_2 T = 0$$

$$\{k_1\ \nabla_1^2 - j\omega\bar{d}\eta_0\}T + \beta T_0(j\omega\eta_1)\ \nabla_1^2\ \phi = 0 \tag{2.64}$$

$$(\nabla_1^2 + \Omega^2)\psi = 0 \tag{2.65}$$

where $\eta_0 = 1 + j\omega\tau_0$, $\eta_1 = 1 + j\omega\delta_{1k}\tau_0$, $\eta_2 = 1 + j\omega\delta_{2k}\tau_1$.

Equation (2.65) in terms of ψ gives a purely transverse wave. In planes perpendicular to the z-axis, this wave is polarized. The following fourth order differential equation is obtained by assuming that the disturbance is time harmonic factor $e^{j\omega t}$, rewriting equation (2.64).

$$(E_9\ \nabla_2^4 + E_{10}\ \nabla_2^2 + E_{11})(\bar{\phi}, \bar{T}) = 0 \tag{2.66}$$

where $E_9 = (2 + \bar{\lambda})k_1$, $E_{10} = \{k_1\Omega^2 - j\omega(2 + \bar{\lambda})\bar{d}\eta_0 + j\omega T_0(2 + \bar{\lambda})\beta\ \eta_1\eta_2\}$,

$$E_{11} = -\left(j\omega\ \Omega^2\ \bar{d}\eta_0\right). \tag{2.67}$$

By solving the partial differential equation (2.66), the solutions are obtained as

$$\bar{\phi} = \sum_{i=1}^{2}[A_i J_\delta(\alpha_i ax) + B_i Y_\delta(\alpha_i ax)]$$

$$\bar{T} = \sum_{i=1}^{2} f_i [A_i J_\delta(\alpha_i ax) + B_i Y_\delta(\alpha_i ax)] \tag{2.68}$$

$$f_i = \{k_1(\alpha_i ax)^4 + (2 + \bar{\lambda})\beta T_0 j\omega\eta_1\eta_2(\alpha_i ax)^2 - (2 + \bar{\lambda})j\omega\bar{d}.$$

Equation (2.65) is a Bessel equation and its solution obtained from the discussion as in equation (2.19) is

$$\bar{\psi} = [A_3 J_\delta(\alpha_3 ax) + Y_\delta B_3(\alpha_3 ax)] \tag{2.69}$$

where $(\alpha_3 a)^2 = \Omega^2$.

2.4.4 Analysis of a rotating cylindrical panel

The deformation of the cylindrical panel in the direction r, θ, z is defined by u, v and w. The cylindrical panel is assumed to be homogenous, isotropic and linearly elastic with a rotational parameter Γ, Young's modulus E, poisson ratio ν and density ρ in an undisturbed state. Here we can take the three-dimensional strain–displacement relation and stress equations of motion in the absence of body force for a linearly elastic medium as defined in equations (2.2) and (2.3). By inserting the rotational parameter Γ in the radial equation defined in the stress equations of motion (2.3)

$$\sigma_{rr,\,r} + r^{-1}\sigma_{r\theta,\,\theta} + \sigma_{rz,\,z} + r^{-1}(\sigma_{rr} - \sigma_{\theta\theta}) + \rho\,\Gamma^2 u = \rho u_{,tt}. \tag{2.70}$$

Substitution of equations (2.1) and (2.2) in equation (2.70) gives the following three displacement equations of motion

$$(\lambda + 2\mu)(u_{,rr} + r^{-1}u_{,r} - r^{-2}u) + \mu r^{-2}u_{,\theta\theta} + \mu u_{,zz} + r^{-1}(\lambda + \mu)v_{,r\theta} - r^{-2}(\lambda + 3\mu)v_{,\theta}$$
$$+ (\lambda + \mu)w_{,rz} + \rho\,\Gamma^2 u = \rho u_{,tt}$$

$$\mu(v_{,rr} + r^{-1}v_{,r} - r^{-2}v) + r^{-2}(\lambda + 2\mu)v_{,\theta\theta} + \mu v_{,zz} + r^{-2}(\lambda + 3\mu)u_{,\theta} + r^{-1}(\lambda + \mu)u_{,r\theta}$$
$$+ r^{-1}(\lambda + \mu)w_{,\theta z} = \rho v_{,tt}$$

$$\begin{aligned}&(\lambda + 2\mu)w_{,zz} + \mu(w_{,rr} + r^{-1}w_{,r} + r^{-2}w_{,\theta\theta}) + (\lambda + \mu)u_{,rz} + r^{-1}(\lambda + \mu)v_{,\theta z}\\ &+ r^{-1}(\lambda + \mu)u_{,z} = \rho w_{,tt}.\end{aligned} \tag{2.71}$$

To solve equation (2.71), we can take the same displacement potential functions from equation (2.5) and using the same in equation (2.71), we find that ϕ, W, T satisfies the following equations

$$((\lambda + 2\mu)\,\nabla_1^2 + \mu\frac{\partial^2}{\partial z^2} - \rho\frac{\partial^2}{\partial t^2} + \rho\Gamma^2)\phi - (\lambda + \mu)\frac{\partial^2 W^2}{\partial z^2} = 0$$

$$\left(\mu\,\nabla_1^2 + (\lambda + 2\mu)\frac{\partial^2}{\partial z^2} - \rho\frac{\partial^2}{\partial t^2}\right)W - (\lambda + \mu)\nabla_1{}^2\phi = 0 \tag{2.72}$$

$$\left(\nabla_1{}^2 + \frac{\partial^2}{\partial z^2} - \frac{\rho}{\mu}\frac{\partial^2}{\partial t^2} - \rho\,\Gamma^2\right)\psi = 0. \tag{2.73}$$

Here also equation (2.73) in ψ gives a purely transverse wave. This wave is polarized in planes perpendicular to the z-axis and the disturbance is time harmonic through the factor $e^{j\omega t}$. To uncouple equation (2.72), we can write three displacement functions which satisfies the simply supported boundary conditions as defined in equation (2.10) and introducing the following dimensionless parameters in equations (2.72) and (2.73)

$$r' = \frac{r}{R},\ z' = \frac{z}{L},\ \delta = \frac{n\pi}{\alpha},\ t_L = \frac{m\pi R}{L},\ \bar{\lambda} = \frac{\lambda}{\mu},\ \epsilon_0 = \frac{1}{2 + \bar{\lambda}}$$

$$C = \frac{\lambda + 2\mu}{\rho},\ \Omega^2 = \frac{\omega^2 R^2}{C^2},\ \varpi = \frac{\rho\,\Gamma^2 R^2}{2 + \bar{\lambda}} \tag{2.74}$$

we obtain the following system of equations

$$(\nabla_2^2 + h_7)\bar{\phi} + h_2\bar{W} = 0$$

$$(\nabla_2^2 + h_3)\bar{W} + (1 + \bar{\lambda})\,\nabla_2^2\,\bar{\phi} = 0 \tag{2.75}$$

and

$$(\nabla_2^2 + k_2{}^2)\bar{\psi} = 0 \tag{2.76}$$

where the constants in the above differential equation are defined as follows

$$h_2 = \epsilon_0(1 + \bar{\lambda})t_L{}^2\ h_3 = (\Omega^2 - \epsilon_0 t_L{}^2) \text{ and } h_7 = h_1 + (2 + \bar{\lambda})\varpi.$$

A non-trivial solution of the algebraic equation (2.75) exists only when the determinant of equation (2.75) is equal to zero

$$\begin{vmatrix} (\nabla_2^2 + h_3) & -h_2 \\ (1 + \bar{\lambda})\nabla_2^2 & (\nabla_2^2 - h_7) \end{vmatrix}(\bar{\phi},\ \bar{W}) = 0. \tag{2.77}$$

Equation (2.77), on simplification, reduces to the following differential equation. In addition, α_1 and α_2 (Re($\alpha_1 \geqslant 0$) and Re($\alpha_2 \geqslant 0$)) are the two roots of the following equation

$$(\nabla_2^4 + F_1\,\nabla_2^2 + F_2)(\bar{\phi},\ \bar{W}) = 0 \tag{2.78}$$

where

$$F_1 = -h_7 + h_2(1 + \bar{\lambda}) + h_3\ F_2 = -h_7 h_3. \tag{2.79}$$

The solution of equation (2.78) has the same term as defined in equation (2.16). Equation (2.76) is a Bessel equation and its possible solution is obtained as discussed in equation (2.19) as

$$\bar{\psi}(r) = A_3 J_\delta(k_2 r) + B_3 Y_\delta(k_2 r) \tag{2.80}$$

where $k_2{}^2 = k_1{}^2 - \varpi$.

2.4.4.1 Elastokinetic for a cylindrical panel

In the present analysis if we take the coupling parameter for rotation $\Gamma = 0$, then the relations will reduce to the classical case in elastokinetic such as

$$\begin{vmatrix} (\nabla_2^2 + h_3) & -h_2 \\ (1 + \bar{\lambda})\nabla_2^2 & (\nabla_2^2 - h_7) \end{vmatrix} (\bar{\phi}, \bar{W}) = 0. \tag{2.81}$$

The above determinant is reduced as

$$(\nabla_2^4 + F_3\, \nabla_2^2 + F_4)(\bar{\phi}, \bar{W}) = 0 \tag{2.82}$$

where

$$F_3 = (1 + \bar{\lambda})h_2 + h_3 + h_7 \tag{2.83}$$

$$F_4 = h_3\, h_7$$

The solution of equation (2.82) is obtained as

$$\bar{\phi}(r) = \sum_{i=1}^{2}[A_i J_\delta(\alpha_i r) + B_i\, Y_\delta(\alpha_i r)]$$

$$\bar{W}(r) = \sum_{i=1}^{2} j_i\, [A_i J_\delta(\alpha_i r) + B_i\, Y_\delta(\alpha_i r)] \tag{2.84}$$

where

$$j_i = \frac{(1 + \bar{\lambda})(\alpha_i r)^2}{(\alpha_i r)^2 + h_3} (i = 1, 2). \tag{2.85}$$

These equations constitute the solution for the homogeneous isotropic cylindrical panel with traction free boundary conditions.

2.4.5 Analysis of rotating thermoelastic cylindrical panel

In cylindrical coordinates the three-dimensional stress equations of motion, strain–displacement relations and heat conduction in the absence of body force for a linearly elastic rotating medium are simplified as in Roychoudhury and Mukhopadhyay (2000)

$$\sigma_{rr,\,r} + r^{-1}\sigma_{r\theta,\,\theta} + \sigma_{rz,\,z} + r^{-1}(\sigma_{rr} - \sigma_{\theta\theta}) + \rho(\vec{\Gamma}\times(\vec{\Gamma}\times\vec{u}) + 2\vec{\Gamma}\vec{u}_{,t}) = \rho u_{,tt}$$

$$\sigma_{r\theta,\,r} + r^{-1}\sigma_{\theta\theta,\,\theta} + \sigma_{,rzz} + \sigma_{\theta z,\,z} + 2r^{-1}\sigma_{r\theta} = \rho v_{,tt}$$

$$\sigma_{rz,\,r} + r^{-1}\sigma_{\theta z,\,\theta} + \sigma_{zz,\,z} + r^{-1}\sigma_{r\theta} + \rho(\vec{\Gamma}\times(\vec{\Gamma}\times\vec{u}) + 2\vec{\Gamma}\vec{u}_{,t}) = \rho w_{,tt}$$

$$K(T_{,rr} + r^{-1}T_{,r} + r^{-2}T_{,\theta\theta} + T_{,zz}) = \rho c_{\nu} T_{,t} + \beta T_0(u_{,rt} + r^{-1}(u_{,t} + v_{,\theta t}) + w_{,tz}) \quad (2.86)$$

where the stress equations of motion have the additional terms with a time dependent centripetal acceleration $\vec{\Gamma}\times(\vec{\Gamma}\times\vec{u})$ and $2\Gamma \times \vec{u}_{,t}$ where, $\vec{u} = (u, 0, w)$ is the displacement vector and $\vec{\Gamma} = (0, \Gamma, 0)$ is a constant, the comma notation used in the subscript denotes the partial differentiation with respect to the respective variables. The stress–strain relations and strain–displacement components have the same meaning as discussed in equations (2.22) and (2.2).

Substituting equations (2.22) and (2.2) in equation (2.86), gives the following three displacement equations of motion and the heat conduction equation

$$\begin{aligned}&(\lambda + 2\mu)(u_{,rr} + r^{-1}u_{,r} - r^{-2}u) + \mu r^{-2}u_{,\theta\theta} + \mu u_{,zz}\\ &+ r^{-1}(\lambda + \mu)v_{,r\theta} - r^{-2}(\lambda + 3\mu)v_{,\theta}\\ &+ (\lambda + \mu)w_{,rz} - \beta T_{,r} + \rho(\Gamma^2 u + 2\Gamma w_{,t}) = \rho u_{,tt}\end{aligned}$$

$$\begin{aligned}&\mu(v_{,rr} + r^{-1}v_{,r} - r^{-2}v) + r^{-2}(\lambda + 2\mu)v_{,\theta\theta}\\ &+ \mu v_{,zz} + r^{-2}(\lambda + 3\mu)u_{,\theta} + r^{-1}(\lambda + \mu)u_{,r\theta}\\ &+ r^{-1}(\lambda + \mu)w_{,\theta z} - \beta T_{,\theta} = \rho v_{,tt}\end{aligned}$$

$$\begin{aligned}&(\lambda + 2\mu)w_{,zz} + \mu(w_{,rr} + r^{-1}w_{,r} + r^{-2}w_{,\theta\theta}) + (\lambda + \mu)u_{,rz} + r^{-1}(\lambda + \mu)v_{,\theta z}\\ &+ r^{-1}(\lambda + \mu)u_{,z} - \beta T_{,z} + \rho(\Gamma^2 w + 2\Gamma u_{,t}) = \rho w_{,tt}\end{aligned}$$

$$K(T_{,rr} + r^{-1}T_{,r} + r^{-2}T_{,\theta\theta} + T_{,zz}) = \rho c_{\nu} T_{,t} + \beta T_0(u_{,tr} + r^{-1}(u_{,t} + v_{,t\theta}) + w_{,tz}). \quad (2.87)$$

To solve equation (2.87), we can take the displacement potential as defined in equation (2.5) and using equation (2.5) in equation (2.87), we find that ϕ, W, T satisfies the following equations

$$\left((\lambda + 2\mu)\,\nabla_1^2 + \mu\frac{\partial^2}{\partial z^2} - \rho\frac{\partial^2}{\partial t^2} + \rho\,\Gamma^2\right)\phi - \left((\lambda + \mu)\frac{\partial^2}{\partial z^2} + 2\Gamma\frac{\partial^2}{\partial z\partial t}\right)W = \beta T$$

$$\left(\mu\,\nabla_1^2 + (\lambda + 2\mu)\frac{\partial^2}{\partial z^2} - \rho\frac{\partial^2}{\partial t^2} + \rho\,\Gamma^2\right)W - \left((\lambda + \mu)\nabla_1{}^2 - 2\Gamma\frac{\partial^2}{\partial r\partial t}\right)\phi = \beta T$$

$$\nabla_1{}^2 T + \frac{\partial^2 T}{\partial z^2} - \frac{1}{K}\frac{\partial T}{\partial t} + \frac{\beta T_0(j\omega)}{\rho c_{\nu} K}\left(\nabla_1{}^2\phi + \frac{\partial^2 W}{\partial z^2}\right) = 0 \quad (2.88)$$

and

$$\left(\nabla_1{}^2 + \frac{\partial^2}{\partial z^2} - \frac{\rho}{\mu}\frac{\partial^2}{\partial t^2} - \rho\,\Gamma^2\right)\psi = 0. \quad (2.89)$$

Again, in terms of ψ, equation (2.89) yields a purely transverse wave that is unaffected by temperature. Equations (2.88) are the three displacement components' coupled partial differential equations. To decouple equation (2.88), we may create three displacement functions that satisfy the simply supported boundary conditions specified in equation (2.10), as well as the temperature displacement, as follows.

$$\psi(r,\ \theta,\ z,\ t) = \bar{\psi}(r)\sin(m\pi z)\cos(n\pi\theta/\alpha)e^{j\omega t}$$

$$\phi(r,\ \theta,\ z,\ t) = \bar{\phi}(r)\sin(m\pi z)\sin(n\pi\theta/\alpha)e^{j\omega t}$$

$$W(r,\ \theta,\ z,\ t) = \bar{\chi}(r)\sin(m\pi z)\sin(n\pi\theta/\alpha)e^{j\omega t}$$

$$T(r,\ \theta,\ z,\ t) = \bar{T}(r,\ \theta,\ z,\ t)\sin(m\pi z)\sin(n\pi\theta/\alpha)e^{j\omega t}. \tag{2.90}$$

Plugging the non-dimensional parameters defined in equations (2.74) and (2.90) into equations (2.88) and (2.89), we obtain the following system of equations

$$(\nabla_2^2 + h_7)\bar{\phi} + h_2\bar{W} - h_4\bar{T} = 0$$

$$(\nabla_2^2 + h_8)\bar{W} + (1+\bar{\lambda})\nabla_2^2\bar{\phi} + (2+\bar{\lambda})h_4\bar{T} = 0$$

$$(\nabla_2{}^2 - t_L{}^2 + \in_2\Omega^2 - j\in_3)\bar{T} + j\in_1\Omega\nabla_2{}^2\bar{\phi} - j\in_1\Omega t_L{}^2\bar{W} = 0 \tag{2.91}$$

and

$$(\nabla_2^2 + k_2{}^2)\bar{\psi} = 0 \tag{2.92}$$

where

$$\in_1 = \frac{T_0R\beta^2}{\rho^2 c_v CK} \quad \in_2 = \frac{C^2}{c_v K} \quad \in_3 = \frac{CR}{K}$$

$$h_2 = \in_4(1+\bar{\lambda})t_L^2 \; h_4 = \frac{\beta T_0 R^2}{\lambda + 2\mu} h_5 = \in_1\Omega.$$

$$h_7 = h_1 + (2+\bar{\lambda})\varpi \; h_8 = h_3 + \varpi.$$

A non-trivial solution of the algebraic equation (2.91) exists only when the determinant of the coefficients of equation (2.91) is equal to zero

$$\begin{vmatrix} (\nabla_2^2 + h_7) & -h_2 & h_4 \\ (1+\bar{\lambda})\nabla_2{}^2 & (\nabla_2{}^2 + h_8) & (2+\bar{\lambda})h_4 \\ jh_5\nabla_2{}^2 & -jh_5 t_L{}^2 & (\nabla_2{}^2 - t_L{}^2 + \in_2\Omega^2 - j\Omega\in_3) \end{vmatrix} (\bar{\phi},\ \bar{W},\ \bar{T}) = 0. \tag{2.93}$$

Equation (2.93), on simplification reduces to the following differential equation

$$(\nabla_2^6 + F_5\nabla_2^4 + F_6\nabla_2^2 + F_7)(\bar{\phi}, \bar{W},\ \bar{T}) = 0 \tag{2.94}$$

where,

$$F_5 = -h_7 + h_2(1 + \bar{\lambda}) + h_3 - h_4h_5jt_L{}^2 + \in_2\Omega^2 - j\in_3\Omega$$

$$F_6 = -h_7h_8 - h_7h_4h_5j - h_2h_4h_5j(2 + \bar{\lambda}) + t_l{}^2(h_7 - h_2 - h_8) + h_4h_5jt_L{}^2 + h_2\Omega^2\in_2(1 + \bar{\lambda}) \\ - h_2j\in_3\Omega(1 + \bar{\lambda}) - h_2t_L{}^2\bar{\lambda} + h_8\Omega(\Omega\in_2 - j\in_3) + h_7\Omega(j\in_3 - \Omega\in_2)$$

$$F_7 = h_7h_8(t_L{}^2 + j\in_3\Omega - \in_2\Omega^2) + jh_3h_4h_5t_L^2(2 + \bar{\lambda}). \tag{2.95}$$

The solutions of equation (2.94) are given as follows

$$\bar{\phi}(r) = \sum_{i=1}^{3}(A_iJ_\delta(\alpha_ir) + B_iY_\delta(\alpha_ir))$$

$$\begin{aligned} \bar{W}(r) &= \sum_{i=1}^{3}m_i(A_iJ_\delta(\alpha_ir) + B_iY_\delta(\alpha_ir)) \\ \bar{T}(r) &= \sum_{i=1}^{3}n_i(A_iJ_\delta(\alpha_ir) + B_iY_\delta(\alpha_ir)). \end{aligned} \tag{2.96}$$

Here, $(\alpha_ir)^2(i = 1, 2, 3)$ are the non-zero roots of the algebraic equation

$$(\alpha_ir)^6 - F_5(\alpha_ir)^4 + F_6(\alpha_ir)^2 - F_7 = 0.$$

The arbitrary constants m_i and $n_i(i = 1, 2, 3)$ are obtained from

$$m_i = \left[\frac{(1 + \bar{\lambda})\alpha_i^2 - (2 + \bar{\lambda})\alpha_i^2h_7}{h_2(2 + \bar{\lambda}) - \alpha_i^2 - h_8}\right]$$

$$n_i = \left(\frac{\bar{\lambda} + 2}{\beta T_0R^2}\right)\left[\frac{\in_0\alpha_i^2 + (\in_0(h_7 + h_8) + \in_0(1 + \bar{\lambda})h_2)\alpha_i^2 + \in_0h_1h_3 + \alpha_i^2) - h_1h_3}{\in_0h_8 + \in_0\alpha_i^2 - h_2}\right]. \tag{2.97}$$

Equation (2.92) is a Bessel equation and its possible solution is obtained by the equation as follows

$$\bar{\psi}(r) = A_4J_\delta(k_2r) + B_4Y_\delta(k_2r) \tag{2.98}$$

where $k_2{}^2 = (2 + \bar{\lambda})\Omega^2 - (t_L{}^2 + \varpi)$.

2.4.5.1 Thermoelasticity for a cylindrical panel

By taking $\Gamma = 0$, the motion corresponding to the rotational mode decouples from the rest of the motion and the various results reduce to the thermoelasticity

$$\phi(r) = \sum_{i=1}^{3}(A_iJ_\delta(\alpha_ir) + B_iY_\delta(\alpha_ir))\sin(m\pi z)\sin(n\pi\theta/\alpha)e^{j\omega t}$$

$$W(r) = \sum_{i=1}^{3}d_i(A_iJ_\delta(\alpha_ir) + B_iY_\delta(\alpha_ir))\sin(m\pi z)\sin(n\pi\theta/\alpha)e^{j\omega t}$$

$$T(r) = \sum_{i=1}^{3} e_i(A_i J_\delta(\alpha_i r) + B_i Y_\delta(\alpha_i r))\sin(m\pi z)\sin(n\pi\theta/\alpha)e^{j\omega t}$$

$$\psi(r) = (A_4 J_\delta(k_1 r) + B_4 Y_\delta(k_1 r))\sin(m\pi z)\cos(n\pi\theta/\alpha)e^{j\omega t} \tag{2.99}$$

with

$$k_1{}^2 = (2 + \bar{\lambda})\Omega^2 - t_L{}^2. \tag{2.100}$$

Equations (2.99) and (2.100) constitute the solution for the homogeneous isotropic cylindrical panel with traction free boundary conditions.

2.4.6 Analysis of a transversely isotropic magneto-thermoelastic cylindrical panel

The stress equations of motion in the absence of body force specified in equation (2.3), as well as the accompanying heat conduction equation, are used to explain the analytical formulation of a transversely isotropic magneto-thermoelastic cylindrical panel.

For a transversely isotropic medium, the steady-state Fourier's heat conduction equation is as follows:

$$\begin{aligned} &K_1(T_{,rr} + r^{-1}T_{,r} + r^{-2}T_{,\theta\theta}) + K_3 T_{,zz} \\ &= \rho c_v T_{,t} + T_0\frac{\partial}{\partial t}\left(\beta_1(e_{rr} + e_{\theta\theta}) + \beta_3 e_{zz} - p_3 H_{,z}\right). \end{aligned} \tag{2.101}$$

The simplified Maxwell charge equilibrium equation for magnetic field is given by Buchanan (2003)

$$\frac{1}{r}\frac{\partial}{\partial r}(rB_r) + \frac{1}{r}\frac{\partial}{\partial \theta}(B_\theta) + \frac{\partial}{\partial z}(B_z) = 0. \tag{2.102}$$

The stress–strain relations for transversely isotropic material by generalized Hooke's law are given by

$$\sigma_{rr} = c_{11}e_{rr} + c_{12}e_{\theta\theta} + c_{13}e_{zz} - \beta_1 T + d_{31}H_{,z}$$

$$\sigma_{\theta\theta} = c_{12}e_{rr} + c_{11}e_{\theta\theta} + c_{13}e_{zz} - \beta_1 T + d_{31}H_{,z}$$

$$\sigma_{zz} = c_{13}e_{rr} + c_{13}e_{\theta\theta} + c_{33}e_{zz} - \beta_3 T + d_{33}H_{,z}$$

$$\sigma_{r\theta} = c_{66}e_{r\theta}\ \sigma_{\theta z} = c_{44}e_{\theta z} + r^{-1}d_{15}H_{,\theta}\ \sigma_{rz} = c_{44}e_{rz} + d_{15}H_{,r} \tag{2.103}$$

$$B_r = d_{15}e_{rz} - \mu_{11}H_{,r}\ \beta_1 = (c_{11} + c_{12})\gamma_1 + c_{13}\gamma_3\quad \beta_3 = 2c_{13}\gamma_1 + c_{13}\gamma_3$$

$$B_\theta = d_{15}e_{z\theta} - r^{-1}\mu_{11}H_{,\theta}\ B_z = d_{31}(e_{rr} + e_{\theta\theta}) + d_{33}e_{zz} - \mu_{33}H_{,z} + p_3 T. \tag{2.104}$$

Substitution of the strain-mechanical displacement relation given in equation (2.2) and the equation (2.103) into equations (2.101) and (2.102) gives the following four equations of motion

$$c_{11}(u_{,rr} + r^{-1}u_{,r} - r^{-2}u) - r^{-2}(c_{11} + c_{66})v_{,\theta} + r^{-2}c_{66}u_{,\theta\theta}$$
$$+ c_{44}u_{,zz} + (c_{44} + c_{13})w_{,rz} + r^{-1}(c_{66} + c_{12})v_{,r\theta} - \beta_1 T_{,r} + (d_{31} + d_{15})H_{,rz} = \rho u_{,tt}$$

$$r^{-1}(c_{12} + c_{66})u_{,r\theta} + r^{-2}(c_{66} + c_{11})u_{,\theta} + c_{66}(v_{,rr} + r^{-1}v_{,r} - r^{-2}v)$$
$$+ r^{-2}c_{11}v_{,\theta\theta} + c_{44}v_{,zz} + r^{-1}(c_{44} + c_{13})w_{,\theta z} - \beta_1 T_{,\theta} + r^{-1}(d_{31} + d_{15})H_{,\theta z} = \rho v_{,tt}$$

$$c_{44}(w_{,rr} + r^{-1}w_{,r} + r^{-2}w_{,\theta\theta}) + r^{-1}(c_{44} + c_{13})(u_{,z} + v_{,\theta z})$$
$$+ (c_{44} + c_{13})u_{,rz} + c_{33}w_{,zz} - \beta_3 T_{,z} + d_{33}H_{,zz} + (H_{,rr} + r^{-1}H_{,r} + r^{-2}H_{,\theta\theta}) = \rho w_{,tt}$$

$$K_1(T_{,rr} + r^{-1}T_{,r} + r^{-2}T_{,\theta\theta}) + K_3 T_{,zz} - \rho c_v T_{,t}$$
$$- T_0\frac{\partial}{\partial t}(\beta_1(u_{,r} + r^{-1}u + r^{-1}v_{,\theta}))(u_{,z} + v_{,\theta z}) \tag{2.105}$$
$$+ \beta_3 w_{,z} + p_3 T = 0.$$

The comma in the subscripts denotes the partial differentiation with respect to the respective variables. Assume the following decoupling factors from Sharma *et al* (2004)

$$u = r^{-1}\psi_{,\theta} - G_{,r}\ v = -r^{-1}G_{,\theta} - \psi_{,r}\ \ w = W_{,z}\ \ H = \chi_{,z}\ . \tag{2.106}$$

Substituting equation (2.106) into equation (2.105) yields the following second order partial differential equation with constant coefficients

$$\left(c_{11}\nabla_1^2 + c_{44}\frac{\partial^2}{\partial z^2} - \rho\frac{\partial^2}{\partial t^2}\right)G - (c_{44} + c_{13})\frac{\partial^2 W}{\partial z^2} - (d_{31} + d_{15})\frac{\partial^2 \chi}{\partial z^2} + \beta_1 T = 0$$

$$\left(c_{44}\nabla_1^2 + c_{33}\frac{\partial^2}{\partial z^2} - \rho\frac{\partial^2}{\partial t^2}\right)W - (c_{13} + c_{14})\nabla_1^2 G + \left(d_{15}\nabla_1^2 + d_{33}\frac{\partial^2}{\partial z^2}\right)\chi - \beta_3 T = 0$$

$$(d_{31} + d_{15})\nabla_1^2 G - \left(d_{15}\nabla_1^2 + d_{33}\frac{\partial^2}{\partial z^2}\right)W + \left(\mu_{11}\nabla_1^2 + \mu_{33}\frac{\partial^2}{\partial z^2} - \rho\frac{\partial^2}{\partial t^2}\right)\chi - p_3 T = 0$$

$$\left(K_1\nabla_1^2 + K_3\frac{\partial^2}{\partial z^2} - \rho c_v\frac{\partial}{\partial t}\right)T + T_0\frac{\partial}{\partial t}\left(\beta_1\nabla_1^2 G - \beta_3\frac{\partial^2}{\partial z^2}W + p_3\frac{\partial^2}{\partial z^2}\chi\right) = 0 \tag{2.107}$$

$$\left(c_{66}\nabla_1^2 + c_{44}\frac{\partial^2}{\partial z^2} - \rho\frac{\partial^2}{\partial t^2}\right)\psi = 0. \tag{2.108}$$

Equations (2.107) are coupled partial differential equations of the three displacement components. To uncouple equations (2.107), we can write three displacement components, temperature and magnetic potential which satisfy the stress-free boundary conditions followed by Sharma *et al* (2004)

$$G(r, \theta, z, t) = \bar{G}(r)\sin(m\pi z)\cos(n\pi\theta/\alpha)e^{j\omega t}$$

$$W(r,\,\theta,\,z,\,t) = \bar{W}(r)\sin(m\pi z)\sin(n\pi\theta/\alpha)e^{j\omega t}$$

$$\chi(r,\,\theta,\,z,\,t) = \bar{\chi}(r)\sin(m\pi z)\cos(n\pi\theta/\alpha)e^{j\omega t}$$

$$T(r,\,\theta,\,z,\,t) = \bar{T}(r)\sin(m\pi z)\cos(n\pi\theta/\alpha)e^{j\omega t} \tag{2.109}$$

and

$$\psi(r,\,\theta,\,z,\,t) = \bar{\psi}(r)\sin(m\pi z)\cos(n\pi\theta/\alpha)e^{j\omega t}. \tag{2.110}$$

It is obvious that $n = 0$ means the axisymmetric vibration. In such a case, the circumferential displacement v vanishes and the radial displacement u, axial displacement w and magnetic potential χ are independent of θ coordinate. Rotating the symmetry axes by $\pi/2$, another set of free vibration modes can be obtained, which corresponds to an interchange of $\cos(n\pi\theta/\alpha)$ by $\sin(n\pi\theta/\alpha)$ in equations (2.109) and (2.110). However, in such cases, $n = 0$ represents the so-called torsional vibration. By introducing the following dimensionless quantities such as

$$r' = \frac{r}{R},\ z' = \frac{z}{L},\ \bar{T} = \frac{T}{T_0},\ \delta = \frac{n\pi}{\alpha},\ t_L = \frac{m\pi R}{L},\ \Omega^2 = \frac{\rho^2\omega^2R^2}{c_{44}},\ \chi^* = \frac{\omega}{\omega^*},\ \omega^* = \frac{c_\nu c_{11}}{K_1}$$

$$R = \frac{a+b}{2},\ c_1 = \frac{c_{33}}{c_{11}},\ c_2 = \frac{c_{44}}{c_{11}},\ c_3 = \frac{c_{13}+c_{44}}{c_{11}},\ c_4 = \frac{c_{66}}{c_{11}}$$

$$d_1 = \frac{d_{15}+d_{31}}{d_{33}},\ d_2 = \frac{d_{15}}{d_{33}},\ \bar{\beta} = \frac{\beta_3}{\beta_1},\ \chi' = \frac{\chi}{\phi_0},\ p = \frac{p_3c_{11}}{\beta_1d_{33}} \tag{2.111}$$

$$\bar{K} = \frac{K_3}{K_1},\ \eta_3 = \frac{\mu_{11}c_{11}}{d_{33}^2},\ \mu_p = \frac{d_{33}\phi_0}{c_{11}},\ \beta^* = \frac{\beta_1T_0R^2}{c_{11}}$$

we can rewrite equations (2.109) and (2.110) in a convenient form of equations as

$$(\nabla_2^2 + g_1)\bar{G} + g_2\bar{W} - \mu_pg_5\bar{\chi} + \beta^*\bar{T} = 0$$

$$c_2(\nabla_2^2 + g_3)\bar{W} - c_3\nabla_2^2\bar{G} + d_2\mu_p(\nabla_2^2 - g_6)\bar{\chi} - \beta^*\bar{\beta}\bar{T} = 0$$

$$d_1\nabla_2^2\bar{G} - d_2(\nabla_2^2 - g_6)\bar{W} + \eta_3\mu_p(\nabla_2^2 + g_7)\bar{\chi} - \beta^*p\bar{T} = 0$$

$$\frac{\in_1c_2\Omega^2}{jx^*}(\nabla_2^2\bar{G} + \bar{\beta}t_L^2\bar{W} - p\mu_pt_L^2\bar{\chi}) + \beta^*(\nabla_2^2 + g_4)\bar{T} = 0 \tag{2.112}$$

and

$$(\nabla_2^2 + k_3{}^2)\bar{\psi} = 0 \tag{2.113}$$

where $g_1 = c_2(\Omega^2 - t_L^2)g_2 = c_3t_L{}^2$

$$g_3 = \Omega^2 - \frac{c_1}{c_2}t_L{}^2\, g_4 = -\frac{jc_2}{\chi^*}\left(\Omega^2 - \frac{j\bar{K}t_L{}^2}{c_2}\chi^*\right)$$

$$g_5 = d_1 t_L^2 \; g_6 = \frac{t_L^2}{d_2} \; g_7 = -\bar{\epsilon}\, t_L^2.$$

For a non-trivial solution of the algebraic equation (2.112), we obtain the frequency equations in the determinant form as

$$\begin{vmatrix} (\nabla_2^2 + g_1) & g_2 & -\mu_p g_5 & \beta^* \\ -c_3\nabla_2^2 & c_2(\nabla_2^2 + g_3) & d_2\mu_p(\nabla_2^2 - g_6) & -\beta^*\bar{\beta} \\ d_1\nabla_2^2 & -d_2(\nabla_2^2 - g_6) & \eta_3\mu_p(\nabla_2^2 + g_7) & -\beta^* p \\ \frac{c_{11}c_2\Omega^2}{j\chi^*}\nabla_2^2 & \bar{\beta}t_L^2 & -p\mu_p t_L^2 & \beta^*(\nabla_2^2 + g_4) \end{vmatrix} (\bar{G}\ \bar{W}\ \bar{\chi}\ \bar{T}) = 0. \quad (2.114)$$

Equation (2.114), on simplification reduces to the following differential equation.

$$(\nabla_2^8 + H_1\nabla_2^6 + H_2\nabla_2^4 + H_3\nabla_2^2 + H_4)(\bar{G}\ \bar{W}\ \bar{\chi}\ \bar{T}) = 0 \qquad (2.115)$$

where

$$H_1 = \frac{1}{\Delta}(c_2\eta_3 g_1 + c_2\eta_3 g_4 - c_2\eta_3 g_3 + d_2^2 g_4 - 2d_2^2 g_6 + c_2\eta_3 g_1 g_7 + c_2\eta_3 g_1 g_3 + d_2^2 g_4 g_1 - d_2^2 g_1 g_6 - \frac{c_2^2\eta_3 \in_1 \Omega^2}{j\chi^*})$$

$$\begin{aligned} H_2 = \tfrac{1}{\Delta}(&c_2\eta_3 g_3 g_7 - c_2\eta_3 p^2 t_L^2 + c_2\eta_3 g_3 g_4 - 2d_2^2 g_4 g_6 - d_2\bar{\beta}pt_L^2 + d_2^2 g_4 g_6 + d_2^2 g_6^2 \\ &d_2\bar{\beta}p + \bar{\beta}^2\eta_3 t_L^2 - c_2 g_1 p^2 t_L^2 + c_2\eta_3 g_1 g_3 g_7 + c_2\eta_3 g_1 g_3 g_4 - d_2^2 g_1 g_4 g_6 \\ &- d_2 p g_1\bar{\beta}t_L^2 - d_2^2 g_1 g_6 - d_2 c_3 g_4 g_5 + d_2 c_3 g_4 g_6 + d_2 c_3 g_4 g_5 g_6 + c_3 g_5\bar{\beta}t_L^2 \\ &+ c_2 d_1 g_4 g_5^2 + \frac{c_2^2 p \in_1 \Omega^2 g_5}{j\chi^*} + c_2 d_1 g_3 g_5 + c_2 d_2 \in_1 \Omega^2\bar{\beta}g_5 + c_2 d_3 p t_L^2\eta_3 \\ &- d_1 c_2 p t_L^2 - \frac{c_2^2\eta_3 \in_1 \Omega^2 g_1}{j\chi^*} - c_2^2 \in_1 \Omega^2\eta_3 g_3/j\chi^* - d_1 d_2\bar{\beta}t_L^2 + d_2^2 c_2 \in_1 \Omega^2 g_6/j\chi^*)/\Delta \end{aligned}$$

$$\begin{aligned} H_3 = \tfrac{1}{\Delta}(&c_2\eta_3 g_4 g_7 + c_2\eta_3 g_3 g_4 g_7 - c_2 p^2 t_L^2 - d_2^2 g_4 g_6^2 + 2d_2\bar{\beta}pt_L^2 g_6 + \bar{\beta}^2\eta_3 t_L^2 g_7 + d_2\eta_3 g_1 g_4 g_7 \\ &- d_2\bar{\beta}p + \bar{\beta}^2\eta_3 t_L^2 + c_2\eta_3 g_1 g_3 g_4 g_7 - c_2 g_1 p^2 t_L^2 - d_2^2 g_1 g_4 g_6 + d_2^2 g_1 g_6^2 - d_2 p g_1\bar{\beta} - d_2 c_3 g_4 g_5 \\ &+ \bar{\beta}g_1\eta_3 t_L^2 - d_2 g_5 c_3 + c_2^2 p \in_1 \Omega^2 g_5 g_3/j\chi^* + d_1 g_5^2\bar{\beta}t_L^2 - d_2 c_2 g_5 g_6\bar{\beta} \in_1 \Omega^2/j\chi^* \\ &- c_3 d_2 t_L^2 g_6 - c_3\bar{\beta}\eta_3 t_L^2 g_7 + c_2 d_1 t_L^2 + c_2 d_2 \in_1 \Omega^2\bar{\beta}g_5 \\ &- c_2^2\eta_3 \in_1 \Omega^2 g_3 g_7/j\chi^* + d_2^2 \in_1 \Omega^2 g_6/j\chi^* + d_1 d_2\bar{\beta}\, t_L^2 g_6) \end{aligned}$$

$$H_4 = \frac{1}{\Delta}(d_2^2 g_1 g_4 g_6^2 + 2d_2\bar{\beta}pt_L^2 g_1 g_6 + \bar{\beta}^2\eta_3 t_L^2 g_1 g_7 - d_2^2 \in_1 c_2\Omega^2 g_6^2/j\chi^*)$$

where

$$\Delta = \left(c_2\eta_3 + d_2^2 + c_2\eta_3 g_1 + d_2^2 g_1\right). \qquad (2.116)$$

The solution of equation (2.115) is

$$\bar{G}(r) = \sum_{i=1}^{4} [A_i J_\delta(\alpha_i r) + B_i Y_\delta(\alpha_i r)]$$

$$\bar{W}(r) = \sum_{i=1}^{4} L_i\, [A_i J_\delta(\alpha_i r) + B_i Y_\delta(\alpha_i r)]$$

$$\bar{\chi}(r) = \sum_{i=1}^{4} M_i\, [A_i J_\delta(\alpha_i r) + B_i Y_\delta(\alpha_i r)]$$

$$\bar{T}(r) = \sum_{i=1}^{4} N_i\, [A_i J_\delta(\alpha_i r) + B_i Y_\delta(\alpha_i r)]. \tag{2.117}$$

Here, $(\alpha_i r)^2 (i = 1, 2, 3, 4)$ are the non-zero roots of the algebraic equation

$$(\alpha_i r)^8 + H_1(\alpha_i r)^6 - H_2(\alpha_i r)^4 + H_3(\alpha_i r)^2 - H_4 = 0. \tag{2.118}$$

The arbitrary constants L_i, M_i, N_i $(i = 1, 2, 3, 4)$ are obtained from

$$L_i = \frac{1}{\Delta_1}\begin{vmatrix} \alpha_i^2 + g_1 & g_5 & -1 \\ c_3\alpha_i^2 & d_2(\alpha_i^2 - g_6) & -\bar{\beta} \\ -d_1\alpha_i^2 & \eta_1(\alpha_i^2 + g_7) & -p \end{vmatrix}$$

$$M_i = \frac{1}{\Delta_1 \mu_p}\begin{vmatrix} g_2 & (\alpha_i^2 + g_1) & -1 \\ c_2(\alpha_i^2 + g_3) & c_3\alpha_i^2 & -\bar{\beta} \\ -d_2(\alpha_i^2 - g_6) & -d_1\alpha_i^2 & -p \end{vmatrix}$$

$$N_i = \frac{1}{\Delta_1 \beta^*}\begin{vmatrix} g_2 & g_5 & (\alpha_i^2 + g_1) \\ c_2(\alpha_i^2 + g_3) & d_2(\alpha_i^2 - g_6) & c_3\alpha_i^2 \\ -d_2(\alpha_i^2 - g_6) & \eta_1(\alpha_i^2 + g_7) & -d_1\alpha_i^2 \end{vmatrix}$$

where

$$\Delta_1 = \begin{vmatrix} g_2 & g_5\mu_p & -1 \\ c_2(\alpha_i^2 + g_3) & d_2\mu_p(\alpha_i^2 - g_6) & \bar{\beta} \\ -d_2(\alpha_i^2 - g_6) & \eta_1\mu_p(\alpha_i^2 + g_7) & -p \end{vmatrix}. \tag{2.119}$$

Equation (2.113) is a Bessel equation and its possible solutions are obtained by the relation in equation (2.19) as

$$\bar{\psi}(r) = A_5 J_\delta(k_3 r) + B_5 Y_\delta(k_3 r) \tag{2.120}$$

where ${k_3}^2 = \frac{c_{44}}{c_{66}}(\Omega^2 - {t_L}^2)$.

2.4.6.1 Magneto elasticity for a transversely isotropic cylindrical panel

By taking $\beta * = p = 0$ the motion corresponding to the thermal wave decouples from the rest of the motion and the various results reduce to those of the magnetic field

$$\begin{vmatrix} (\nabla_2^2 + g_1) & g_2 & -\mu_p g_5 \\ -c_3 \nabla_2^2 & c_2(\nabla_2^2 + g_3) & d_2 \mu_p(\nabla_2^2 - g_6) \\ d_1 \nabla_2^2 & -d_2(\nabla_2^2 - g_6) & \eta_3 \mu_p(\nabla_2^2 + g_7) \end{vmatrix} (\bar{G}\ \bar{W}\ \bar{\chi}) = 0. \tag{2.121}$$

The equation (2.121) will reduce to

$$(\nabla_2^6 + H_5\ \nabla_2^4 + H_6\ \nabla_2^2 + H_7)(\bar{G}\ \bar{W}\ \bar{\chi}) = 0 \tag{2.122}$$

where $H_5,\ H_6,\ H_7$ are defined as

$$H_5 = \left(\begin{array}{l} c_2\eta_3\mu_p g_7 + c_2\eta_3\mu_p g_3 + 2d_2\mu_p g_6 + c_2\eta_3\mu_p g_1 - d_2\mu_p g_1 \\ -c_2\eta_3\mu_p c_3 - d_1 d_2 \mu_p - \eta_3 c_3 g_5 \mu_p + d_1 c_2 g_5 \mu_p \end{array} \right) \Big/ c_2\eta_3\mu_p - d_2\mu_p$$

$$H_6 = \left(\begin{array}{l} c_2\eta_3\mu_p g_3 g_7 - c_2\mu_p g_6^2 + c_2\eta_3\mu_p g_1 g_7 + c_2\eta_3\mu_p g_1 g_3 - g_2\eta_3\mu_p c_3 \\ + d_1 d_2 g_6 + 2d_2\mu_p g_1 g_6 - d_2\ \mu_p c_3 g_1 g_6 + d_1\ \mu_p c_2 g_5 g_6 \end{array} \right) \Big/ c_2\eta_1\mu_p - d_2\mu_p$$

$$H_7 = \left(c_2\eta_3\mu_p g_1 g_3 g_7 - d_2\mu_p g_1 g_6^2\right)/c_2\eta_3\mu_p - d_2\mu_p. \tag{2.123}$$

The solution of equation (2.122) is obtained as

$$\bar{G}(r) = \sum_{i=1}^{3} [A_i J_\delta(\alpha_i r) + B_i\, Y_\delta(\alpha_i r)]$$

$$\bar{W}(r) = \sum_{i=1}^{3} q_i\ [A_i J_\delta(\alpha_i r) + B_i\, Y_\delta(\alpha_i r)]$$

$$\bar{\chi}(r) = \sum_{i=1}^{3} r_i [A_i J_\delta(\alpha_i r) + B_i\, Y_\delta(\alpha_i r)]\ \bar{\psi}(r) = A_4 J_\delta(k_3 r) + B_4\, Y_\delta(k_3 r) \tag{2.124}$$

where $q_i,\ r_i (i = 1, 2, 3)$ are given by

$$q_i = \left(\frac{\mu_p g_5}{g_2}\right) \left(\frac{\begin{array}{l} \alpha_i^4\left(c_2 - \mu_p c_2 g_5 - 2\mu_p d_2 g_2\right) + \alpha_i^2\left(g_1 + g_3 + c_3 g_2 - \mu_p c_2 g_5^2 - \mu_p c_2 g_5 g_1\right) \\ + \mu_p d_2 g_2 g_6 - \mu_p c_2 g_5^2 g_1 + \mu_p d_2 g_2 g_6 g_1 \end{array}}{\mu_p c_2 g_5(\alpha_i^2 + g_5) + \mu_p d_2 g_2(\alpha_i^2 - g_6)} \right)$$

$$r_i = \left(\frac{\alpha_i^4 c_2 + \alpha_i^2 (g_1 + g_3 + c_3 g_2) + g_1 g_3}{\mu_p c_2 g_5(\alpha_i^2 + g_5) + \mu_p d_2 g_2 (\alpha_i^2 - g_6)} \right). \tag{2.125}$$

Equations (2.121)–(2.125) are the solutions for frequency analysis of a magneto-elastic cylindrical panel with traction free boundary conditions.

2.4.6.2 Thermoelasticity for a transversely isotropic cylindrical panel

By taking $\mu_p = p = 0$ the motion corresponding to the magnetic wave decouples from the rest of the motion and the various results reduce to those of a thermoelastic field and equation (2.114) is simplified to

$$\begin{vmatrix} (\nabla_2^2 + g_1) & g_2 & \beta_* \\ -c_3\nabla_2^2 & c_2(\nabla_2^2 + g_3) & -\beta_*\bar{\beta} \\ \dfrac{\epsilon_1 c_2 \Omega^2 \nabla_2^2}{j\chi^*} & \beta t_L^2 & \beta_*(\nabla_2^2 + g_4) \end{vmatrix} (\bar{G}\ \bar{W}\ \bar{T}) = 0. \tag{2.126}$$

Equation (2.126) is reduced to

$$(\nabla_2^6 + H_8\,\nabla_2^4 + H_9\,\nabla_2^2 + H_{10})\,(\bar{G}\ \overline{W}\ \bar{T}) = 0 \tag{2.127}$$

where H_8, H_9, H_{10} are given by

$$H_8 = g_1 + g_3 + g_4 + \frac{c_3}{c_2} g_5 - \frac{c_2\,\epsilon_1\,\Omega^2}{j\chi^*}$$

$$H_9 = g_1 g_3 + g_1 g_4 + g_3 g_4 + \frac{c_3}{c_2} g_2 g_4 + \frac{\epsilon_1 \Omega^2 \bar{\beta}}{j\chi^*}\left(g_2 - \frac{c_2 g_3}{\bar{\beta}} + (c_3 + \bar{\beta}) t_L^2\right)$$

$$H_{10} = g_1 g_3 g_4 + \frac{\epsilon_1 \Omega^2 \bar{\beta} t_L^2 g_1}{j\chi^*}. \tag{2.128}$$

The solutions of equation (2.127) are obtained as follows

$$\bar{G}(r) = \sum_{i=1}^{3} [A_i J_\delta(\alpha_i r) + B_i\, Y_\delta(\alpha_i r)]$$

$$\bar{W}(r) = \sum_{i=1}^{3} k_i [A_i J_\delta(\alpha_i r) + B_i\, Y_\delta(\alpha_i r)]$$

$$\bar{T}(r) = \sum_{i=1}^{3} l_i [A_i J_\delta(\alpha_i r) + B_i\, Y_\delta(\alpha_i r)]\ \bar{\psi}(r) = A_4 J_\delta(k_3 r) + B_4\, Y_\delta(k_3 r) \tag{2.129}$$

where k_i, $l_i (i = 1, 2, 3)$ are

$$\begin{aligned} k_i &= \left(\frac{\alpha_i^2 (c_3 + \bar{\beta}) + \bar{\beta} g_1}{c_2 \alpha_i^2 + c_2 g_3 - \bar{\beta} g_2}\right) \\ l_i &= \left(\frac{c_{11}}{\beta_1 T_0 R^2}\right)\left(\frac{\alpha_i^4 c_2 + \alpha_i^2 (c_3 g_2 + c_2 (g_1 + g_3)) + c_2 g_1 g_3}{c_2 \alpha_i^2 + c_2 g_3 - \bar{\beta} g_2}\right). \end{aligned} \tag{2.130}$$

Equations (2.126)–(2.130) are the solutions for the frequency waves in a thermoelastic cylindrical panel with stress-free, thermally insulated boundary conditions.

2.4.6.3 Elastokinetics for a transversely isotropic cylindrical panel

In the present analysis if we take the coupling parameter for magnetic and thermal field $\mu_p = \beta* = p = 0$ equal to zero, then equations (2.121)–(2.130) reduce to the classical case in elastokinetics

$$\begin{vmatrix} (\nabla_2^2 + g_3) & -g_2 \\ (1+\bar{\lambda})\nabla_2^2 & (\nabla_2^2 - g_1) \end{vmatrix} (\bar{G}, \bar{W}) = 0. \tag{2.131}$$

The simplified form of equation (2.131) is obtained as

$$(\nabla_2^4 + H_{12}\,\nabla_2^2 + H_{13})(\bar{G}, \bar{W}) = 0. \tag{2.132}$$

The constants in the coefficient of the above equation are defined as follows

$$H_{12} = g_1 + (c_3/c_2)g_2 + g_3,\ \ H_{13} = g_1 g_3. \tag{2.133}$$

The solutions of equation (2.132) are as follows

$$\bar{G}(r) = \sum_{i=1}^{2} [A_i J_\delta(\alpha_i r) + B_i Y_\delta(\alpha_i r)]$$

$$\bar{W}(r) = \sum_{i=1}^{2} v_i\, [A_i J_\delta(\alpha_i r) + B_i Y_\delta(\alpha_i r)] \tag{2.134}$$

where $v_i = \frac{(c_3/c_2)(\alpha_i r)^2}{(\alpha_i r)^2 + g_3}, (i = 1, 2)$.

Equations (2.131)–(2.134) constitute the solution for the homogenous transversely isotropic cylindrical panel with traction free boundary conditions.

2.4.7 Analysis of generalized magneto-thermoelastic waves in a transversely isotropic cylindrical panel

The three-dimensional stress equations of motion in the absence of body force in a cylindrical coordinate (r, θ, z) are taken from equation (2.3) in chapter 2. The heat conduction equation for a homogeneous transversely isotropic generalized thermoelastic cylindrical panel is investigated by introducing the thermal relaxation times τ_0 and τ_1 in the heat conduction equation (2.101) and considering the same Maxwell's equation defined in equation (2.102). The heat conduction equation in the context of a generalized theory of thermoelasticity is given by

$$K_1(T_{,rr} + r^{-1}T_{,r} + r^{-2}T_{,\theta\theta}) + K_3 T_{,zz} = \rho c_v[T + \tau_0 T_{,tt}] + T_0\left(\frac{\partial}{\partial t} + \delta_{1k}\tau_0 \frac{\partial^2}{\partial t^2}\right) \tag{2.135}$$

$$(\beta_1(e_{rr} + e_{\theta\theta}) + \beta_3 e_{zz} - p_3 H_{,z}).$$

The modified stress–strain relations for transversely isotropic material by generalized Hooke's law are given by

$$\sigma_{rr} = c_{11}e_{rr} + c_{12}e_{\theta\theta} + c_{13}e_{zz} - \beta_1(T + \delta_{2k}\tau_1 T_{,t}) + d_{31}H_{,z}$$

$$\sigma_{\theta\theta} = c_{12}e_{rr} + c_{11}e_{\theta\theta} + c_{13}e_{zz} - \beta_1(T + \delta_{2k}\tau_1 T_{,t}) + d_{31}H_{,z}$$

$$\sigma_{zz} = c_{13}e_{rr} + c_{13}e_{\theta\theta} + c_{33}e_{zz} - \beta_3(T + \delta_{2k}\tau_1 T_{,t}) + d_{33}H_{,z}. \tag{2.136}$$

The remaining relations which describe the strain–displacement have the same definitions as in equation (2.2).

2.4.7.1 Lord–Shulman (L–S) theory for a magneto-thermoelastic cylindrical panel

The three-dimensional rate-dependent temperature with one relaxation time called the Lord–Shulman theory of thermoelasticity is obtained by replacing $k = 1$ in equations (2.135) and (2.136).

The heat conduction equation is simplified as

$$K_1(T_{,rr} + r^{-1}T_{,r} + r^{-2}T_{,\theta\theta}) + K_3 T_{,zz} = \rho c_v(T + \tau_0 T_{,tt}) + T_0\left(\frac{\partial}{\partial t} + \tau_0\frac{\partial^2}{\partial t^2}\right) \tag{2.137}$$
$$(\beta_1(e_{rr} + e_{\theta\theta}) + \beta_3 e_{zz} - p_3 H_{,z}).$$

Then the stress–strain relations are

$$\sigma_{rr} = c_{11}e_{rr} + c_{12}e_{\theta\theta} + c_{13}e_{zz} - \beta_1 T + d_{31}H_{,z}$$

$$\sigma_{\theta\theta} = c_{12}e_{rr} + c_{11}e_{\theta\theta} + c_{13}e_{zz} - \beta_1 T + d_{31}H_{,z}$$
$$\sigma_{zz} = c_{13}e_{rr} + c_{13}e_{\theta\theta} + c_{33}e_{zz} - \beta_3 T + d_{33}H_{,z}. \tag{2.138}$$

Upon using these relations in equation (2.137) along with equations (2.103) and (2.104), we can get the following displacement equations of motion and the heat conduction equation

$$c_{11}(u_{,rr} + r^{-1}u_{,r} - r^{-2}u) - r^{-2}(c_{11} + c_{66})v_{,\theta} + r^{-2}c_{66}u_{,\theta\theta}$$
$$+ c_{44}u_{,zz} + (c_{44} + c_{13})w_{,rz} + r^{-1}(c_{66} + c_{12})v_{,r\theta} - \beta_1 T_{,r} + (d_{31} + d_{15})H_{,rz} = \rho u_{,tt}$$

$$r^{-1}(c_{12} + c_{66})u_{,r\theta} + r^{-2}(c_{66} + c_{11})u_{,\theta} + c_{66}(v_{,rr} + r^{-1}v_{,r} - r^{-2}v)$$
$$+ r^{-2}c_{11}v_{,\theta\theta} + c_{44}v_{,zz} + r^{-1}(c_{44} + c_{13})w_{,\theta z} - \beta_3 T_{,\theta} + r^{-1}(d_{31} + d_{15})H_{,\theta z} = \rho v_{,tt}$$

$$c_{44}(w_{,rr} + r^{-1}w_{,r} + r^{-2}w_{,\theta\theta}) + r^{-1}(c_{44} + c_{13})(u_{,z} + v_{,\theta z})$$
$$+ (c_{44} + c_{13})u_{,rz} + c_{33}w_{,zz} - \beta_3 T_{,z} + d_{33}H_{,zz} + (H_{,rr} + r^{-1}H_{,r} + r^{-2}H_{,\theta\theta}) = \rho w_{,tt}$$

$$K_1(T_{,rr} + r^{-1}T_{,r} + r^{-2}T_{,\theta\theta}) + K_3 T_{,zz} - \rho c_v(T_{,t} + \tau_0 T_{tt}) - T_0\left(\frac{\partial}{\partial t} + \tau_0\frac{\partial^2}{\partial t^2}\right) \tag{2.139}$$
$$(\beta_1(u_{,r} + r^{-1}u + r^{-1}v_{,\theta}))(u_{,z} + v_{,\theta z}) + \beta_3 w_{,z} + p_3 T = 0$$

where the symbols and notations have the same meaning as defined in earlier sections. Since the heat conduction equation in this theory is of a hyperbolic wave type, it automatically ensures the finite speeds of propagation for heat and elastic waves.

2.4.7.2 Green–Lindsay (G–L) theory for a magneto-thermoelastic cylindrical panel

The second generalization to the coupled thermoelasticity with two relaxation times called Green–Lindsay theory of thermoelasticity is obtained by setting $k = 2$ in the equations (2.135) and (2.136), the heat conduction equation is simplified as

$$K_1(T_{,rr} + r^{-1}T_{,r} + r^{-2}T_{,\theta\theta}) + K_3 T_{,zz} = \rho c_v(T + \tau_0 T_{,tt}) + T_0 \frac{\partial}{\partial t}\left(\beta_1(e_{rr} + e_{\theta\theta}) + \beta_3 e_{zz} - p_3 H_{,z}\right) \tag{2.140}$$

then the stress–strain relations are

$$\begin{aligned}\sigma_{rr} &= c_{11}e_{rr} + c_{12}e_{\theta\theta} + c_{13}e_{zz} - \beta_1(T + \tau_1 T_{,t}) + d_{31}H_{,z}\\ \sigma_{\theta\theta} &= c_{12}e_{rr} + c_{11}e_{\theta\theta} + c_{13}e_{zz} - \beta_1(T + \tau_1 T_{,t}) + d_{31}H_{,z}\\ \sigma_{zz} &= c_{13}e_{rr} + c_{13}e_{\theta\theta} + c_{33}e_{zz} - \beta_3(T + \tau_1 T_{,t}) + d_{33}H_{,z}.\end{aligned} \tag{2.141}$$

Substituting these relations in equation (2.140), the displacement equations are reduced to

$$c_{11}(u_{,rr} + r^{-1}u_{,r} - r^{-2}u) - r^{-2}(c_{11} + c_{66})v_{,\theta} + r^{-2}c_{66}u_{,\theta\theta} + c_{44}u_{,zz} + (c_{44} + c_{13})w_{,rz} + r^{-1}(c_{66} + c_{12})v_{,r\theta} - \beta_1(T_{,r} + \tau_1 T_{,rt}) + (d_{31} + d_{15})H_{,rz} = \rho u_{,tt}$$

$$r^{-1}(c_{12} + c_{66})u_{,r\theta} + r^{-2}(c_{66} + c_{11})u_{,\theta} + c_{66}(v_{,rr} + r^{-1}v_{,r} - r^{-2}v) + r^{-2}c_{11}v_{,\theta\theta} + c_{44}v_{,zz} + r^{-1}(c_{44} + c_{13})w_{,\theta z} - \beta_1(T_{,\theta} + \tau_1 T_{,\theta t}) + r^{-1}(d_{31} + d_{15})H_{,\theta z} = \rho v_{,tt}$$

$$c_{44}(w_{,rr} + r^{-1}w_{,r} + r^{-2}w_{,\theta\theta}) + r^{-1}(c_{44} + c_{13})(u_{,z} + v_{,\theta z}) + (c_{44} + c_{13})u_{,rz} + c_{33}w_{,zz} - \beta_3(T_{,z} + \tau_1 T_{,zt}) + d_{33}H_{,zz} + (H_{,rr} + r^{-1}H_{,r} + r^{-2}H_{,\theta\theta}) = \rho w_{,tt}$$

$$K_1(T_{,rr} + r^{-1}T_{,r} + r^{-2}T_{,\theta\theta}) + K_3 T_{,zz} - \rho c_v(T_{,t} + \tau_0 T_{tt}) - T_0\frac{\partial}{\partial t} (\beta_1(u_{,r} + r^{-1}u + r^{-1}v_{,\theta}))(u_{,z} + v_{,\theta z}) + \beta_3 w_{,z} + p_3 T = 0. \tag{2.142}$$

In the preceding part, the symbols and notations were specified. The generalized thermoelasticity theories are supposed to be more realistic than the conventional theory in dealing with practical problems involving very large heat fluxes and/or short time intervals, such as those found in laser units and energy channels, due to available experimental evidence in favour of the finiteness of heat propagation speeds. The following four equations of motion are obtained by substituting equations (2.136) and (2.2) into equation (2.135).

$$c_{11}(u_{,rr} + r^{-1}u_{,r} - r^{-2}u) - r^{-2}(c_{11} + c_{66})v_{,\theta} + r^{-2}c_{66}u_{,\theta\theta}$$
$$+ c_{44}u_{,zz} + (c_{44} + c_{13})w_{,rz} + r^{-1}(c_{66} + c_{12})v_{,r\theta} - \beta_1(T_{,r} + \tau_1\delta_{2k}T_{,rt})$$
$$+ (d_{31} + d_{15})H_{,rz} = \rho u_{,tt}$$

$$r^{-1}(c_{12} + c_{66})u_{,r\theta} + r^{-2}(c_{66} + c_{11})u_{,\theta} + c_{66}(v_{,rr} + r^{-1}v_{,r} - r^{-2}v)$$
$$+ r^{-2}c_{11}v_{,\theta\theta} + c_{44}v_{,zz} + r^{-1}(c_{44} + c_{13})w_{,\theta z} - \beta_1(T_{,\theta} + \tau_1\delta_{2k}T_{,\theta t})$$
$$+ r^{-1}(d_{31} + d_{15})H_{,\theta z} = \rho v_{,tt}$$

$$c_{44}(w_{,rr} + r^{-1}w_{,r} + r^{-2}w_{,\theta\theta}) + r^{-1}(c_{44} + c_{13})(u_{,z} + v_{,\theta z})$$
$$+ (c_{44} + c_{13})u_{,rz} + c_{33}w_{,zz} - \beta_3(T_{,z} + \tau_1\delta_{2k}T_{,zt}) + d_{33}H_{,zz}$$
$$+ (H_{,rr} + r^{-1}H_{,r} + r^{-2}H_{,\theta\theta}) = \rho w_{,tt}$$

$$K_1(T_{,rr} + r^{-1}T_{,r} + r^{-2}T_{,\theta\theta}) + K_3T_{,zz} - \rho c_v(T_{,t} + \tau_0 T_{tt})$$
$$- T_0\left(\frac{\partial}{\partial t} + \delta_{1k}\tau_0\frac{\partial^2}{\partial t^2}\right)(\beta_1(u_{,r} + r^{-1}u + r^{-1}v_{,\theta}))(u_{,z} + v_{,\theta z}) \tag{2.143}$$
$$+ \beta_3 w_{,z} + p_3 T = 0.$$

To uncouple equation (2.143), the mechanical displacement u, v, w along the radial, circumferential and axial directions is taken as defined in equation (2.106) and substituting this equation in equation (2.143) yields the following second order partial differential equation with constant coefficients

$$\left(c_{11}\nabla_1^2 + c_{44}\frac{\partial^2}{\partial z^2} - \rho\frac{\partial^2}{\partial t^2}\right)G - (c_{44} + c_{13})\frac{\partial^2 W}{\partial z^2} - (d_{31} + d_{15})\frac{\partial^2\chi}{\partial z^2} = \beta_1(T + \tau_1\delta_{2k}T_{,t})$$

$$\left(c_{44}\nabla_1^2 + c_{33}\frac{\partial^2}{\partial z^2} - \rho\frac{\partial^2}{\partial t^2}\right)W - (c_{13} + c_{14})\nabla_1^2 G + \left(d_{15}\nabla_1^2 + d_{33}\frac{\partial^2}{\partial z^2}\right)\chi = \beta_3(T + \tau_1\delta_{2k}T_{,t})$$

$$(d_{31} + d_{15})\nabla_1^2 G - \left(d_{15}\nabla_1^2 + d_{33}\frac{\partial^2}{\partial z^2}\right)W + \left(\mu_{11}\nabla_1^2 + \mu_{33}\frac{\partial^2}{\partial z^2} - \rho\frac{\partial^2}{\partial t^2}\right)\chi - p_3 T = 0$$

$$\left(K_1\nabla_1^2 + K_3\frac{\partial^2}{\partial z^2} - \rho c_v j\omega\eta_0\frac{\partial}{\partial t}\right)T + T_0(j\omega\eta_1)\left[\frac{\partial}{\partial t} + \tau_0\frac{\partial^2}{\partial t^2}\right]$$
$$\left(\beta_1\nabla_1^2 G - \beta_3\frac{\partial^2}{\partial z^2}W + p_3\frac{\partial^2}{\partial z^2}\chi\right) = 0 \tag{2.144}$$

$$\left(c_{66}\nabla_1^2 + c_{44}\frac{\partial^2}{\partial z^2} - \rho\frac{\partial^2}{\partial t^2}\right)\psi = 0 \tag{2.145}$$

where $\eta_0 = 1 + j\omega\tau_0$, $\eta_1 = 1 + j\omega\delta_{1k}\tau_0$, $\eta_2 = 1 + j\omega\delta_{2k}\tau_1$.

Equation (2.145) in ψ gives a purely transverse wave, this wave is polarized in planes perpendicular to the z-axis. We assume that the disturbance is time harmonic through the factor $e^{j\omega t}$. Equations (2.144) are coupled partial differential equations of the three displacement components. To uncouple equations (2.144), we can write three displacement components, temperature and magnetic potential which satisfies the stress-free boundary conditions followed by Sharma *et al* (2004)

$$G(r,\theta,z,t)=\bar{G}(r)\sin(m\pi z)\cos(n\pi\theta/\alpha)e^{j\omega t}$$

$$W(r,\theta,z,t)=\bar{W}(r)\sin(m\pi z)\sin(n\pi\theta/\alpha)e^{j\omega t}$$

$$\chi(r,\theta,z,t)=\bar{\chi}(r)\sin(m\pi z)\sin(n\pi\theta/\alpha)e^{j\omega t}$$

$$T(r,\theta,z,t)=\bar{T}(r)\sin(m\pi z)\sin(n\pi\theta/\alpha)e^{j\omega t} \tag{2.146}$$

$$\psi(r,\theta,z,t)=\bar{\psi}(r)\sin(m\pi z)\sin(n\pi\theta/\alpha)e^{j\omega t} \tag{2.147}$$

where n is the circumferential mode and m is the axial mode. By introducing the dimensionless quantities defined in equation (2.111), we can rewrite equations (2.146) and (2.147) in a convenient form of equations as follows

$$(\nabla_2^2+g_1)\bar{G}+g_2\bar{W}-\mu_p g_5\bar{\chi}+\beta^*\eta_2\bar{T}=0$$

$$c_2(\nabla_2^2+g_3)\bar{W}-c_3\nabla_2^2\bar{G}+d_2\mu_p(\nabla_2^2-g_6)\bar{\chi}-\beta^*\eta_2\bar{\beta}\,\bar{T}=0$$

$$d_1\nabla_2^2\bar{G}-d_2(\nabla_2^2-g_6)\bar{W}+\eta_3\mu_p(\nabla_2^2+g_7)\bar{\chi}-\beta^*\eta_2 p\,\bar{T}=0$$

$$\frac{\in_1\eta_1 c_2\Omega^2}{jx^*}(\nabla_2^2\bar{G}+\bar{\beta}\,t_L^2\bar{W}-p\mu_p t_L^2\bar{\chi})+\beta^*(\nabla_2^2+g_4)\bar{T}=0 \tag{2.148}$$

$$(\nabla_2^2+k_3{}^2)\bar{\psi}=0 \tag{2.149}$$

where

$$g_2=c_3t_L{}^2\; g_2=c_3t_L{}^2\; g_3=\Omega^2-\frac{c_1}{c_2}t_L{}^2$$

$$g_4=-\frac{j\eta_1c_2}{\chi^*}\left(\Omega^2-\frac{j\bar{K}t_L{}^2}{c_2\eta_1}\chi^*\right)g_5=d_1t_L^2\; g_6=\frac{t_L^2}{d_2}\; g_7=-\bar{\in}\,t_L^2.$$

A non-trivial solution of the algebraic equation (2.148) exists only when the determinant of the coefficients of equation (2.148) is equal to zero

$$\begin{vmatrix} (\nabla_2^2+g_1) & g_2 & -\mu_p g_5 & \beta^*\eta_2 \\ -c_3\nabla_2^2 & c_2(\nabla_2^2+g_3) & d_2\mu_p(\nabla_2^2-g_6) & -\beta^*\eta_2\bar{\beta} \\ d_1\nabla_2^2 & -d_2(\nabla_2^2-g_6) & \eta_3\mu_p(\nabla_2^2+g_7) & -\beta^*\eta_2 p \\ \frac{c_{11}\eta_1c_2\Omega^2}{jx^*}\nabla_2^2 & \bar{\beta}t_L^2 & -p\mu_p t_L^2 & \beta^*(\nabla_2^2+g_4) \end{vmatrix}(\bar{G}\;\bar{W}\;\bar{\chi}\;\bar{T})=0. \tag{2.150}$$

Equation (2.150), on simplification reduces to the following differential equation

$$(\nabla_2^8 + G_1\,\nabla_2^6 + G_2\,\nabla_2^4 + G_3\,\nabla_2^2 + G_4)(\bar{G}\ \bar{W}\ \chi\ \bar{T}) = 0 \tag{2.151}$$

where the constants in the coefficients are defined as

$$G_1 = (c_2\eta_3 g_1 + c_2\eta_3 g_4 - c_2\eta_3 g_3 + d_2^2 g_4 - 2d_2^2 g_6 + c_2\eta_3 g_1 g_7 + c_2\eta_3 g_1 g_3 + d_2^2 g_4 g_1 - d_2^2 g_1 g_6 - \frac{c_2^2\eta_3 \in_1 \eta_1 \Omega^2}{j\chi^*})$$

$$G_2 = \frac{1}{\Delta_2}(c_2\eta_3 g_3 g_7 - c_2\eta_3 p^2 t_L^2 + c_2\eta_3 g_3 g_4 - 2d_2^2 g_4 g_6 - d_2\overline{\beta}\eta_2 p t_L^2 + d_2^2 g_4 g_6 + d_2^2 g_6^2$$
$$- d_2\overline{\beta}\eta_2 p + \overline{\beta}^2\eta_2\eta_3 t_L^2 - c_2 g_1 p^2 t_L^2 + c_2\eta_3 g_1 g_3 g_7 + c_2\eta_3 g_1 g_3 g_4 - d_2^2 g_1 g_4 g_6$$
$$- d_2 p g_1\overline{\beta}\eta_2 t_L^2 - d_2^2 g_1 g_6 - d_2 c_3 g_4 g_5 + d_2 c_3 g_4 g_6 + d_2 c_3 g_4 g_5 g_6 + c_3 g_5\overline{\beta}\eta_2 t_L^2$$
$$+ c_2 d_1 g_4 g_5^2 + \frac{c_2^2 p\eta_1 \in_1 \Omega^2 g_5}{j\chi^*} + c_2 d_1 g_3 g_5 + c_2 d_2 \in_1 \Omega^2\overline{\beta}\eta_2 g_5 + c_2 d_3 p t_L^2\eta_3$$
$$- d_1 c_2 p t_L^2 - \frac{c_2^2\eta \in_1 \eta_1\Omega^2 g_1}{j\chi^*} - c_2^2 \in_1 \Omega^2\eta_1 g_3/j\chi^* - d_1 d_2\overline{\beta}\eta_2 t_L^2 + d_2^2 c_2 \in_1 \eta_1\Omega^2 g_6/j\chi^*)$$

$$G_3 = \frac{1}{\Delta_2}(c_2\eta_3 g_4 g_7 + c_2\eta_3 g_3 g_4 g_7 - c_2 p^2 t_L^2 - d_2^2 g_4 g_6^2 + 2d_2\overline{\beta}\eta_2 p t_L^2 g_6 + \overline{\beta}^2\eta_2\eta_3 t_L^2 g_7$$
$$+ d_2\eta_3 g_1 g_4 g_7 - d_2\overline{\beta}\eta_2 p + \overline{\beta}^2\eta_2\eta t_L^2 + c_2\eta_3 g_1 g_3 g_4 g_7 - c_2 g_1 p^2 t_L^2 - d_2^2 g_1 g_4 g_6$$
$$+ d_2^2 g_1 g_6^2 - d_2 p g_1\eta_2\overline{\beta} - d_2 c_3 g_4 g_5 + \overline{\beta} g_1\eta_3\eta_2 t_L^2 - d_2 g_5 c_3 + c_2^2 p \in_1 \eta_1\Omega^2 g_5 g_3/j\chi^*$$
$$+ d_1 g_5^2\overline{\beta}\eta_2 t_L^2 - d_2 c_2 g_5 g_6\overline{\beta} \in_1 \Omega^2/j\chi^* - c_3 d_2 t_L^2 g_6 - c_3\overline{\beta}\eta_3\eta_2 t_L^2 g_7 + c_2 d_1 t_L^2$$
$$+ c_2 d_2 \in_1 \eta_1\Omega^2\overline{\beta} g_5 - c_2^2\eta \in_1 \eta_1\Omega^2 g_3 g_7/j\chi^* + d_2^2 \in_1 \eta_1\Omega^2 g_6/j\chi^* + d_1 d_2\overline{\beta}\eta_2 t_L^2 g_6$$

$$G_4 = \frac{1}{\Delta_2}(d_2^2 g_1 g_4 g_6^2 + 2d_2\overline{\beta}\eta_2 p t_L^2 g_1 g_6 + \overline{\beta}^2\eta_3\eta_2 t_L^2 g_1 g_7 - d_2^2 \in_1 \eta_1 c_2\Omega^2 g_6^2/j\chi^*)$$

where

$$\Delta_2 = \left(c_2\eta_3 + d_2^2 + c_2\eta_3 g_1 + d_2^2 g_1\right). \tag{2.152}$$

The solution of equation (2.151) is obtained as defined in equation (2.117), where the arbitrary constants U_i, V_i, $W_i (i = 1, 2, 3, 4)$ are defined by

$$U_i = \frac{1}{\Delta_3}\begin{vmatrix} \alpha_i^2 + g_1 & g_5 & -1 \\ c_3\alpha_i^2 & d_2(\alpha_i^2 - g_6) & -\overline{\beta} \\ -d_1\alpha_i^2 & \eta_1(\alpha_i^2 + g_7) & -p \end{vmatrix}$$

$$V_i = \frac{1}{\Delta_3\mu_p}\begin{vmatrix} g_2 & (\alpha_i^2 + g_1) & -1 \\ c_2(\alpha_i^2 + g_3) & c_3\alpha_i^2 & -\overline{\beta} \\ -d_2(\alpha_i^2 - g_6) & -d_1\alpha_i^2 & -p \end{vmatrix}$$

$$W_i = \frac{1}{\Delta_3 \beta*} \begin{vmatrix} g_2 & g_5 & (\alpha_i^2 + g_1) \\ c_2(\alpha_i^2 + g_3) & d_2(\alpha_i^2 - g_6) & c_3\alpha_i^2 \\ -d_2(\alpha_i^2 - g_6) & \eta_1(\alpha_i^2 + g_7) & -d_1\alpha_i^2 \end{vmatrix}$$

$$\Delta_3 = \begin{vmatrix} g_2 & g_5\mu_p & -1 \\ c_2(\alpha_i^2 + g_3) & d_2\mu_p(\alpha_i^2 - g_6) & \bar{\beta} \\ -d_2(\alpha_i^2 - g_6) & \eta_1\mu_p(\alpha_i^2 + g_7) & -p \end{vmatrix}.$$

Equation (2.149) is a Bessel equation and its solution is obtained from equation (2.19)

$$\bar{\psi}(r) = A_5 J_\delta(k_3 r) + B_5 Y_\delta(k_3 r) \tag{2.153}$$

where $k_3{}^2 = \frac{c_{44}}{c_{66}}(\Omega^2 - t_L{}^2)$.

The waves are attenuated in space because the algebraic equation (2.151) contains all of the information regarding wave speed and angular frequency, and the roots are complex for all wave number values considered. We compose

$$c^{-1} = s^{-1} + i\omega^{-1}q \tag{2.154}$$

so that $\delta = R + iq$, where $R = \omega/s$ and the wave speed (s) and the attenuation coefficient (q) are real numbers.

References

Berliner J and Solecki R 1996 Wave propagation in a fluid-loaded transversely isotropic cylinder. Part I. Analytical formulation. Part II: numerical results *J. Acoust. Soc. Am.* **99** 1841–53

Buchanan G R 2003 Free vibration of an infinite magneto-electro-elastic cylinder *J. Sound Vib.* **268** 413–26

Ponnusamy P 2007 Wave propagation in a generalized thermo elastic solid cylinder of arbitrary cross-section *Int. J. Solids Struct.* **44** 5336–48

Roychoudhury S K and Mukhopadhyay S 2000 Effect of rotation and relaxation times on plane waves in generalized thermo visco elasticity *Int. J. Math. Math. Sci.* **23** 497–505

Selvamani R and Ponnusamy P 2012 Damping of generalized thermo elastic waves in a homogeneous isotropic plate *Mater. Phys. Mech.* **14** 64–73

Sharma J N 2001 Three dimensional vibration of a homogenous transversely isotropic thermo elastic cylindrical panel *J. Acoust. Soc. Am.* **110** 648–53

Sharma J N, Pal M and Chand D 2004 Three dimensional vibration analysis of a piezo-thermo elastic cylindrical panel *Int. J. Eng. Sci.* **42** 1655–73

IOP Publishing

Mathematical Modelling and Characterization of Cylindrical Structures

Farzad Ebrahimi and Rajendran Selvamani

Chapter 3

Wave propagation in a homogeneous isotropic thermoelastic cylindrical panel

3.1 Introduction

In this chapter, in the context of the linear theory of thermoelasticity, investigation is conducted into the boundary conditions and frequency equations of three-dimensional wave propagation of homogeneous isotropic thermoelastic, generalized thermoelastic cylindrical panel and plate. For the material zinc, frequency equations are numerically calculated, and for varying cylindrical panel specifications the dispersion curves are plotted.

3.2 Boundary conditions and frequency equations

We derive the secular equation for the three-dimensional vibrations of a thermoelastic cylindrical panel subjected to traction free boundary conditions at the upper and lower surfaces of the panel at $r = a, b$

$$\sigma_{rr} = \sigma_{r\theta} = \sigma_{rz} = 0 \quad T_{,r} = 0. \tag{3.1}$$

Using these boundary conditions in equations (2.22) and (2.2) with the corresponding solution yields a system of eight simultaneous equations in A_i and B_i, which will have a non-trivial solution if the determinant of their coefficients vanishes. We have

$$|M_{k\,l}| = 0 \quad k, l = 1, 2,\ldots,8 \tag{3.2}$$

The coefficients in the determinant of equation (3.2) are

$$M_{11} = (2 + \bar{\lambda})\Big((\delta J_\delta(\alpha_1 t_1)/t_1^2 - \frac{\alpha_1}{t_1} J_{\delta+1}(\alpha_1 t_1)) - ((\alpha_1 t_1)^2 R^2 - \delta^2) J_\delta(\alpha_1 t_1)/t_1{}^2\Big)$$
$$+ \bar{\lambda}\Big(\delta(\delta - 1) J_\delta(\alpha_1 t_1)/t_1{}^2 - \frac{\alpha_1}{t_1} J_{\delta+1}(\alpha_1 t_1)\Big) + \bar{\lambda} b_1 t_L{}^2 J_\delta(\alpha_1 t_1) - \beta T_0 R^2 c_1 \bar{\lambda}$$

doi:10.1088/978-0-7503-5638-1ch3

$$M_{13} = (2+\bar{\lambda})\left((\delta J_\delta(\alpha_2 t_1)/t_1{}^2 - \tfrac{\alpha_2}{t_2} J_{\delta+1}(\alpha_2 t_1)) - ((\alpha_2 t_1)^2 R^2 - \delta^2) J_\delta(\alpha_2 t_1)/t_1{}^2\right)$$
$$+\bar{\lambda}\left(\delta(\delta-1)J_\delta(\alpha_2 t_1)/t_1{}^2 - \tfrac{\alpha_2}{t_1} J_{\delta+1}(\alpha_2 t_1)\right) + \bar{\lambda} b_2 t_L{}^2 J_\delta(\alpha_2 t_1) - \beta T_0 R^2 c_2 \bar{\lambda}$$

$$M_{15} = (2+\bar{\lambda})\left((\delta J_\delta(\alpha_3 t_1)/t_1{}^2 - \tfrac{\alpha_2}{t_2} J_{\delta+1}(\alpha_3 t_1)) - ((\alpha_3 t_1)^2 R^2 - \delta^2) J_\delta(\alpha_3 t_1)/t_1{}^2\right)$$
$$+\bar{\lambda}\left(\delta(\delta-1)J_\delta(\alpha_3 t_1)/t_1{}^2 - \tfrac{\alpha_2}{t_1} J_{\delta+1}(\alpha_3 t_1)\right) + \bar{\lambda} b_3 t_L{}^2 J_\delta(\alpha_3 t_1) - \beta T_0 R^2 c_3 \bar{\lambda}$$

$$M_{17} = (2+\bar{\lambda})\left((\tfrac{k_1\delta}{t_1} J_{\delta+1}(k_1 t_1) - \delta(\delta-1)J_\delta(k_1 t_1)/t_1{}^2\right)$$
$$+\bar{\lambda}\left(\delta(\delta-1)J_\delta(k_1 t_1)/t_1{}^2 - \tfrac{k_1\delta}{t_1} J_{\delta+1}(k_1 t_1)\right)$$

$$M_{21} = 2\delta((\alpha_1/t_1)J_{\delta+1}(\alpha_1 t_1) - \delta(\delta-1)J_\delta(\alpha_1 t_1))$$

$$M_{23} = 2\delta((\alpha_2/t_1)J_{\delta+1}(\alpha_2 t_1) - \delta(\delta-1)J_\delta(\alpha_2 t_1))$$

$$M_{25} = 2\delta((\alpha_3/t_1)J_{\delta+1}(\alpha_3 t_1) - \delta(\delta-1)J_\delta(\alpha_3 t_1))$$

$$M_{27} = (k_1 t_1)^2 R^2 J_\delta(k_1 t_1) - 2\delta(\delta-1)J_\delta(k_1 t_1)/t_1{}^2 + k_1/t_1 J_{\delta+1}(k_1 t_1)$$

$$M_{31} = -t_L(1+b_1)(\delta/t_1 J_\delta(\alpha_1 t_1) - \alpha_1 J_{\delta+1}(\alpha_1 t_1))$$

$$M_{33} = -t_L(1+b_2)(\delta/t_1 J_\delta(\alpha_2 t_1) - \alpha_2 J_{\delta+1}(\alpha_2 t_1))$$

$$M_{35} = -t_L(1+b_3)(\delta/t_1 J_\delta(\alpha_3 t_1) - \alpha_2 J_{\delta+1}(\alpha_3 t_1))$$

$$M_{37} = -t_L(\delta/t_1)J_\delta(k_1 t_1)$$

$$M_{41} = c_1[(\delta/t_1)J_\delta(\alpha_1 t_1) - (\alpha_1)J_{\delta+1}(\alpha_1 t_1)]$$

$$M_{43} = c_2[(\delta/t_1)J_\delta(\alpha_2 t_1) - (\alpha_2)J_{\delta+1}(\alpha_2 t_1)]$$

$$M_{45} = c_3[(\delta/t_1)J_\delta(\alpha_3 t_1) - (\alpha_3)J_{\delta+1}(\alpha_3 t_1)],\ M_{47} = 0$$

in which $t_1 = a/R = 1 - t_*/2$, $t_2 = b/R = 1 + t_*/2$ and $t_* = b - a/R$ is the thickness-to-mean radius ratio of the panel. Obviously $M_{kl}(l = 2, 4, 6, 8)$ can be obtained by just replacing the modified Bessel function of the first kind in $M_{kl}(k = 1, 3, 5, 7)$ with the ones of the second kind, respectively, while $M_{kl}(k = 5, 6, 7, 8)$ can be obtained by just replacing t_1 in $M_{kl}(k = 1, 2, 3, 4)$ with t_2.

3.2.1 Wave propagation in a homogeneous isotropic generalized thermoelastic cylindrical panel

In this section we derive the boundary condition and frequency equation for the three-dimensional vibration of a generalized thermoelastic cylindrical panel subjected to traction free boundary conditions at the upper and lower surfaces taken from equation (3.1) at $r = a, b$ then the frequency equation is obtained for a generalized thermoelastic cylindrical panel as

$$|N_{k\,l}| = 0 \quad k, l = 1, 2, \ldots, 8. \tag{3.3}$$

The elements of the determinant in equation (3.3) are obtained from the relations

$$N_{11} = (2 + \bar{\lambda})\Big((\delta J_\delta(\alpha_1 t_1)/t_1^2 - \tfrac{\alpha 1}{t_1} J_{\delta+1}(\alpha_1 t_1)) - ((\alpha_1 t_1)^2 R^2 - \delta^2) J_\delta(\alpha_1 t_1)/t_1{}^2\Big)$$
$$+ \bar{\lambda}\Big(\delta(\delta - 1) J_\delta(\alpha_1 t_1)/t_1{}^2 - \tfrac{\alpha 1}{t_1} J_{\delta+1}(\alpha_1 t_1)\Big) + \bar{\lambda} d_1 t_L{}^2 J_\delta(\alpha_1 t_1) - \beta T_0 R^2 e_1 \bar{\lambda}$$

$$N_{13} = (2 + \bar{\lambda})\Big((\delta J_\delta(\alpha_2 t_1)/t_1{}^2 - \tfrac{\alpha 2}{t_2} J_{\delta+1}(\alpha_2 t_1)) - ((\alpha_2 t_1)^2 R^2 - \delta^2) J_\delta(\alpha_2 t_1)/t_1{}^2\Big)$$
$$+ \bar{\lambda}\Big(\delta(\delta - 1) J_\delta(\alpha_2 t_1)/t_1{}^2 - \tfrac{\alpha 2}{t_1} J_{\delta+1}(\alpha_2 t_1)\Big) + \bar{\lambda} d_2 t_L{}^2 J_\delta(\alpha_2 t_1) - \beta T_0 R^2 e_2 \bar{\lambda}$$

$$N_{15} = (2 + \bar{\lambda})\Big((\delta J_\delta(\alpha_3 t_1)/t_1{}^2 - \tfrac{\alpha 2}{t_2} J_{\delta+1}(\alpha_3 t_1)) - ((\alpha_3 t_1)^2 R^2 - \delta^2) J_\delta(\alpha_3 t_1)/t_1{}^2\Big)$$
$$+ \bar{\lambda}\Big(\delta(\delta - 1) J_\delta(\alpha_3 t_1)/t_1{}^2 - \tfrac{\alpha 2}{t_1} J_{\delta+1}(\alpha_3 t_1)\Big) + \bar{\lambda} d_3 t_L{}^2 J_\delta(\alpha_3 t_1) - \beta T_0 R^2 e_3 \bar{\lambda}$$

$$N_{17} = (2 + \bar{\lambda})\left(\left(\frac{k_1 \delta}{t_1}\right) J_{\delta+1}(k_1 t_1) - \delta(\delta - 1) J_\delta(k_1 t_1)/t_1{}^2\right) + \bar{\lambda}\left(\delta(\delta - 1) J_\delta(k_1 t_1)/t_1{}^2 - \frac{k_1 \delta}{t_1} J_{\delta+1}(k_1 t_1)\right)$$

$$N_{21} = 2\delta((\alpha_1/t_1) J_{\delta+1}(\alpha_1 t_1) - \delta(\delta - 1) J_\delta(\alpha_1 t_1))$$

$$N_{23} = 2\delta((\alpha_2/t_1) J_{\delta+1}(\alpha_2 t_1) - \delta(\delta - 1) J_\delta(\alpha_2 t_1))$$

$$N_{25} = 2\delta((\alpha_3/t_1) J_{\delta+1}(\alpha_3 t_1) - \delta(\delta - 1) J_\delta(\alpha_3 t_1))$$

$$N_{27} = (k_1 t_1)^2 R^2 J_\delta(k_1 t_1) - 2\delta(\delta - 1) J_\delta(k_1 t_1)/t_1{}^2 + k_1/t_1 J_{\delta+1}(k_1 t_1)$$

$$N_{31} = -t_L(1 + d_1)(\delta/t_1 J_\delta(\alpha_1 t_1)) - \alpha_1 J_{\delta+1}(\alpha_1 t_1)$$

$$N_{33} = -t_L(1 + d_2)(\delta/t_1 J_\delta(\alpha_2 t_1)) - \alpha_2 J_{\delta+1}(\alpha_2 t_1)$$

$$N_{35} = -t_L(1 + d_3)(\delta/t_1 J_\delta(\alpha_3 t_1)) - \alpha_3 J_{\delta+1}(\alpha_3 t_1)$$

$$N_{37} = -t_L(\delta/t_1) J_\delta(k_1 t_1)$$

$$N_{41} = e_1(\delta/t_1) J_\delta(\alpha_1 t_1) - \alpha_1 J_{\delta+1}(\alpha_1 t_1)$$

$$N_{43} = e_2(\delta/t_1) J_\delta(\alpha_2 t_1) - \alpha_2 J_{\delta+1}(\alpha_2 t_1)$$

$$N_{45} = e_3(\delta/t_1) J_\delta(\alpha_3 t_1) - \alpha_3 J_{\delta+1}(\alpha_3 t_1)$$

$$N_{47} = 0$$

in which $t_1 = a/R = 1 - t^*/2$, $t_2 = b/R = 1 + t^*/2$ and $t^* = b - a/R$ is the thickness-to-mean radius ratio of the panel. Obviously, $N_{kl}(l = 2, 4, 6, 8)$ can be obtained by just replacing modified Bessel function of the first kind in $N_{kl}(k = 1, 3, 5, 7)$ with the ones of the second kind, respectively, while $N_{kl}(k = 5, 6, 7, 8)$ can be obtained by just replacing t_1 in $N_{kl}(k = 1, 2, 3, 4)$ with t_2.

3.2.2 Damping of generalized thermoelastic waves in a homogeneous isotropic plate

The coupled partial differential equation (2.59) is subjected to the following non-dimensional boundary conditions at the surfaces $r = a,\ b$.

(i) Stress-free boundary (free edge)

$$\sigma_{rr} = \sigma_{r\theta} = 0. \tag{3.4}$$

(ii) Rigidly fixed boundary (clamped edge)

$$u = v = 0. \tag{3.5}$$

(iii) Thermal boundary

$$T_{,r} + hT = 0 \tag{3.6}$$

where h is the surface heat transfer coefficient. Here, $h \to 0$ corresponds to a thermally insulated surface and $h \to \infty$ refers to an isothermal one. Substituting the expressions in equations (2.51) and (2.2) into equations (3.4)–(3.6), we can get the frequency equation for free vibration as follows:

$$|P_{k\,l}| = 0\ k,\ l = 1,\ 2,\ldots,6. \tag{3.7}$$

The elements of the determinant in equation (3.7) are considered as given below

$$\begin{aligned}P_{11} = {} & (2 + \bar{\lambda})((\delta J_\delta(\alpha_1 ax) + (\alpha_1 ax)J_{\delta+1}(\alpha_1 ax)) - ((\alpha_1 ax)^2 R^2 - \delta^2)J_\delta(\alpha_1 ax)) \\ & + \bar{\lambda}(\delta(\delta - 1)(J_\delta(\alpha_1 ax) - (\alpha_1 ax)J_{\delta+1}(\alpha_1 ax))) - \beta T(j\omega)\eta_2 f_1(\alpha_1 ax)^2\end{aligned}$$

$$\begin{aligned}P_{13} = {} & (2 + \bar{\lambda})((\delta J_\delta(\alpha_2 ax) + (\alpha_2 ax)J_{\delta+1}(\alpha_2 ax)) - ((\alpha_2 ax)^2 R^2 - \delta^2)J_\delta(\alpha_2 ax)) \\ & + \bar{\lambda}(\delta(\delta - 1)(J_\delta(\alpha_2 ax) - (\alpha_2 ax)J_{\delta+1}(\alpha_2 ax))) - \beta T(j\omega)\eta_2 f_2(\alpha_2 ax)^2\end{aligned}$$

$$\begin{aligned}P_{15} = {} & (2 + \bar{\lambda})((\delta(\delta - 1)J_\delta(\alpha_3 ax) - (\alpha_3 ax)J_{\delta+1}(\alpha_3 ax)) \\ & + \bar{\lambda}(\delta(\delta - 1)J_\delta(\alpha_3 ax) - (\alpha_3 ax)J_{\delta+1}(\alpha_3 ax))\end{aligned}$$

$$P_{21} = 2\delta(\delta - 1)J_\delta(\alpha_1 ax) - 2\delta(\alpha_1 ax)J_{\delta+1}(\alpha_1 ax)$$

$$P_{23} = 2\delta(\delta - 1)J_\delta(\alpha_2 ax) - 2\delta(\alpha_2 ax)J_{\delta+1}(\alpha_2 ax)$$

$$P_{25} = 2\delta(\delta - 1)J_\delta(\alpha_3 ax) - 2(\alpha_3 ax)J_{\delta+1}(\alpha_3 ax) + ((\alpha_3 ax)^2 - \delta^2)J_\delta(\alpha_3 ax))$$

$$P_{31} = f_1(\delta J_\delta(\alpha_1 ax) - (\alpha_1 ax)J_{\delta+1}(\alpha_1 ax) + hJ_\delta(\alpha_1 ax))$$

$$P_{33} = f_2(\delta J_\delta(\alpha_2 ax) - (\alpha_2 ax)J_{\delta+1}(\alpha_2 ax) + hJ_\delta(\alpha_2 ax)),\ P_{35} = 0.$$

Obviously, $P_{kl}(l = 2, 4, 6)$ can be obtained by just replacing the Bessel functions of the first kind in $P_{kl}(k = 1, 3, 5)$ with those of the second kind, respectively, while $P_{kl}(k = 4, 5, 6)$ can be obtained by just replacing a in $P_{kl}(k = 1, 2, 3)$ with b.

3.3 Numerical results and discussion

The frequency equation (3.2) is numerically solved for zinc material. For the purpose of numerical computation, we consider a closed circular cylindrical shell with the centre angle $\alpha = 2\pi$ and the integer n must be even since the shell vibrates in a circumferential full wave. The frequency equation for a closed cylindrical shell can be obtained by setting $\delta = N(N = 1, 2, 3, \ldots)$ where N is the circumferential wave number in equation (3.2). The material properties of zinc for an isotropic material are, given as in Sharma (2001)

$$\lambda = 0.385 \times 10^{11}\,\mathrm{Nm}^{-2},\quad \mu = 0.508 \times 10^{11}\,\mathrm{Nm}^{-2},\quad \rho = 7.14 \times 10^{3}\,\mathrm{kgm}^{-3}$$

$$T_0 = 296\,\mathrm{K},\quad \beta = 1,\quad K = 1,\quad \nu = 0.3,\quad c_\nu = 3.9 \times 10^{2}\,\mathrm{J\,kg}^{-1}\,\mathrm{deg}^{-1}. \qquad (3.8)$$

A combination of the Birge–Vita and Newton–Raphson methods is used to determine the roots of the algebraic equation (2.31). The simple Birge–Vita method does not work in this case to determine the root of the algebraic equation. The Newton–Raphson method is used to correct the roots of the algebraic equation after they have been obtained using the Birge–Vita method. This combination has solved the problem of obtaining the roots of the algebraic equations of the governing equations.

A dispersion curve is drawn between the non-dimensional length-to-mean radius ratio versus dimensionless frequency for the different thickness parameters $t^* = 0.01$, 0.1 0.25 with the axial wave number $t_L = 1$ and $t_L = 2$ and the circumferential wave number $\delta = 1, 2$, as shown in figures 3.1 and 3.2, respectively. From figures 3.1 and

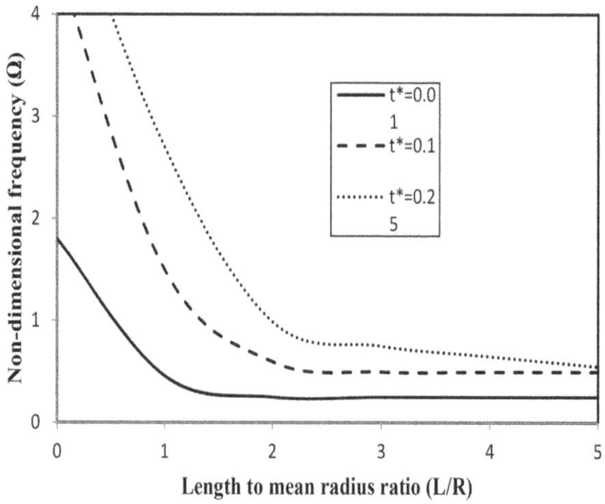

Figure 3.1. Variation of non-dimensional frequency Ω versus length-to-mean radius ratio L/R with different t^* for $t_L = 1$ and $\delta = 1$.

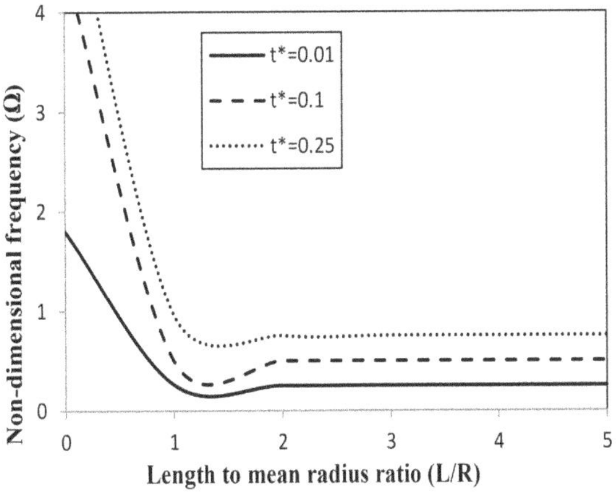

Figure 3.2. Variation of non-dimensional frequency Ω versus length-to-mean radius ratio L/R with different t^* for $t_L = 2$ and $\delta = 1$.

3.2, it is observed that the non-dimensional frequency decreases rapidly and becomes linear at $L/R = 3$ for both $t_L = 1$ and $t_L = 2$.

The dimensionless frequency decreases as the thickness of the cylindrical panel increases. This is a cylindrical panel's proper physical behaviour in relation to its thickness. The comparison of figures 3.1 and 3.2 shows that the non-dimensional frequency decrease exponentially for $L/R < 3$ in the case of the axial wave number $t_L = 1$ and $t_L = 2$ and the circumferential wave number $\delta = 1{,}2$ for all values of t^*, but for the case when $L/R > 3$ the non-dimensional frequency is slow and steady for all values of t^*.

3.3.1 Wave propagation in a homogeneous isotropic generalized thermoelastic cylindrical panel

To solve the frequency equation (3.3), we consider the same discussion as in thermoelastic cylindrical panel with $\alpha = 2\pi$. The frequency equation for a closed cylindrical shell can be obtained by setting $\delta = N(N = 1, 2, 3,\ldots)$ here N is the circumferential wave number in equation (3.3). The material constants of zinc are taken from equation (3.8) together with thermoelastic coupling constants

$$\varepsilon_1 = 0.0221 \; \omega_1^* = 5.01 \times 10^{11}\,\mathrm{s}^{-1}.$$

Figures 3.3 and 3.4 illustrate a dispersion curve drawn between the non-dimensional axial wave number against dimensionless frequency with and without thermal effect for the different thickness parameters $t^* = 0.01,0.1,0.25,0.5$ for the circumferential wave number $\delta = 1$.

From figures 3.3 and 3.4, it is observed that the non-dimensional frequency increases rapidly to become linear for $t_L \leqslant 0.8$ and quite dispersive for $t_L > 0.8$ for all values of t^*. On comparison, the trends of variation of non-dimensional frequency in thermoelastic shell are similar to those in an elastic shell, as can be

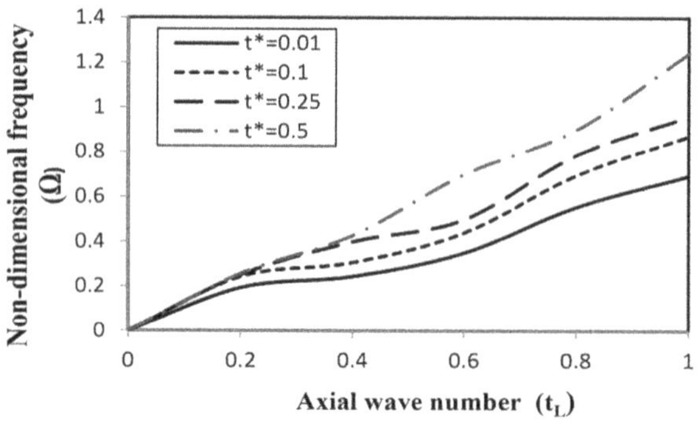

Figure 3.3. Variation of non-dimensional frequency Ω in elastic cylindrical shell with axial wave number t_L.

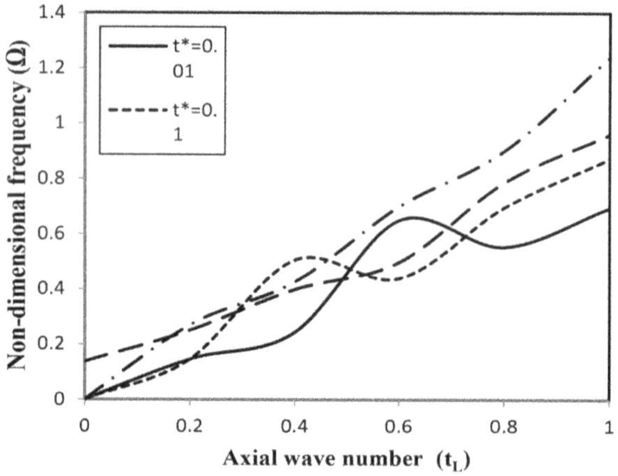

Figure 3.4. Variation of non-dimensional frequency Ω in thermoelastic cylindrical shell with axial wave number t_L.

observed from these figures, but there are significant modifications in their magnitude due to thermal effect for all values of t^* with t_L.

The dimensionless frequency decreases as the thickness of the cylindrical panel increases. This is a cylindrical panel's proper physical behaviour in relation to its thickness.

3.3.2 Damping of generalized thermoelastic waves in a homogeneous isotropic plate

The frequency equation (3.5) for the damping of generalized thermoelastic waves in a simply supported homogenous isotropic cylindrical plate is numerically solved for zinc and the material properties of zinc are taken from equation (3.8). Combining the Birge–Vita and Newton–Raphson methods to calculate the roots of the algebraic equation in equation (2.66), the simple Birge–Vita method does not work in this case to determine the root of the algebraic equation. After obtaining the roots of the algebraic equation using the Birge–Vita method, the roots are corrected for the desired accuracy using the Newton–Raphson method. Such a combination can overcome the difficulties encountered in finding the roots of the algebraic equations of the governing equations. Here the values of the thermal relaxation times are calculated as

$$\tau_0 = 0.75 \times 10^{-13}\,\text{s and } \tau_1 = 0.5 \times 10^{-13}\,\text{s}.$$

Due to the presence of a dissipation term in the heat conduction equation, the frequency equation (2.66) in the general complex transcendental equation provides us with a complex value of frequency. The thermoelastic damping factor is defined by $Q^{-1} = 2\,|\frac{\text{Im}(\omega)}{\text{Re}(\omega)}|$. Tables 3.1 and 3.2, compare the non-dimensional frequencies of the Green–Lindsay (G–L), Lord–Shulman Theory (L–S), and the classical theory (CT) of thermoelasticity for the free and clamped boundaries of the thermally insulated and isothermal circular plate. It is obvious from these tables that when the number of vibration modes rises sequentially, the non-dimensional frequencies increase for both clamped and unclamped cases. Also, due to the effect of thermal relaxation times, the non-dimensional frequency for the L–S theory have higher amplitudes than the G–L and CT theories.

The dispersion of the thermoelastic damping factor with the wave number is explored for both the thermally insulated and isothermal boundaries of the cylindrical plate in different modes of vibration as shown in figures 3.5 and 3.6. It

Table 3.1. Comparison of non-dimensional frequency among the G–L, L–S and CT theories of thermoelasticity for free and clamped boundaries of a thermally insulated circular plate.

	Free edge			Clamped edge		
Mode	LS	GL	CT	LS	GL	CT
1	1.3937	1.3927	1.3295	1.2289	1.2278	1.5565
2	1.6542	1.6533	1.5886	1.4614	1.4604	1.8391
3	1.9176	1.9156	1.8529	1.7009	1.7019	2.1227
4	2.1832	2.1802	2.1204	1.9486	1.9475	2.4048
	2.5840	2.5810	2.5245	2.3381	2.3375	2.8318

Table 3.2. Comparison of non-dimensional frequency among the G–L, L–S and CT of thermoelasticity for free and clamped boundaries of an isothermal circular plate.

	Free edge			Clamped edge		
Mode	LS	GL	CT	LS	GL	CT
1	1.4558	1.4543	1.4153	1.4049	1.4037	1.3588
2	1.7260	1.7251	1.6827	1.6611	1.6602	1.6123
3	1.9967	1.9957	1.9511	1.9182	1.9176	1.8682
4	2.2678	2.2648	2.2213	2.1768	2.1753	2.1264
5	2.6754	2.6732	2.6303	2.5680	2.5670	2.5183

modes of vibration increases exponentially with increasing wave number.

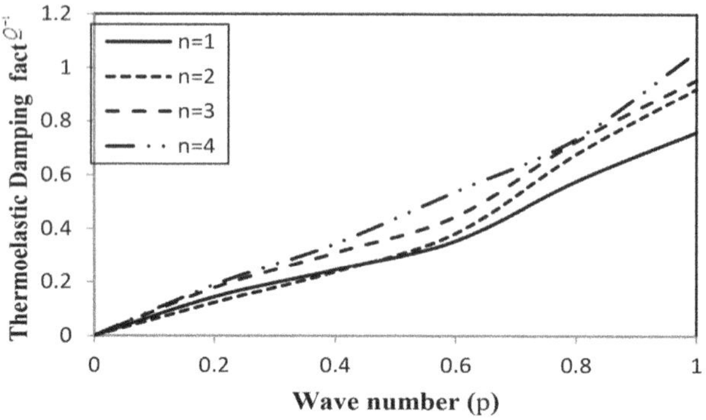

Figure 3.5. Variation of thermoelastic damping factor Q^{-1} of thermally insulated cylindrical plate with wave number p.

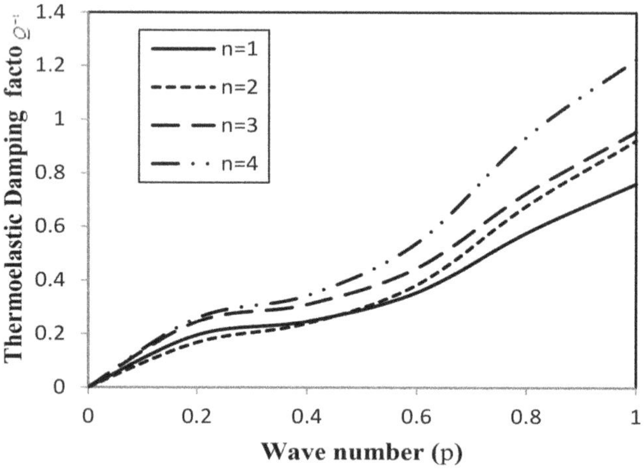

Figure 3.6. Variation of thermoelastic damping factor Q^{-1} of isothermal cylindrical plate with wave number p.

can be seen in figure 3.5 that the damping factor for thermally insulated modes of vibration increases exponentially with increasing wave number.

However, due to the combined effect of damping and insulation, there is smaller dispersion in the damping factor in the current range of wave numbers in figure 3.6 for the isothermal mode. Because of the combined effect of thermal relaxation durations and mechanical field, the consequences of stress-free thermally insulated and isothermal plate borders are extremely relevant, as shown in figures 3.5 and 3.6.

Reference

Sharma J N 2001 Three dimensional vibration of a homogenous transversely isotropic thermo elastic cylindrical panel *J. Acoust. Soc. Am.* **110** 648–53

IOP Publishing

Mathematical Modelling and Characterization of Cylindrical Structures

Farzad Ebrahimi and Rajendran Selvamani

Chapter 4

Mathematical modelling of waves in a homogeneous isotropic rotating cylindrical panel

4.1 Introduction

As the rotation rate will explain the speed of the disturbed waves, this study will help us to design high-speed steam and gas turbines and rotation rate sensors. Through copper and zinc materials the frequency equations are arrived at via stressless boundary conditions, and are evaluated numerically. With the aid of MATLAB software, dispersion curves have been drawn for non-dimensional frequencies

4.2 Boundary conditions and frequency equations

The three-dimensional vibrations of a rotating cylindrical panel subjected to traction-free boundary conditions at the upper and lower surfaces at $r = a, b$ are defined in the following form

$$u = (-\bar{\phi}' - r^{-1}\delta\bar{\psi})\sin(m\pi z)\sin(\delta\theta)e^{j\omega t}$$

$$v = (-\bar{\psi}' - r^{-1}\delta\bar{\phi})\sin(m\pi z)\cos(\delta\theta)e^{j\omega t}$$

$$w = \bar{\chi}\ t_L \cos(m\pi z)\sin(\delta\theta)e^{j\omega t} \tag{4.1}$$

$$\bar{\sigma}_{rr} = \begin{bmatrix} (2+\bar{\lambda})\delta\left(\frac{\bar{\psi}'}{r} - \frac{\bar{\psi}}{r^2}\right) + (2+\bar{\lambda})\left(\frac{1}{r}\bar{\phi}' + \left(\alpha_i^2 - \frac{\delta^2}{r^2}\bar{\phi}\right)\right) \\ + \bar{\lambda}\left(\frac{\delta}{r^2}\bar{\psi} - \frac{1}{r}\bar{\phi}' - \frac{\delta^2}{r^2}\bar{\phi} - \frac{\delta}{r}\bar{\psi}' - t_L^2\bar{\chi}\right) \end{bmatrix}\sin(m\pi z)\cos(\delta\theta)e^{j\omega t}$$

doi:10.1088/978-0-7503-5638-1ch4

$$\bar{\sigma}_{r\theta} = 2\left(\frac{1}{r}\bar{\psi} + \left(\alpha_i^2 - \frac{\delta^2}{r^2}\right)\bar{\psi} - \frac{2\delta}{r}\bar{\phi}' + \frac{2\delta}{r^2}\bar{\phi} + \frac{\bar{\psi}'}{r} - \frac{\delta^2}{r^2}\bar{\psi}\right)\sin(m\pi z)\cos(\delta\theta)e^{j\omega t}$$

$$\bar{\sigma}_{rz} = 2t_L\left(-\bar{\phi}' - \frac{\delta}{r}\bar{\psi} + \bar{\chi}'\right)\cos(m\pi z)\sin(\delta\theta)\, e^{j\omega t} \tag{4.2}$$

where the prime denotes the differentiation with respect to r, $\bar{u}_i = u_i/R$, $(i = r, \theta, z)$ are three non-dimensional displacements and $\bar{\sigma}_{rr} = \sigma_{rr}/\mu$, $\bar{\sigma}_{r\theta} = \sigma_{r\theta}/\mu$, $\bar{\sigma}_{rz} = \sigma_{rz}/\mu$ are three non-dimensional stresses. In this case, both convex and concave surfaces of the panel are traction-free and the boundary conditions are the same as defined in equation (3.1) without the temperature boundary condition.

Using the result obtained in equations (2.2) and (2.3) in equation (3.1) we can get the frequency equation in the following determinant form

$$|Q_{kl}| = 0 \quad k, l = 1, 2, \ldots, 6. \tag{4.3}$$

The coefficients of the above determinant are obtained as

$$Q_{11} = (2 + \bar{\lambda})\left((\delta J_\delta(\alpha_1 t_1)/t_1^2 - \tfrac{\alpha 1}{t_1}J_{\delta+1}(\alpha_1 t_1)) - ((\alpha_1 t_1)^2R^2 - \delta^2)J_\delta(\alpha_1 t_1)/t_1^2\right)$$
$$+ \bar{\lambda}\left(\delta(\delta - 1)J_\delta(\alpha_1 t_1)/t_1^2 - \tfrac{\alpha 1}{t_1}J_{\delta+1}(\alpha_1 t_1)\right) + \bar{\lambda}a_1 t_L{}^2 J_\delta(\alpha_1 t_1)$$

$$Q_{13} = (2 + \bar{\lambda})\left((\delta J_\delta(\alpha_2 t_1)/t_1^2 - \tfrac{\alpha 2}{t2}J_{\delta+1}(\alpha_2 t_1)) - ((\alpha_2 t_1)^2R^2 - \delta^2)J_\delta(\alpha_2 t_1)/t_1^2\right)$$
$$+ \bar{\lambda}\left(\delta(\delta - 1)J_\delta(\alpha_2 t_1)/t_1^2 - \tfrac{\alpha 2}{t_1}J_{\delta+1}(\alpha_2 t_1)\right) + \bar{\lambda}a_2 t_L{}^2 J_\delta(\alpha_2 t_1)$$

$$Q_{15} = (2 + \bar{\lambda})\left((\tfrac{k_1\delta}{t_1}J_{\delta+1}(k_2 t_1) - \delta(\delta - 1)J_\delta(k_2 t_1)/t_1^2\right)$$
$$+ \bar{\lambda}\left(\delta(\delta - 1)J_\delta(k_2 t_1)/t_1^2 - \tfrac{k_1\delta}{t_1}J_{\delta+1}(k_2 t_1)\right)$$

$$Q_{21} = 2\delta((\alpha_1/t_1)J_{\delta+1}(\alpha_1 t_1) - \delta(\delta - 1)J_\delta(\alpha_1 t_1))$$

$$Q_{23} = 2\delta((\alpha_2/t_1)J_{\delta+1}(\alpha_2 t_1) - \delta(\delta - 1)J_\delta(\alpha_2 t_1))$$

$$Q_{25} = (k_2 t_1)^2R^2J_\delta(k_2 t_1) - 2\delta(\delta - 1)J_\delta(k_2 t_1)/t_1^2 + k_1/t_1 J_{\delta+1}(k_2 t_1)$$

$$Q_{31} = -t_L(1 + a_1)(\delta/t_1 J_\delta(\alpha_1 t_1) - \alpha_1 J_{\delta+1}(\alpha_1 t_1))$$

$$Q_{33} = -t_L(1 + a_2)(\delta/t_1 J_\delta(\alpha_2 t_1) - \alpha_2 J_{\delta+1}(\alpha_2 t_1))\ Q_{35} = -t_L(\delta/t_1)J_\delta(k_1 t_1)$$

in which $t_1 = a/R = 1 - t*/2$, $t_2 = b/R = 1 + t*/2$ and $t* = b - a/R$ is the thickness-to-mean radius ratio of the panel. Obviously $Q_{kl}(l = 2, 4, 6)$ can be obtained by just replacing the modified Bessel function of the first kind in $Q_{kl}(k = 1, 3, 5)$ with the ones of the second kind, respectively, while $Q_{kl}(k = 4, 5, 6)$ can be obtained by just replacing t_1 in $Q_{kl}(k = 1, 2, 3)$ with t_2.

4.2.1 Wave propagation in a homogeneous isotropic thermoelastic rotating cylindrical panel

Using zero traction boundary conditions, the frequency equations of three-dimensional wave propagation in a homogeneous isotropic rotating thermoelastic cylindrical panel are studied. For a zinc material, the frequency equation is demonstrated numerically. With the temperature change, the displacement functions are as described in equations (4.1) and (4.2).

$$T(r,\theta,z,t) = \bar{T}(r,\theta,z,t)\sin(m\pi z)\sin(n\pi\theta/\alpha)e^{j\omega t} \tag{4.4}$$

$\bar{u}_i = u_i/R$, $(i = r, \theta, z)$ are three non-dimensional displacements and $\bar{\sigma}_{rr} = \sigma_{rr}/\mu$, $\bar{\sigma}_{r\theta} = \sigma_{r\theta}/\mu$, $\bar{\sigma}_{rz} = \sigma_{rz}/\mu$ are three non-dimensional stresses. The coupled partial differential equations defined in equation (2.87) are also subjected to the following non-dimensional boundary conditions at surfaces $r = a,\ b$ as:

(i) The traction-free non-dimensional mechanical boundary conditions for a stress-free edge are given by

$$\sigma_{rr} = \sigma_{r\theta} = \sigma_{rz} = 0, \tag{4.5}$$

(ii) The non-dimensional insulated or isothermal thermal boundary condition is given as

$$T_{,r} + hT = 0 \tag{4.6}$$

where h is the surface heat transfer coefficient. Here, $h \to 0$ corresponds to a thermally insulated surface and $h \to \infty$ refers to an isothermal one. Invoking the boundary conditions in equations (4.5) and (4.6) in equations (2.22) and (2.2), we can get the frequency equation in the vanishing determinant form as follows

$$|R_{kl}| = 0 \quad k, l = 1, 2, \ldots 8. \tag{4.7}$$

The values of R_{kl} are given by

$$R_{11} = (2+\bar{\lambda})\Big((\delta J_\delta(\alpha_1 t_1)/t_1^2 - \tfrac{\alpha 1}{t1}J_{\delta+1}(\alpha_1 t_1)) - ((\alpha_1 t_1)^2R^2 - \delta^2)J_\delta(\alpha_1 t_1)/t_1{}^2\Big)$$
$$+\bar{\lambda}\Big(\delta(\delta-1)J_\delta(\alpha_1 t_1)/t_1{}^2 - \tfrac{\alpha 1}{t1}J_{\delta+1}(\alpha_1 t_1)\Big) + \bar{\lambda} m_1 t_L{}^2 J_\delta(\alpha_1 t_1) - \beta T_0 R^2 n_1 \bar{\lambda}$$

$$R_{13} = (2+\bar{\lambda})\Big((\delta J_\delta(\alpha_2 t_1)/t_1{}^2 - \tfrac{\alpha 2}{t2}J_{\delta+1}(\alpha_2 t_1)) - ((\alpha_2 t_1)^2R^2 - \delta^2)J_\delta(\alpha_2 t_1)/t_1{}^2\Big)$$
$$+\bar{\lambda}\Big(\delta(\delta-1)J_\delta(\alpha_2 t_1)/t_1{}^2 - \tfrac{\alpha 2}{t1}J_{\delta+1}(\alpha_2 t_1)\Big) + \bar{\lambda} m_2 t_L{}^2 J_\delta(\alpha_2 t_1) - \beta T_0 R^2 n_2 \bar{\lambda}$$

$$R_{15} = (2+\bar{\lambda})\Big((\delta J_\delta(\alpha_3 t_1)/t_1{}^2 - \tfrac{\alpha 2}{t2}J_{\delta+1}(\alpha_3 t_1)) - ((\alpha_3 t_1)^2R^2 - \delta^2)J_\delta(\alpha_3 t_1)/t_1{}^2\Big)$$
$$+\bar{\lambda}\Big(\delta(\delta-1)J_\delta(\alpha_3 t_1)/t_1{}^2 - \tfrac{\alpha 2}{t1}J_{\delta+1}(\alpha_3 t_1)\Big) + \bar{\lambda} m_3 t_L{}^2 J_\delta(\alpha_3 t_1) - \beta T_0 R^2 n_3 \bar{\lambda}$$

$$R_{17} = (2+\bar{\lambda})\left(\left(\frac{k_1\delta}{t_1}J_{\delta+1}(k_2t_1) - \delta(\delta-1)J_\delta(k_2t_1)/t_1{}^2\right) + \bar{\lambda}\left(\delta(\delta-1)J_\delta(k_2t_1)/t_1{}^2 - \frac{k_1\delta}{t_1}J_{\delta+1}(k_2t_1)\right)\right.$$

$$R_{21} = 2\delta((\alpha_1/t_1)J_{\delta+1}(\alpha_1t_1) - \delta(\delta-1)J_\delta(\alpha_1t_1))$$

$$R_{23} = 2\delta((\alpha_2/t_1)J_{\delta+1}(\alpha_2t_1) - \delta(\delta-1)J_\delta(\alpha_2t_1))$$

$$R_{25} = 2\delta((\alpha_3/t_1)J_{\delta+1}(\alpha_3t_1) - \delta(\delta-1)J_\delta(\alpha_3t_1))$$

$$R_{27} = (k_2t_1)^2R^2J_\delta(k_2t_1) - 2\delta(\delta-1)J_\delta(k_2t_1)/t_1{}^2 + k_1/t_1J_{\delta+1}(k_2t_1)$$

$$R_{31} = -t_L(1+m_1)(\delta/t_1J_\delta(\alpha_1t_1) - \alpha_1J_{\delta+1}(\alpha_1t_1))$$

$$R_{33} = -t_L(1+m_2)(\delta/t_1J_\delta(\alpha_2t_1) - \alpha_2J_{\delta+1}(\alpha_2t_1))$$

$$R_{35} = -t_L(1+m_3)(\delta/t_1J_\delta(\alpha_3t_1) - \alpha_2J_{\delta+1}(\alpha_3t_1))$$

$$R_{37} = -t_L(\delta/t_1)J_\delta(k_2t_1)$$

$$R_{41} = n_1((\delta/t_1)J_\delta(\alpha_1t_1) - (\alpha_1)J_{\delta+1}(\alpha_1t_1) + hJ_\delta(\alpha_1t_1))$$

$$R_{43} = n_2((\delta/t_1)J_\delta(\alpha_2t_1) - (\alpha_2)J_{\delta+1}(\alpha_2t_1) + hJ_\delta(\alpha_2t_1))$$

$$R_{45} = n_3((\delta/t_1)J_\delta(\alpha_3t_1) - (\alpha_3)J_{\delta+1}(\alpha_3t_1) + hJ_\delta(\alpha_3t_1)),\ R_{47} = 0$$

in which $t_1 = a/R = 1 - t_*/2$, $t_2 = b/R = 1 + t_*/2$ and $t_* = b - a/R$ is the thickness-to-mean radius ratio of the panel. Obviously, $R_{kl}(l = 2, 4, 6, 8)$ can be obtained by just replacing the modified Bessel function of the first kind in $R_{kl}(k = 1, 3, 5, 7)$ with the ones of the second kind, respectively, while $R_{kl}(k = 5, 6, 7, 8)$ can be obtained by just replacing t_1 in $R_{kl}(k = 1, 2, 3, 4)$ with t_2.

4.3 Numerical results and discussion

For copper material, the frequency equation (4.3) is numerically solved. For the purpose of numerical computation we consider the closed circular cylindrical shell with the centre angle $\alpha = 2\pi$ and the frequency equation for a closed cylindrical shell can be obtained by setting $\delta = N(N = 1, 2, 3,\ldots)$ and N is the circumferential wave number in equation (4.3). The material properties of copper are given by

$$\lambda = 8.20 \times 10^{11}\ \text{kg m s}^{-2}\ \mu = 4.20 \times 10^{11}\ \text{kg m s}^{-2}\ \rho = 8.96 \times 10^{3}\ \text{kg m}^{-3}$$

and

$$\nu = 0.3,\ E = 2.139 \times 10^{11}\ \text{N m}^{-2}.$$

The dispersion curves are drawn between the non-dimensional circumferential wave number versus frequency of the cylindrical shell with respect to different rotating speeds $\Gamma = 0.2, 0.4, 0.6, 0.8$ as shown in figures 4.1 and 4.2. The frequency parameter rises monotonically with increasing rotational speed in the current range

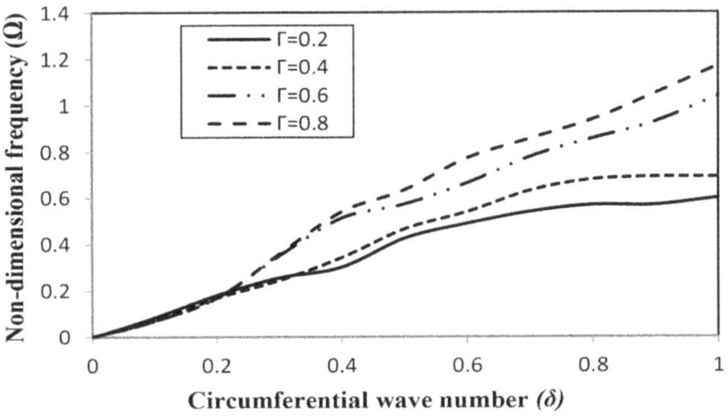

Figure 4.1. Variation of non-dimensional frequency Γ with circumferential wave number δ for $t_L = 1$.

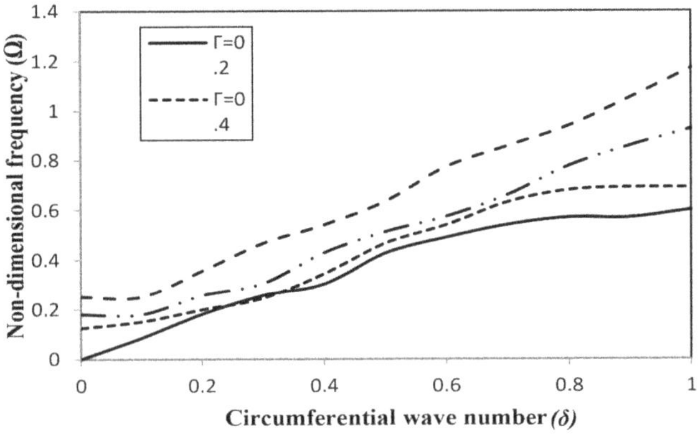

Figure 4.2. Variation of non-dimensional frequency Γ with circumferential wave number δ for $t_L = 2$.

of circumferential wave number. It shows the cylindrical shell rotating in high speed, the effect of the rotating speed on the frequency characteristics is more significant.

From figures 4.1 and 4.2, it is observed that the non-dimensional frequency increases rapidly to become dispersive at higher values of wave number for both $t_L = 1$ and $t_L = 2$. The frequency of increasing value of Γ of rotating shell is observed to increase from zero wave number and becomes quite dispersive at higher values of wave number for both $t_L = 1$ and $t_L = 2$.

When the rotational speed of the cylindrical panel is increased, the dimensionless frequency also increases for both $t_L = 1$ and $t_L = 2$. The comparison of figures 4.1 and 4.2 shows that the non-dimensional frequency increases exponentially for smaller wave number in the case of the axial wave number $t_L = 1$ and $t_L = 2$ for all values of Γ, but in the case of higher wave number the non-dimensional frequency is dispersive and slow for all values of the rotational speed Γ.

4.3.1 Wave propagation in a homogeneous isotropic thermoelastic rotating cylindrical panel

For the purpose of numerical computation of the frequency equation defined in (4.7), we consider a closed zinc circular cylindrical shell as defined in the case of a rotating cylindrical panel. The material properties of zinc are taken from equation (3.8) together with rotational speed $\Gamma = 0.3$ rotation per second. To validate the present analysis a comparative study is presented in table 4.1 for different values of thickness-to-inner radius ratio ($h/b = 0.1, 0.2, 0.3$) and centre angle $\alpha = 30^\circ, 60^\circ, 90^\circ$ of a cylindrical panel in the absence of thermal and rotational effect. A comparison is made between the non-dimensional frequencies of thermally insulated and isothermal boundaries of a rotating and non-rotating cylindrical shell with respect to different rotational speed as shown in tables 4.2 and 4.3, respectively, for the first three modes $n = 1$, $n = 2$ and $n = 3$. From tables 4.2 and 4.3 it is clear that as the rotational speed increases, the non-dimensional frequencies also increase in both rotating and non-rotating cases. As the rotation of the cylindrical shell increases, the coupling effect of various interacting fields also increases resulting in higher frequency.

A dispersion curve is drawn between the non-dimensional circumferential wave number versus dimensionless phase velocity in the case of rotating and non-rotating thermally insulated cylindrical shells with respect to different thickness parameters

Table 4.1. The lowest natural frequency of a zinc cylindrical panel with respect to thickness-to-inner radius ratio.

h/b	(α)	Loy and Lam (1995)	Bhimraddi (1984)	Present
0.1	30	0.7207	0.7207	0.7190
	60	0.8262	0.8257	0.8192
	90	0.9697	0.9680	0.9533
0.2	30	1.3448	1.3429	1.3325
	60	1.3118	1.3055	1.1990
	90	1.3015	1.2901	1.2877
0.3	30	1.9803	1.9706	1.9690
	60	1.8362	1.8099	1.8135
	90	1.6937	1.6552	1.6743

Table 4.2. Comparison between the non-dimensional frequency of rotating and non-rotating thermoelastic cylindrical shell for thermally insulated boundary in the first three modes of vibration.

	Rotating			Non-rotating		
Γ	$n = 1$	$n = 2$	$n = 3$	$n = 1$	$n = 2$	$n = 3$
0.1	0.1033	0.1159	0.1462	0.0899	0.1059	0.1259
0.3	0.3721	0.4821	0.5250	0.2897	0.2707	0.3779
0.5	0.5285	0.6221	0.6614	0.5406	0.5241	0.6327
0.7	0.9898	0.9053	0.7999	0.7840	0.9005	0.8945
1.0	1.3144	1.3728	1.4663	1.1353	1.2064	1.3977

Table 4.3. Comparison between the non-dimensional frequency of rotating and non-rotating thermoelastic cylindrical shell for isothermal boundary in the first three modes of vibration.

	Rotating			Non-rotating		
Γ	$n=1$	$n=2$	$n=3$	$n=1$	$n=2$	$n=3$
0.1	0.1026	0.1215	0.1413	0.0741	0.1078	0.1214
0.3	0.4443	0.4549	0.5245	0.2243	0.3550	0.4247
0.5	0.6077	0.7075	0.7378	0.5922	0.7071	0.7077
0.7	0.9196	0.8200	0.9044	0.9094	0.9909	0.9909
1.0	1.4149	1.4256	1.4644	1.4142	1.4156	1.4142

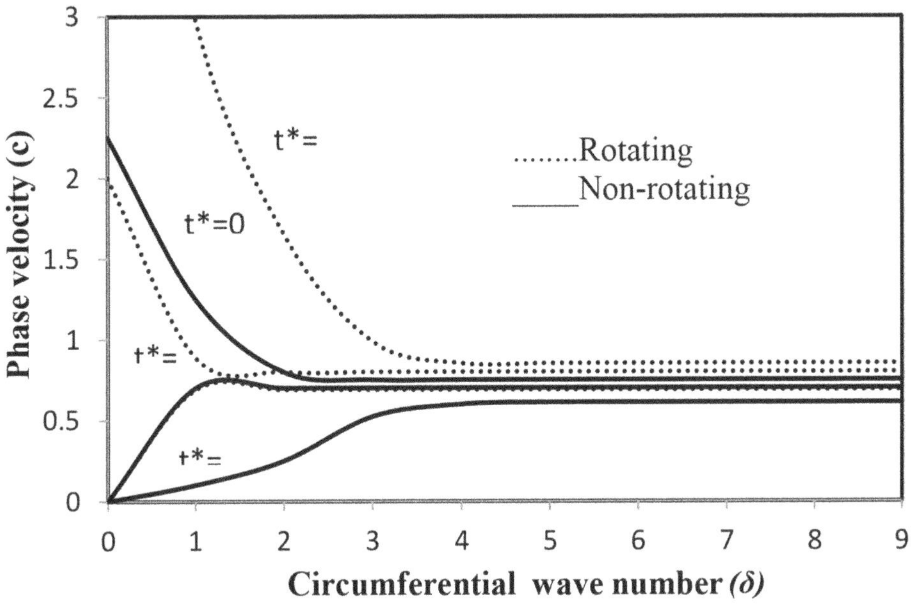

Figure 4.3. Variation of phase velocity c with circumferential wave number δ for $t_L = 1$.

$t_* = b - a/R = 0.1, 0.25, 0.5$ for $t_L = 1, 2$ as shown in figures 4.3 and 4.4, respectively. The solid line curves correspond to a rotating thermoelastic cylindrical shell and the dotted line curves to that of a non-rotating shell. From figures 4.3 and 4.4, it is observed that the non-dimensional phase velocity decreases rapidly to become linear at higher values of circumferential wave number for both $t_L = 1$ and $t_L = 2$. The phase velocity of lower value of t_* in the case of a non-rotating shell is observed to increase from zero wave number and become stable at higher values of wave number for both $t_L = 1$ and $t_L = 2$.

The phase velocity at higher value of t_* attains quite large values at the vanishing wave number and are non-dispersive due to rotation.

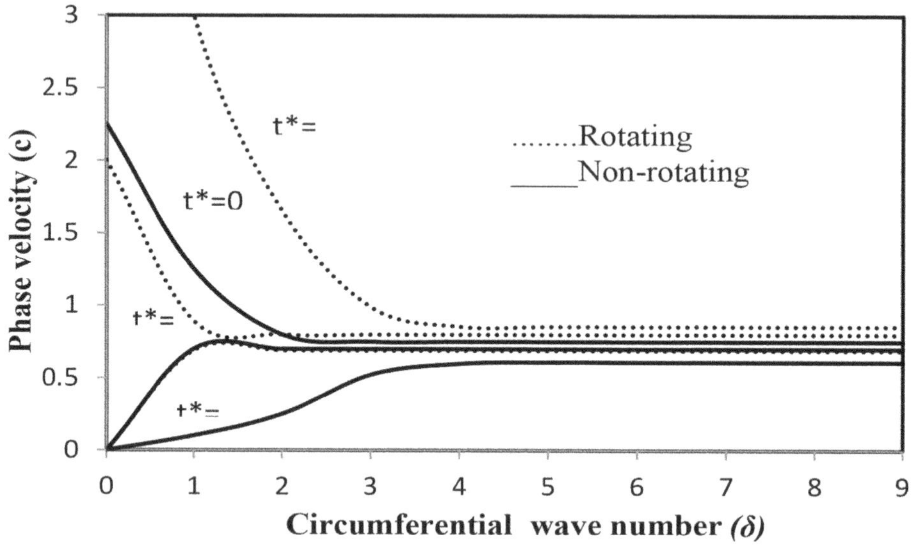

Figure 4.4. Variation of phase velocity c with circumferential wave number δ for $t_L = 2$.

When the thickness parameter of the cylindrical panel is increased, the dimensionless phase velocity decreases for both a rotating and non-rotating cylindrical shell. The comparison of figures 4.3 and 4.4 shows that the non-dimensional phase velocity decreases exponentially for a smaller wave number in the case of the axial wave number $t_L = 1$ and $t_L = 2$ for all values of t^*, but in the case of higher wave number the non-dimensional phase velocity is steady and slow for all values of t^*.

References

Bhimraddi A A 1984 A higher order theory for free vibration analysis of circular cylindrical shell *Int. J. Solids Struct.* **20** 623–30

Loy C T and Lam K Y 1995 Vibration of rotating thin cylindrical panels *J. Appl. Acoust.* **46** 327–43

Selvamani R and Ponnusamy P 2013 Dispersion of thermo elastic waves in a rotating cylindrical panel *Int. J. Adv. Appl. Sci.* **2** 41–50

IOP Publishing

Mathematical Modelling and Characterization of Cylindrical Structures

Farzad Ebrahimi and Rajendran Selvamani

Chapter 5

Wave propagation in a transversely isotropic magneto-thermoelastic cylindrical panel

5.1 Introduction

The study of free vibration modes and fundamental frequency can also help with structural design of a transversely isotropic material under dynamic loads. In the design of sensors and surface acoustic damping wave filters, the coupling effect between thermal, magnetic, and elastic in magneto-thermoelastic materials provides a mechanism for sensing thermomechanical disturbance from measurements of induced magnetic potentials, and for altering structural responses through applied magnetic fields. The boundary conditions and frequency equations of three-dimensional wave propagation of homogeneous transversely isotropic magneto-thermoelastic and generalized magneto-thermo-elastic cylindrical panels are investigated in this chapter.

5.2 Boundary conditions and frequency equations

The secular equation for the boundary conditions and frequency equations of three-dimensional vibrations of a magneto-thermoelastic cylindrical panel exposed to zero traction has been derived. We may combine the upper and lower boundary conditions from equation (3.1) with the magnetic boundary condition on the upper and lower surfaces of a cylindrical panel $r = a, b$

$$B_{,r} = 0 \tag{5.1}$$

Incorporating the result obtained in equations (2.103) and (2.2) into equation (5.1) we can get the frequency equation of coupled free vibration as

$$|S_{kl}| = 0 \quad (k, l = 1, 2, \ldots 10). \tag{5.2}$$

doi:10.1088/978-0-7503-5638-1ch5

Here the elements in the determinant are obtained by

$$S_{11}=c_{11}\left((\delta J_\delta(\alpha_1 t_1)/t_1^2-\tfrac{\alpha 1}{t_1}J_{\delta+1}(\alpha_1 t_1))-((\alpha_1 t_1)^2R^2-\delta^2)J_\delta(\alpha_1 t_1)/t_1{}^2\right)$$
$$+c_{12}\left(\delta(\delta-1)J_\delta(\alpha_1 t_1)/t_1{}^2-\tfrac{\alpha 1}{t_1}J_{\delta+1}(\alpha_1 t_1)\right)+(t_L^2L_1-\beta_1R^2N_1-d_{31}t_L^2M_1)J_\delta(\alpha_1 t_1)$$

$$S_{13}=c_{11}\left((\delta J_\delta(\alpha_2 t_1)/t_1^2-\tfrac{\alpha 2}{t_1}J_{\delta+1}(\alpha_2 t_1))-((\alpha_2 t_1)^2R^2-\delta^2)J_\delta(\alpha_2 t_1)/t_1{}^2\right)$$
$$+c_{12}\left(\delta(\delta-1)J_\delta(\alpha_2 t_1)/t_1{}^2-\tfrac{\alpha 2}{t_1}J_{\delta+1}(\alpha_2 t_1)\right)+(t_L^2L_2-\beta_1R^2N_2-d_{31}t_L^2M_2)J_\delta(\alpha_2 t_1)$$

$$S_{15}=c_{11}\left((\delta J_\delta(\alpha_3 t_1)/t_1^2-\tfrac{\alpha 1}{t_1}J_{\delta+1}(\alpha_3 t_1))-((\alpha_3 t_1)^2R^2-\delta^2)J_\delta(\alpha_3 t_1)/t_1{}^2\right)$$
$$+c_{12}\left(\delta(\delta-1)J_\delta(\alpha_3 t_1)/t_1{}^2-\tfrac{\alpha 3}{t_1}J_{\delta+1}(\alpha_3 t_3)\right)+(t_L^2L_3-\beta_1R^2N_3-d_{31}t_L^2M_3)J_\delta(\alpha_3 t_1)$$

$$S_{17}=c_{11}\left((\delta J_\delta(\alpha_4 t_1)/t_1^2-\tfrac{\alpha 1}{t_1}J_{\delta+1}(\alpha_4 t_1))-((\alpha_4 t_1)^2R^2-\delta^2)J_\delta(\alpha_4 t_1)/t_1{}^2\right)$$
$$+c_{12}\left(\delta(\delta-1)J_\delta(\alpha_4 t_1)/t_1{}^2-\tfrac{\alpha 1}{t_1}J_{\delta+1}(\alpha_4 t_1)\right)+(t_L^2L_4-\beta_1R^2N_4-d_{31}t_L^2M_4)J_\delta(\alpha_4 t_1)$$

$$S_{19}=c_{11}((k_3\delta J_{\delta+1}(k_3 t_1)/t_1)-\delta(\delta-1)J_\delta(k_3 t_1)/t_1{}^2)$$
$$+c_{12}(\delta(\delta-1)J_\delta(k_3 t_1)/t_1{}^2-k_1\delta J_{\delta+1}(k_3 t_1)/t_1)$$

$$S_{21}=2\delta\alpha_1 J_{\delta+1}(\alpha_1 t_1)/t_1-2\delta(\delta-1)J_\delta(\alpha_1 t_1)/t_1{}^2$$

$$S_{23}=2\delta\alpha_2 J_{\delta+1}(\alpha_2 t_1)/t_1-2\delta(\delta-1)J_\delta(\alpha_2 t_1)/t_1{}^2$$

$$S_{25}=2\delta\alpha_3 J_{\delta+1}(\alpha_3 t_1)/t_1-2\delta(\delta-1)J_\delta(\alpha_3 t_1)/t_1{}^2$$

$$S_{27}=2\delta\alpha_4 J_{\delta+1}(\alpha_4 t_1)/t_1-2\delta(\delta-1)J_\delta(\alpha_4 t_1)/t_1{}^2$$

$$S_{29}=k_3R^2t_1^2J_\delta(k_3 t_1)/t_1^2-2\delta(\delta-1)J_\delta(k_3 t_1)/t_1{}^2+k_1/t_1J_{\delta+1}(k_3 t_1)$$

$$S_{31}=t_L(1-L_1)(\delta J_\delta(\alpha_1 t_1)/t_1-\alpha_1 J_{\delta+1}(\alpha_1 t_1))$$

$$S_{33}=t_L(1-L_2)(\delta J_\delta(\alpha_2 t_1)/t_1-\alpha_2 J_{\delta+1}(\alpha_2 t_1))$$

$$S_{35}=t_L(1-L_3)(\delta J_\delta(\alpha_3 t_1)/t_1-\alpha_3 J_{\delta+1}(\alpha_3 t_1))$$

$$S_{37}=t_L(1-L_4)(\delta J_\delta(\alpha_4 t_1)/t_1-\alpha_4 J_{\delta+1}(\alpha_4 t_1))$$

$$S_{39}=t_L\delta J_\delta(k_3 t_1)/t_1$$

$$S_{41}=M_1(\delta J_\delta(\alpha_1 t_1)/t_1-\alpha_1 J_{\delta+1}(\alpha_1 t_1)+hJ_\delta(\alpha_1 t_1))$$

$$S_{43}=M_2(\delta J_\delta(\alpha_2 t_1)/t_1-\alpha_2 J_{\delta+1}(\alpha_2 t_1)+hJ_\delta(\alpha_2 t_1))$$

$$S_{45} = M_3(\delta J_\delta(\alpha_3 t_1)/t_1 - \alpha_3 J_{\delta+1}(\alpha_3 t_1) + hJ_\delta(\alpha_3 t_1))$$

$$S_{47} = M_4(\delta J_\delta(\alpha_4 t_1)/t_1 - \alpha_4 J_{\delta+1}(\alpha_4 t_1) + hJ_\delta(\alpha_4 t_1))$$

$$S_{49} = 0$$

$$S_{51} = t_L\left(1 + \mu_{11} N_1 - L_1\right)[\delta J_\delta(\alpha_1 t_1)/t_1 - \alpha_1 J_{\delta+1}(\alpha_1 t_1)]$$

$$S_{53} = t_L\left(1 + \mu_{11} N_2 - L_2\right)[\delta J_\delta(\alpha_2 t_1)/t_1 - \alpha_2 J_{\delta+1}(\alpha_2 t_1)]$$

$$S_{55} = t_L\left(1 + \mu_{11} N_3 - L_3\right)[\delta J_\delta(\alpha_3 t_1)/t_1 - \alpha_3 J_{\delta+1}(\alpha_3 t_1)]$$

$$S_{57} = t_L\left(1 + \mu_{11} N_4 - L_4\right)[\delta J_\delta(\alpha_4 t_1)/t_1 - \alpha_4 J_{\delta+1}(\alpha_4 t_1)]$$

$$S_{59} = t_L J_\delta(k_3 t_1)/t_1$$

in which $t_1 = a/R = 1 - t*/2$, $t_2 = b/R = 1 + t*/2$ and $t* = b - a/R$ is the thickness-to-mean radius ratio of the panel. Obviously, $S_{kl}(l = 2, 4, 6, 8, 10)$ can be obtained by just replacing modified Bessel function of the first kind in $S_{kl}(k = 1, 3, 5, 7, 9)$ with the ones of the second kind, respectively, while $S_{kl}(k = 6, 7, 8, 9, 10)$ can be obtained by just replacing t_1 in $S_{kl}(k = 1, 2, 3, 4, 5)$ with t_2.

5.2.1 Dispersion analysis of generalized magneto-thermoelastic waves in a transversely isotropic cylindrical panel

The frequency equation of three-dimensional vibration of a generalized magneto-thermoelastic cylindrical panel is obtained by using the boundary conditions in equations (4.6) and (5.1) in equations (2.103) and (2.2), we get

$$|T_{kl}| = 0 \quad (k, l = 1, 2, \ldots, 10) \tag{5.3}$$

where the elements in the determinant are given by

$$T_{11} = c_{11}\left((\delta J_\delta(\alpha_1 t_1)/t_1^2 - \frac{\alpha_1}{t_1} J_{\delta+1}(\alpha_1 t_1)) - ((\alpha_1 t_1)^2 R^2 - \delta^2) J_\delta(\alpha_1 t_1)/{t_1}^2\right)$$
$$+ c_{12}\left(\delta(\delta - 1) J_\delta(\alpha_1 t_1)/{t_1}^2 - \frac{\alpha_1}{t_1} J_{\delta+1}(\alpha_1 t_1)\right) + (t_L^2 U_1 - \beta_1 R^2 \eta_2 V_1 - d_{31} t_L^2 W_1) J_\delta(\alpha_1 t_1)$$

$$T_{13} = c_{11}\left((\delta J_\delta(\alpha_2 t_1)/t_1^2 - \frac{\alpha_2}{t_1} J_{\delta+1}(\alpha_2 t_1)) - ((\alpha_2 t_1)^2 R^2 - \delta^2) J_\delta(\alpha_2 t_1)/{t_1}^2\right)$$
$$+ c_{12}\left(\delta(\delta - 1) J_\delta(\alpha_2 t_1)/{t_1}^2 - \frac{\alpha_2}{t_1} J_{\delta+1}(\alpha_2 t_1)\right) + (t_L^2 U_2 - \beta_1 R^2 \eta_2 V_2 - d_{31} t_L^2 W_2) J_\delta(\alpha_2 t_1)$$

$$T_{15} = c_{11}\left((\delta J_\delta(\alpha_3 t_1)/t_1^2 - \frac{\alpha_1}{t_1} J_{\delta+1}(\alpha_3 t_1)) - ((\alpha_3 t_1)^2 R^2 - \delta^2) J_\delta(\alpha_3 t_1)/{t_1}^2\right)$$
$$+ c_{12}\left(\delta(\delta - 1) J_\delta(\alpha_3 t_1)/{t_1}^2 - \frac{\alpha_3}{t_1} J_{\delta+1}(\alpha_3 t_3)\right) + (t_L^2 U_3 - \beta_1 R^2 \eta_2 V_3 - d_{31} t_L^2 W_3) J_\delta(\alpha_3 t_1)$$

$$T_{17} = c_{11}\Big((\delta J_\delta(\alpha_4 t_1)/t_1^2 - \tfrac{\alpha 1}{t_1} J_{\delta+1}(\alpha_4 t_1)) - ((\alpha_4 t_1)^2 R^2 - \delta^2) J_\delta(\alpha_4 t_1)/t_1{}^2\Big)$$
$$+ c_{12}\Big(\delta(\delta - 1)J_\delta(\alpha_4 t_1)/t_1{}^2 - \tfrac{\alpha 1}{t_1} J_{\delta+1}(\alpha_4 t_1)\Big) + (t_L^2 U_4 - \beta_1 R^2 \eta_2 V_4 - d_{31} t_L^2 W_4) J_\delta(\alpha_4 t_1)$$

$$T_{19} = c_{11}((k_3 \delta J_{\delta+1}(k_3 t_1)/t_1) - \delta(\delta - 1)J_\delta(k_3 t_1)/t_1{}^2)$$
$$+ c_{12}(\delta(\delta - 1)J_\delta(k_3 t_1)/t_1{}^2 - k_1 \delta J_{\delta+1}(k_3 t_1)/t_1)$$

$$T_{21} = 2\delta\alpha_1 J_{\delta+1}(\alpha_1 t_1)/t_1 - 2\delta(\delta - 1)J_\delta(\alpha_1 t_1)/t_1{}^2$$

$$T_{23} = 2\delta\alpha_2 J_{\delta+1}(\alpha_2 t_1)/t_1 - 2\delta(\delta - 1)J_\delta(\alpha_2 t_1)/t_1{}^2$$

$$T_{25} = 2\delta\alpha_3 J_{\delta+1}(\alpha_3 t_1)/t_1 - 2\delta(\delta - 1)J_\delta(\alpha_3 t_1)/t_1{}^2$$

$$T_{27} = 2\delta\alpha_4 J_{\delta+1}(\alpha_4 t_1)/t_1 - 2\delta(\delta - 1)J_\delta(\alpha_4 t_1)/t_1{}^2$$

$$T_{29} = k_3 R^2 t_1{}^2 J_\delta(k_3 t_1)/t_1^2 - 2\delta(\delta - 1)J_\delta(k_3 t_1)/t_1{}^2 + k_1/t_1 J_{\delta+1}(k_3 t_1)$$

$$T_{31} = t_L(1 - U_1)(\delta J_\delta(\alpha_1 t_1)/t_1 - \alpha_1 J_{\delta+1}(\alpha_1 t_1))$$

$$T_{33} = t_L(1 - U_2)(\delta J_\delta(\alpha_2 t_1)/t_1 - \alpha_2 J_{\delta+1}(\alpha_2 t_1))$$

$$T_{35} = t_L(1 - U_3)(\delta J_\delta(\alpha_3 t_1)/t_1 - \alpha_3 J_{\delta+1}(\alpha_3 t_1))$$

$$T_{37} = t_L(1 - U_4)(\delta J_\delta(\alpha_4 t_1)/t_1 - \alpha_4 J_{\delta+1}(\alpha_4 t_1))$$

$$T_{39} = t_L \delta J_\delta(k_3 t_1)/t_1$$

$$T_{41} = W_1(\delta J_\delta(\alpha_1 t_1)/t_1 - \alpha_1 J_{\delta+1}(\alpha_1 t_1) + h J_\delta(\alpha_1 t_1))$$

$$T_{43} = W_2(\delta J_\delta(\alpha_2 t_1)/t_1 - \alpha_2 J_{\delta+1}(\alpha_2 t_1) + h J_\delta(\alpha_2 t_1))$$

$$T_{45} = W_3(\delta J_\delta(\alpha_3 t_1)/t_1 - \alpha_3 J_{\delta+1}(\alpha_3 t_1) + h J_\delta(\alpha_3 t_1))$$

$$T_{47} = W_4(\delta J_\delta(\alpha_4 t_1)/t_1 - \alpha_4 J_{\delta+1}(\alpha_4 t_1) + h J_\delta(\alpha_4 t_1)) \; T_{49} = 0$$

$$T_{51} = t_L\big(1 + \mu_{11} V_1 - U_1\big)(\delta J_\delta(\alpha_1 t_1)/t_1 - \alpha_1 J_{\delta+1}(\alpha_1 t_1))$$

$$T_{53} = t_L\big(1 + \mu_{11} V_2 - U_2\big)(\delta J_\delta(\alpha_2 t_1)/t_1 - \alpha_2 J_{\delta+1}(\alpha_2 t_1))$$

$$T_{55} = t_L\big(1 + \mu_{11} V_3 - U_3\big)(\delta J_\delta(\alpha_3 t_1)/t_1 - \alpha_3 J_{\delta+1}(\alpha_3 t_1))$$

$$T_{57} = t_L\big(1 + \mu_{11} V_4 - U_4\big)(\delta J_\delta(\alpha_4 t_1)/t_1 - \alpha_4 J_{\delta+1}(\alpha_4 t_1))$$

$$T_{59} = t_L J_\delta(k_3 t_1)/t_1.$$

in which $t_1 = a/R = 1 - t*/2$, $t_2 = b/R = 1 + t*/2$ and $t* = b - a/R$ is the thickness-to-mean radius ratio of the panel. Obviously $T_{kl}(l = 2, 4, 6, 8, 10)$ can be obtained by just replacing the modified Bessel function of the first kind in $T_{kl}(k = 1, 3, 5, 7, 9)$ with ones of the second kind, respectively, while $T_{kl}(k = 6, 7, 8, 9, 10)$ can be obtained by just replacing t_1 in $T_{kl}(k = 1, 2, 3, 4, 5)$ with t_2.

5.3 Numerical results and discussion

This section has been reproduced from Ponnusamy and Selvamani (2013).

The frequency equation (5.2) is numerically solved for cobalt iron oxide ($CoFe_2O_4$) material. For the purpose of numerical computation we consider the closed circular cylindrical shell with the centre angle $\alpha = 2\pi$ and the integer n must be even since the shell vibrates in a circumferential full wave. In fact, the frequency equation for a closed cylindrical shell can be obtained by setting $\delta = N(N = 1, 2, 3, \ldots)$ and N is the circumferential wave number in equation (5.2). For a transversely isotropic material the elastic constants of $CoFe_2O_4$ are taken from Buchanan (2003)

$$\begin{aligned}
c_{11} &= 6.31 \times 10^9\ \mathrm{N\,m^{-2}} & \mu_{11} &= -344.66 \times 10^{-6}\ \mathrm{N\,s^2\,C^{-2}}\\
c_{12} &= 3.81 \times 10^9\ \mathrm{N\,m^{-2}} & \mu_{33} &= 91.72 \times 10^{-6}\mathrm{N\,s^2\,C^{-2}}\\
c_{13} &= 3.75 \times 10^9\ \mathrm{N\,m^{-2}} & d_{31} &= 8.2 \times 10^{-1}\ \mathrm{N\,A^{-1}\,m^{-1}}\\
c_{33} &= 5.94 \times 10^9\ \mathrm{N\,m^{-2}} & d_{33} &= 1.0\ \mathrm{N\,A^{-1}\,m^{-1}}\\
c_{44} &= 1.00 \times 10^9\ \mathrm{N\,m^{-2}} & d_{51} &= 7.810^{-1}\ \mathrm{N\,A^{-1}m^{-1}}\\
c_{66} &= 1.24 \times 10^9\ \mathrm{N\,m^{-2}} & c_\nu &= 420\ \mathrm{J\,kg^{-1}\,K^{-1}}\\
T_0 &= 298\ \mathrm{K} & \beta_1 &= 1.52 \times 10^6\ \mathrm{N\,K^{-1}\,m^{-2}}\\
\rho &= 5.3 \times 10^3\ \mathrm{kg\,m^{-3}} & \beta_3 &= 1.53 \times 10^6\ \mathrm{N\,K^{-1}\,m^{-2}}\\
p_3 &= -452 \times 10^{-6}\ \mathrm{C\,K^{-1}\,m^{-2}} & K_1 &= K_3 = 1.5\ \mathrm{W\,m^{-1}\,K^{-1}}.
\end{aligned} \tag{5.4}$$

A comparison of the non-dimensional frequency modes S1, S2, S3, S4 and S5 of symmetric and anti-symmetric vibration among the longitudinal, flexural, torsional modes of thermally insulated and isothermal boundary conditions for a transversely isotropic cylindrical panel with $a/b = 0.25$ are shown in the tables 5.1 and 5.2, respectively. From these tables it is clear that as the first five vibration modes increase, the non-dimensional frequencies also increase in both symmetric and anti-symmetric modes of vibration and also it is clear that the non-dimensional frequency profiles exhibit high amplitude for anti-symmetric mode compared with symmetric modes of vibration.

In the other spectrum of the non-dimensional frequencies of symmetric and anti-symmetric modes of longitudinal, flexural, torsional vibrations of thermally insulated and isothermal boundary conditions for a transversely isotropic cylindrical panel with $a/b = 0.5$ are presented in tables 5.3 and 5.4. From tables 5.3 and 5.4, it is observed that as the modes increase, the non-dimensional frequencies also increase, whereas the dispersion of longitudinal and flexural modes are almost the

Table 5.1. Non-dimensional frequency for first five symmetric modes of longitudinal, flexural and torsional vibration of transversely isotropic cylindrical panel with $a/b = 0.25$.

	Thermally insulated			Isothermal		
Mode	Longitudinal mode	Flexural mode	Torsional mode	Longitudinal mode	Flexural mode	Torsional mode
S1	1.3937	1.3927	1.5565	1.2289	1.2278	1.4295
S2	1.6542	1.6533	1.8391	1.4614	1.4604	1.7886
S3	1.9176	1.9156	2.1227	1.7009	1.7019	2.0529
S4	2.1832	2.1802	2.4048	1.9486	1.9475	2.2504
S5	2.5840	2.5810	2.8318	2.3381	2.3375	3.5245

Table 5.2. Non-dimensional frequency for first five anti-symmetric modes of longitudinal, flexural and torsional vibration of transversely isotropic cylindrical panel with $a/b = 0.25$.

	Thermally insulated			Isothermal		
Mode	Longitudinal mode	Flexural mode	Torsional mode	Longitudinal mode	Flexural mode	Torsional mode
S1	1.4069	1.3405	1.2414	1.4049	1.3237	1.3588
S2	1.6702	1.6039	1.4868	1.5611	1.6502	1.6123
S3	1.9360	1.8722	1.7452	1.9182	1.9076	1.8682
S4	2.2408	2.1458	2.0174	2.0768	2.1753	2.1264
S5	2.6137	2.5641	2.4496	2.5680	2.6670	2.5183

Table 5.3. Non-dimensional frequency for first five symmetric modes of longitudinal, flexural, and torsional vibration of transversely isotropic cylindrical panel with $a/b = 0.5$.

	Thermally insulated			Isothermal		
Mode	Longitudinal mode	Flexural mode	Torsional mode	Longitudinal mode	Flexural mode	Torsional mode
S1	1.4558	1.4543	1.5153	1.4049	1.4037	1.5588
S2	1.7260	1.7251	1.8827	1.6611	1.5602	1.7123
S3	1.9967	1.9057	2.0511	1.9182	1.9076	1.9682
S4	2.1778	2.2648	3.0213	2.1768	2.1753	2.5264
S5	2.6754	2.6732	3.8303	2.5680	2.5670	2.6183

same and the torsional mode gets dominant in symmetric and anti-symmetric cases. The amplitude of all modes of vibrations is increased in magnitude with respect to aspect ratios also.

Table 5.4. Non-dimensional frequency for first five anti-symmetric modes of longitudinal, flexural and torsional vibration of transversely isotropic cylindrical panel with $a/b = 0.5$.

	Thermally insulated			Isothermal		
Mode	Longitudinal mode	Flexural mode	Torsional mode	Longitudinal mode	Flexural mode	Torsional mode
S1	1.4237	1.5028	1.6295	1.4125	1.4337	1.4768
S2	1.6842	1.6842	1.7986	1.5611	1.6602	1.6323
S3	1.9576	1.9239	1.9529	1.8180	1.9176	1.9682
S4	2.3232	2.1832	2.7204	2.2768	2.1753	2.6264
S5	2.6840	2.5840	3.9245	2.6443	2.5870	3.5183

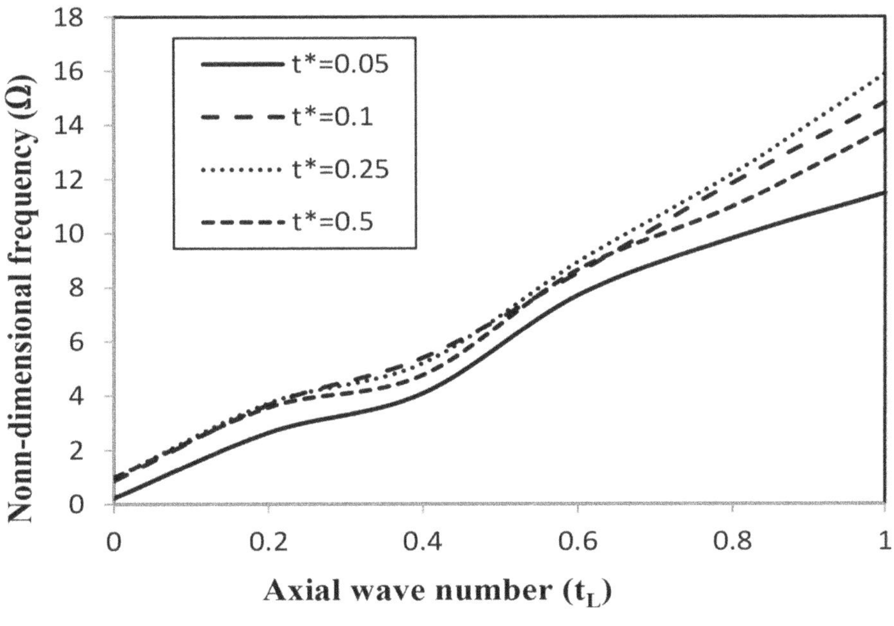

Figure 5.1. Dispersion curves for non-dimensional frequency Ω versus the axial wave number t_L with different t_* for isothermal cylindrical shell.

Figures 5.1 and 5.2 represent the variations of the non-dimensional frequency Ω of an elastic cylindrical shell with respect to the parameter $t_L = m\pi R/L$ shown for different values of the thickness-to-mean radius of the shell $t* = 0.05$, 0.1, 0.25 and 0.5 for isothermal and thermally insulated boundary conditions, respectively. From figure 5.1 it is observed that the non-dimensional frequency of the shell becomes dispersive with respect to t_L whenever the thickness parameter $t*$ increases.

But in figure 5.2 some dispersions are observed from the linear behaviour of frequency with respect to t_L for different values of $t*$ due to the insulated nature of

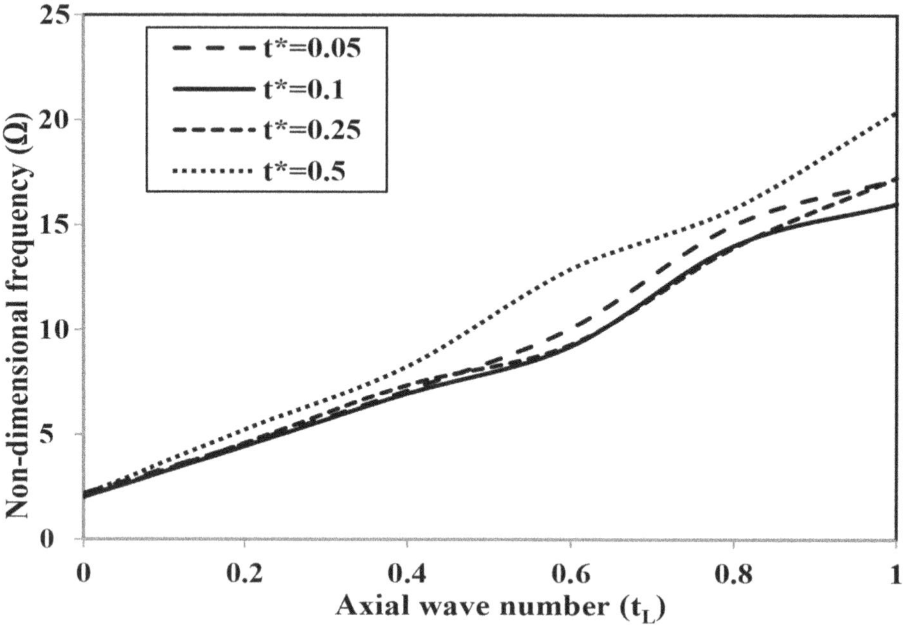

Figure 5.2. Dispersion curves for non-dimensional frequency Ω versus the axial wave number t_L with different t_* for thermally insulated cylindrical shell.

the boundary. In both the cases at small values of the parameter t_L in the range $0 \leqslant t_L \leqslant 1$ the values of frequency are almost steady for different values of the thickness parameter $t_* = 0.05,\ 0.1,\ 0.25$ and 0.5, whereas for higher values of t_L the frequency is large and starting to be dispersive for increasing values of t_*.

The variations of the non-dimensional phase velocity with respect to the wave number for isothermal and thermally insulated cylindrical shell are shown in figures 5.3 and 5.4, respectively. From these figures, it is seen that the phase velocity curves are dispersive only for small values of wave number in the range $0 \leqslant \delta \leqslant 0.4$, but for higher values of wave number, these become non-dispersive for both isothermal and insulated boundary conditions.

There is a small deviation in the magnitude of frequency in an insulated boundary which might happen because of the dissipation of energy and random behaviour of molecules due to thermal and magnetic waves and strong alignment of molecules. The phase velocity of higher modes of propagation attains quite large values at vanishing wave number which sharply slashes down to become steady and asymptotic with increasing wave number due to the coupling effect of magnetic and thermal fields. In figure 5.5, the variation of attenuation coefficient q with respect to circumferential wave number δ of an isothermal cylindrical shell is discussed for different thickness parameters t_*. The magnitude of the attenuation coefficient increases monotonically to attain maximum value in $0.4 \leqslant \delta \leqslant 0.8$ and slashes down to became asymptotically linear in the remaining range of circumferential wave number.

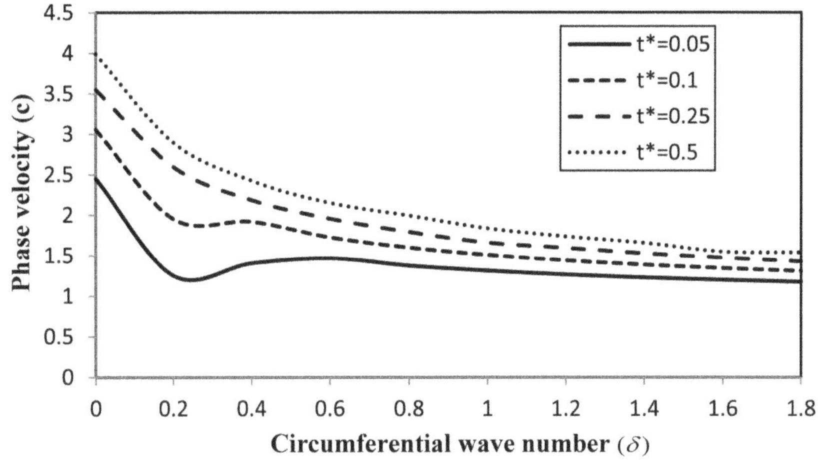

Figure 5.3. Dispersion curves for phase velocity c versus circumferential wave number δ with different t_* for isothermal cylindrical shell.

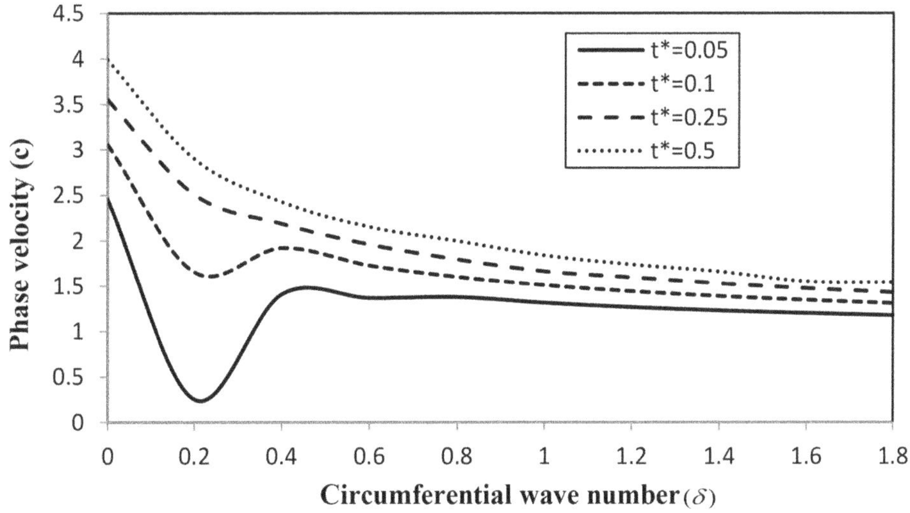

Figure 5.4. Dispersion curves for phase velocity c versus circumferential wave number δ with different t_* for thermally insulated cylindrical shell.

The variations of attenuation coefficient q with respect to circumferential wave number δ of a thermally insulated cylindrical shell are discussed for different thickness parameters t_* are shown in figure 5.6. Here the attenuation coefficient attains maximum value in $0.2 \leqslant \delta \leqslant 0.6$ for $t_* = 0.5,\ 0.1$ and $0.6 \leqslant \delta \leqslant 1$ for $t_* = 0.25,\ 0.5$ and slashes down to become linear due to the insulation. From figures 5.5 and 5.6, it is clear that the attenuation profiles exhibit oscillating nature

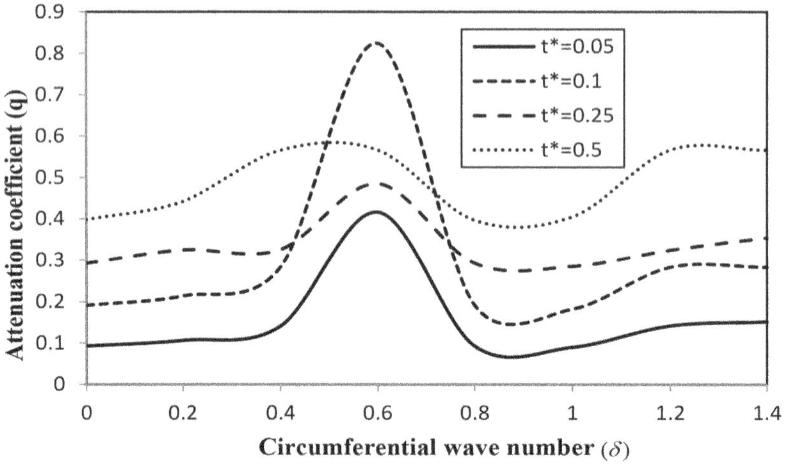

Figure 5.5. Dispersion curves for attenuation coefficient q versus circumferential wave number δ with different t^* for isothermal cylindrical shell.

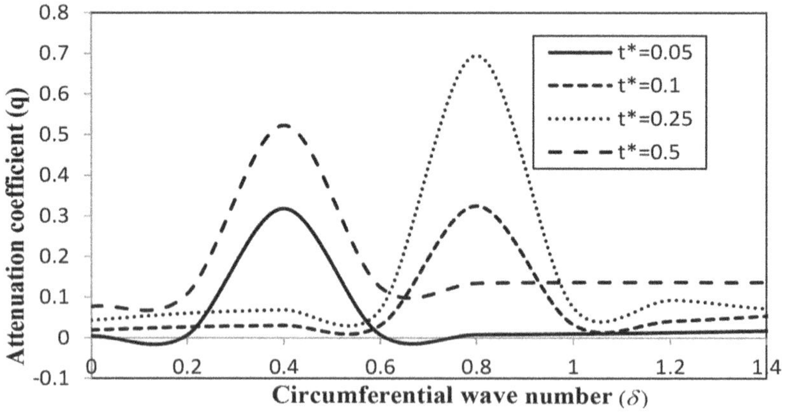

Figure 5.6. Dispersion curves for attenuation coefficient q versus circumferential wave number δ with different t^* for thermally insulated cylindrical shell.

both in an isothermal and a thermally insulated boundary due to the combined effect of thermal and magnetic fields.

5.3.1 Dispersion analysis of generalized magneto-thermoelastic waves in a transversely isotropic cylindrical panel

The numerical example which will demonstrate the dispersion analysis of a generalized thermoelastic cylindrical panel is studied for the material cobalt iron oxide ($CoFe_2O_4$). The computed non-dimensional phase velocity, attenuation coefficient, specific loss and thermomechanical coupling factor are plotted in the form of dispersion curves with the support of Matlab programming. The frequency equation (5.3) is numerically solved for cobalt iron oxide ($CoFe_2O_4$) and its material constantsare taken from equation (5.4) together with the thermal relaxation times

$$\tau_0 = 0.75 \times 10^{-13}\,\text{s and } \tau_1 = 0.5 \times 10^{-13}\,\text{s}.$$

A comparison is made for the natural frequencies among the Green–Lindsay (G–L), Lord–Shulman Theory (L–S) and classical theory (CT) of thermoelasticity for the symmetric and skew-symmetric modes of magneto-thermoelastic cylindrical shell is given in tables 5.5 and 5.6, respectively.

From these tables, it is clear that as the sequential number of the wave number increases, the non-dimensional frequencies also increase for both the symmetric and skew-symmetric modes. Also, it is clear that the non-dimensional frequency exhibits higher amplitudes for the L–S theory compared with the G–L and CT theories due to the combined effect of thermal relaxation times and magnetic field.

Figures 5.7 and 5.8 represent the variations of the non-dimensional phase velocity with wave number of a thermally insulated cylindrical shell in the symmetric and

Table 5.5. Comparison of non-dimensional frequency among the L–S, G–L and CT for symmetric and skew-symmetric modes of the magneto-thermoelastic cylindrical shell in first mode.

	Symmetric			Skew-symmetric		
Wave number (δ)	LS	GL	CT	LS	GL	CT
0.1	0.0779	0.0532	0.1236	0.0259	0.1149	0.1084
0.2	0.1936	0.1801	0.2736	0.1702	0.1213	0.2130
0.6	0.2978	0.2063	0.2928	0.2950	0.2563	0.3220
1.2	0.4998	0.3967	0.3727	0.3837	0.3732	0.3295
1.8	0.6349	0.5010	0.4036	0.5129	0.4831	0.4752
2.4	0.7947	0.6400	0.5308	0.8727	0.6422	0.6349
3.0	1.3191	0.9025	0.7015	0.9308	0.8231	0.9142

Table 5.6. Comparison of non-dimensional frequency among L–S, G–L and CT for symmetric and skew-symmetric modes of magneto-thermoelastic cylindrical shell in second mode.

	Symmetric			Skew-symmetric		
Wave number (δ)	LS	GL	CT	LS	GL	CT
0.1	0.1527	0.1433	0.1140	0.1608	0.1342	0.1152
0.2	0.2498	0.2417	0.1923	0.2255	0.1969	0.1564
0.6	0.3128	0.3057	0.2433	0.5773	0.3248	0.2444
1.2	0.4470	0.4385	0.3489	0.5941	0.5593	0.3487
1.0	0.7113	0.6952	0.5532	0.6303	0.8050	0.6584
2.4	0.9091	0.8714	0.6934	0.7170	0.8512	0.7551
3.0	1.3452	1.1350	0.9650	1.2007	1.0230	0.9038

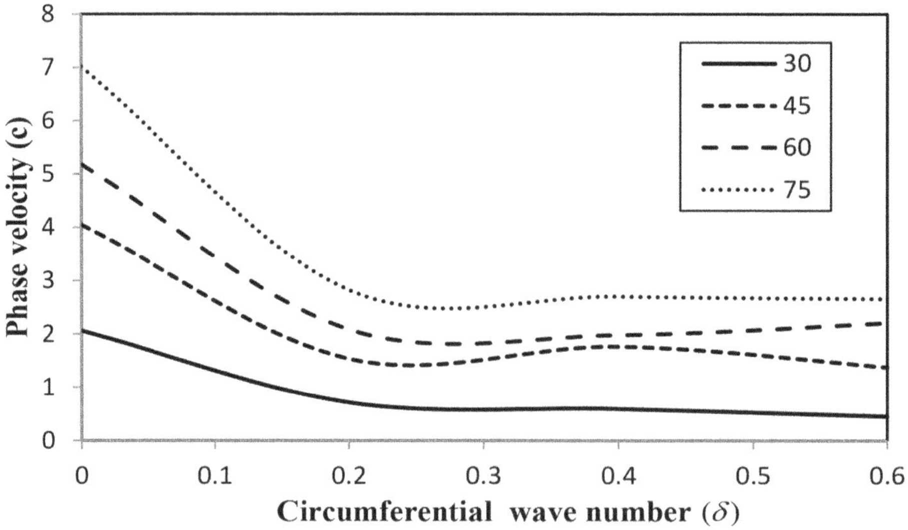

Figure 5.7. Phase velocity profile c of symmetric mode with circumferential wave number δ for thermally insulated cylindrical shell.

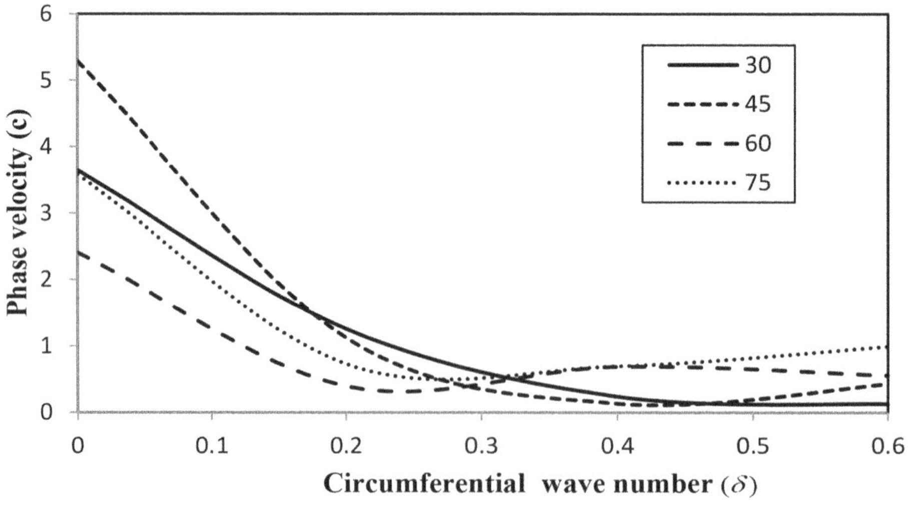

Figure 5.8. Phase velocity profile c of skew-symmetric mode with circumferential wave number δ for thermally insulated cylindrical shell.

skew-symmetric modes of vibrations shown for the different angles $\alpha = 30°$, 45°, 60° and 75°. From figure 5.7, it is clear that the phase velocity slashes down at small wave number and becomes asymptotically linear for $\delta \geqslant 0.2$ in all directions. But in figure 5.8 there is a small dispersion in the linear behaviour along 45° and 75°. From figures 5.7 and 5.8, it is observed that the non-dimensional phase velocity of the thermally insulated cylindrical shell decreases as wave normal inclination progresses from 30° to 75°, it shows the directional dependence of the velocity of the wave motion. Thus, for the thermally insulated cylindrical shell, the phase velocity decreases from its maximum value at minimal wave number and becomes asymptotically linear at higher wave number.

Figures 5.9 and 5.10 explain the variations of attenuation coefficient with respect to the wave number of a thermally insulated cylindrical shell discussed for symmetric and skew-symmetric modes. From figure 5.9, the magnitude of the attenuation coefficient increases monotonically to attain maximum value in $0 \leqslant \delta \leqslant 0.4$ for small wave number and slashes down to became oscillating in nature in the remaining range of wave number for symmetric mode.

The skew-symmetric modes of vibration are oscillating at a higher wavelength in all directions. From figures 5.9 and 5.10, it is clear that the attenuation profiles exhibit oscillation for both symmetric and skew-symmetric modes. From these figures it is clear that the velocity is quite dependent on propagation direction and attains maximum value along 75° and minimum along 30° due to the fact that the wave velocity is maximum near to the free surface and minimum when it penetrates deeper to the medium.

Figures 5.11 and 5.12 represent the variation of the thermomechanical coupling factor with wave number for symmetric and skew-symmetric modes of the cylindrical shell. The magnitude of the factor increases monotonically at small wave number and becomes steady for higher wave number for both symmetric and skew-symmetric modes in all directions because in such situations all the interacting forces become significantly operative. The effect of the thermomechanical coupling factor is dispersive at higher wave number in skew-symmetric mode, which might

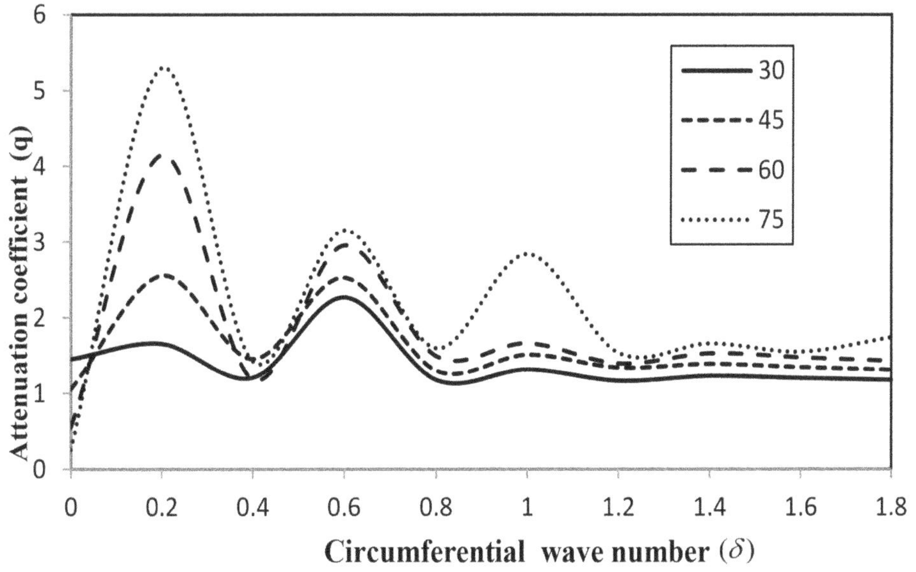

Figure 5.9. Variation of attenuation coefficient q of symmetric mode with circumferential wave number δ for thermally insulated cylindrical shell.

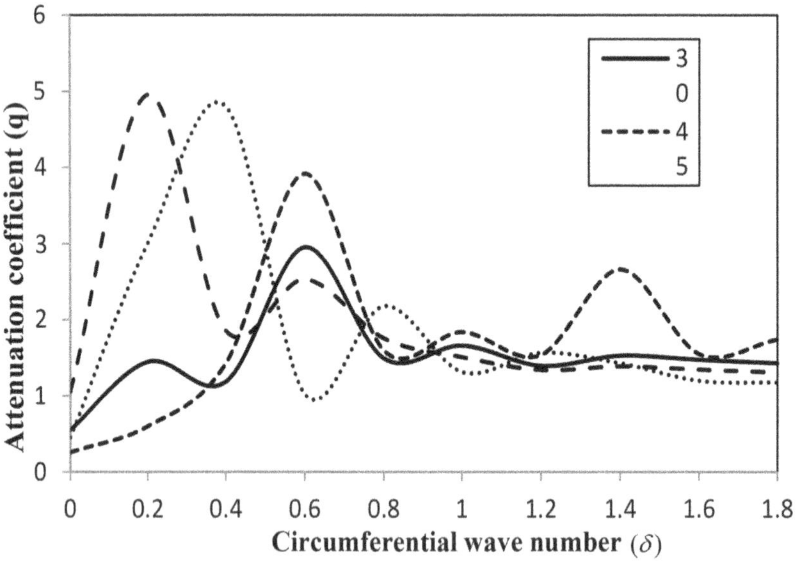

Figure 5.10. Variation of attenuation coefficient q of skew-symmetric mode with circumferential wave number δ for thermally insulated cylindrical shell.

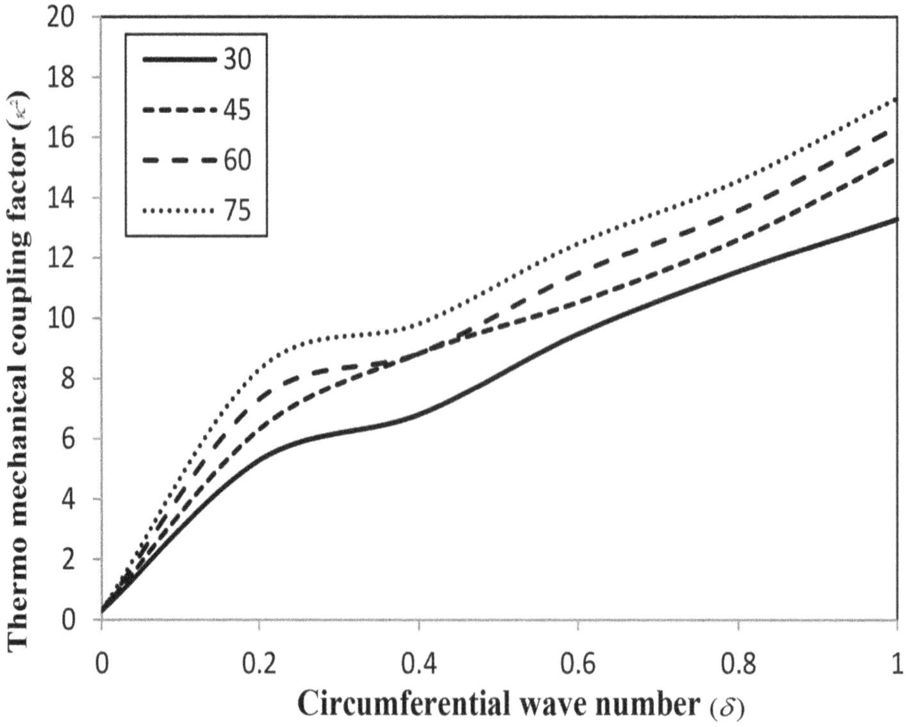

Figure 5.11. Variation of thermomechanical coupling factor κ^2 of symmetric mode with circumferential wave number δ for thermally insulated cylindrical shell.

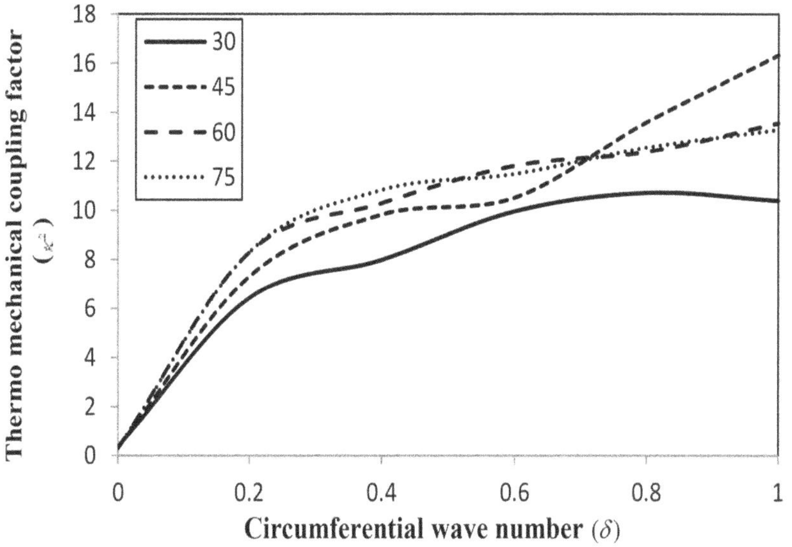

Figure 5.12. Variation of thermomechanical coupling factor κ^2 of skew-symmetric mode with wave number δ for thermally insulated cylindrical shell.

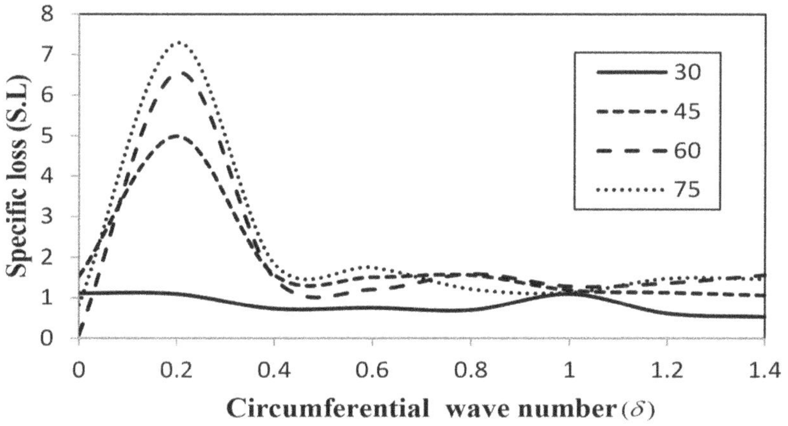

Figure 5.13. Variation of specific loss factor (S.L) of symmetric mode with circumferential wave number δ for thermally insulated cylindrical shell.

happen because of the dissipation of energy and random behaviour of molecules due to thermal waves and the strong alignment of molecules.

Figures 5.13 and 5.14 represents the effect of the specific loss factor for symmetric and skew-symmetric modes of vibrations. The magnitude of the specific loss factor attains maximum value at $0 \leqslant \delta \leqslant 0.4$ for the insulated symmetric mode of the cylindrical shell, as shown in figure 5.13, in all directions except 30°.

For skew-symmetric mode, the specific loss is getting maximum value at the wave number range $0 \leqslant \delta \leqslant 0.6$ due to the specific loss factor attaining maximum along 45°. In comparison of figures 5.13 and 5.14, it is clear that the specific loss factor is

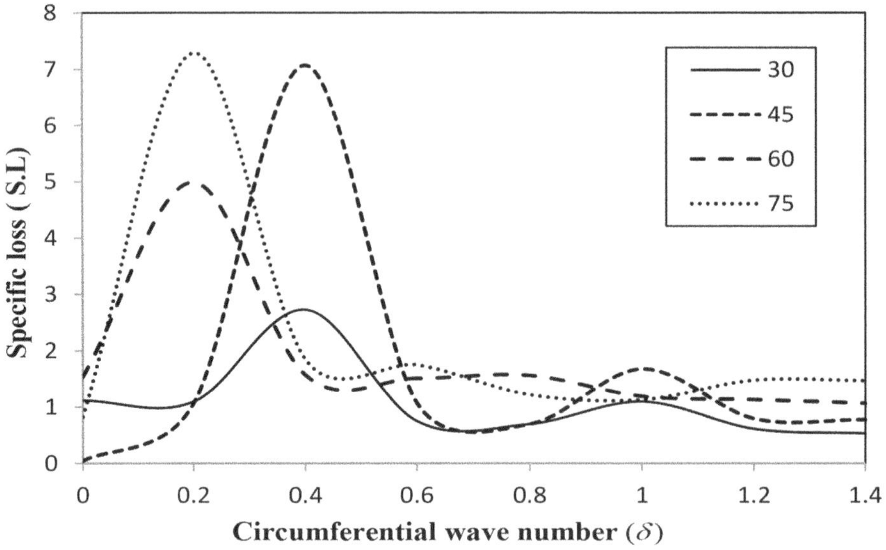

Figure 5.14. Variation of specific loss (S.L) of skew-symmetric mode with circumferential wave number δ for thermally insulated cylindrical shell.

quite high at smaller wave number and decreases asymptotically with increasing wave number due to the combined effect of the magnetic and thermal field.

5.4 Thickness shear wave propagation in a transversely isotropic piezoelectric cylindrical panel

The thickness shear wave propagation in a homogeneous transversely isotropic piezoelectric cylindrical panel is investigated in the context of the linear theory of elasticity. The fundamental equations are simplified and the free vibration solution for a simply supported transversely isotropic cylindrical panel is obtained by using Bessel functions with complex arguments. To clarify the correctness and effectiveness of the developed method, the dispersion curves of different panel parameters are computed and presented for PZT-5A material.

5.4.1 Formulation of the problem

The three-dimensional stress equations of motion in the absence of body force in a cylindrical coordinate (r, θ, z) are taken from equation (2.3) together with the Gauss charge equation given as in Sharma *et al* (2004)

$$D_{r,\,r} + \frac{1}{r}D_r + \frac{1}{r}D_{\theta,\,\theta} + D_{z,\,z} = 0. \tag{5.5}$$

The stress–strain relation for isotropic material by generalized Hooke's law is given by

$$\sigma_{rr} = c_{11}e_{rr} + c_{12}e_{\theta\theta} + c_{13}e_{zz} + e_{31}E_z$$

$$\sigma_{\theta\theta} = c_{12}e_{rr} + c_{11}e_{\theta\theta} + c_{13}e_{zz} + e_{31}E_z$$

$$\sigma_{zz} = c_{13}e_{rr} + c_{13}e_{\theta\theta} + c_{33}e_{zz} + e_{33}E_z \tag{5.6}$$

$$\sigma_{r\theta} = 2c_{66}e_{r\theta},\ \sigma_{\theta z} = 2c_{44}e_{\theta z} + r^{-1}e_{15}E_\theta,\ \sigma_{rz} = 2c_{44}e_{rz} + e_{15}E_r$$

$$\begin{aligned} D_r &= 2e_{15}e_{rz} - \epsilon_{11}E_r,\ D_\theta = 2e_{15}e_{\theta z} - \epsilon_{11}r^{-1}E_\theta, \\ D_z &= e_{31}(e_{rr} + e_{\theta\theta}) + e_{33}e_{zz} - \epsilon_{33}E_z. \end{aligned} \tag{5.7}$$

The strain–displacement relation is given as

$$e_{rr} = u_{,r},\ e_{\theta\theta} = r^{-1}(u + v_{,\theta}),\ e_{zz} = w_{,z} \tag{5.8}$$

$$2e_{r\theta} = v_{,r} - r^{-1}(v - u_{,\theta}),\ 2e_{zr} = (u_{,z} + w_{,r}),\ 2e_{\theta z} = (v_{,z} + r^{-1}w_{,\theta})$$

$$E_r = -\vartheta_{,r}\ E_\theta = -\vartheta_{,\theta}\ E_z = -\vartheta_{,z}. \tag{5.9}$$

Substituting the equations (5.6)–(5.9) into equations (2.3) and (5.5) gives the following three displacement equations of motion and the charge equation

$$\begin{aligned} &c_{11}(u_{,rr} + r^{-1}u_{,r} - r^{-2}u) - r^{-2}(c_{11} + c_{66})v_{,\theta} + r^{-2}c_{66}u_{,\theta\theta} \\ &+ c_{11}u_{,zz} + (c_{44} + c_{13})w_{,rz} + r^{-1}(c_{66} + c_{12})v_{,r\theta} + (e_{31} + e_{15})\vartheta_{,rz} = \rho u_{,tt} \end{aligned}$$

$$\begin{aligned} &r^{-1}(c_{12} + c_{66})u_{,r\theta} + r^{-2}(c_{66} + c_{11})u_{,\theta} + c_{66}(v_{,rr} + r^{-1}v_{,r} - r^{-2}v) \\ &+ r^{-2}c_{11}v_{,\theta\theta} + c_{44}v_{,zz} + r^{-1}(c_{44} + c_{13})w_{,\theta z} + (e_{31} + e_{15})r^{-1}\vartheta_{,\theta z} = \rho v_{,tt} \end{aligned}$$

$$\begin{aligned} &c_{44}(w_{,rr} + r^{-1}w_{,r} + r^{-2}w_{,\theta\theta}) + r^{-1}(c_{44} + c_{13})(u_{,z} + v_{,\theta z}) \\ &+ (c_{44} + c_{13})u_{,rz} + c_{33}w_{,zz} + e_{15}(\vartheta_{,rr} + r^{-1}\vartheta_{,r} + r^{-2}\vartheta_{,\theta\theta}) + e_{33}\vartheta_{,zz} = \rho w_{,tt} \end{aligned}$$

$$\begin{aligned} &(e_{31} + e_{15})(u_{,rz} + r^{-1}u_{,z} + r^{-1}v_{,\theta z}) + e_{15}(w_{,rr} + r^{-1}w_{,r} + r^{-2}w_{,\theta\theta}) + e_{33}w_{,zz} \\ &- \epsilon_{11}(\vartheta_{,rr} + r^{-1}\vartheta_{,r} + r^{-2}\vartheta_{,\theta\theta}) - \epsilon_{33}\vartheta_{,zz} = 0. \end{aligned} \tag{5.10}$$

Thickness shear wave propagation of a cylindrical panel is as follows.

We consider a cylindrical panel, shown in figure 2.1. For the case of thickness shear wave propagation the radial and circumferential displacements u and v vanish and the axial displacement w, electric potential ϑ are independent on the coordinate z. The governing equations can be rewritten as

$$\sigma_{rz,\,r} + r^{-1}\sigma_{rz} = \rho w_{,tt}\ D_{r,\,r} + r^{-1}D_r = 0$$

$$\sigma_{rz} = 2c_{44}e_{rz} + e_{15}E_r\ \sigma_{\theta z} = 2c_{44}e_{\theta z} + r^{-1}e_{15}E_\theta$$

$$D_r = 2e_{15}e_{rz} - \epsilon_{11}E_r\ D_\theta = 2e_{15}e_{\theta z} - \epsilon_{11}r^{-1}E_\theta \tag{5.11}$$

$$2e_{zr} = w_{,r}\ E_r = -\vartheta_{,r}.$$

The displacement equations become

$$c_{44}(w_{,rr} + r^{-1}w_{,r}) + e_{15}(\vartheta_{,rr} + r^{-1}\vartheta_{,r}) = \rho w_{,tt}$$

$$e_{15}(w_{,rr} + r^{-1}w_{,r}) - \epsilon_{11}(\vartheta_{,rr} + r^{-1}\vartheta_{,r}) = 0. \tag{5.12}$$

For the free time harmonic thickness shear vibration in the axial direction, we seek the solutions of the form

$$w(r, \theta, z, t) = \overline{w}(r)\sin(m\pi z)\sin(n\pi\theta/\alpha)e^{j\omega t}$$

$$\vartheta(r, \theta, z, t) = \overline{\vartheta}(r)\sin(m\pi z)\sin(n\pi\theta/\alpha)e^{j\omega t}. \tag{5.13}$$

By dropping the tildes superimposed upon w and ϑ

$$c_{44}(\overline{w}_{,rr} + r^{-1}\overline{w}_{,r}) - e_{15}(\overline{\vartheta}_{,rr} + r^{-1}\overline{\vartheta}_{,r}) + \rho\omega^2 = 0 \tag{5.14}$$

$$e_{15}(\overline{w}_{,rr} + r^{-1}\overline{w}_{,r}) + \epsilon_{11}(\overline{\vartheta}_{,rr} + r^{-1}\overline{\vartheta}_{,r}) = 0. \tag{5.15}$$

Eliminating ϑ from the equations (5.13) and (5.15), we obtain the following equation

$$\overline{w}_{,rr} + r^{-1}\overline{w}_{,r} + k^2\overline{w} = 0 \tag{5.16}$$

$$k^2 = \rho\omega^2/\bar{c}_{44} \text{ and } \bar{c}_{44} = c_{44} + e_{15}^2/\epsilon_{11}.$$

Solving equation (5.16) for w and ϑ we obtain

$$\overline{w} = A\, J_0(kr) + B\, Y_0(kr) \tag{5.17}$$

$$\overline{\vartheta} = \frac{e_{15}}{\epsilon_{11}}w + \frac{e_{15}}{\epsilon_{11}}(C \ln r + D) \tag{5.18}$$

where J_0 and Y_0 are Bessel functions of first and second kind of order zero, respectively, and the constants A, B, C and D are to be determined from the boundary conditions. The requirements that function w and ϑ satisfy the boundary condition $\sigma_{rz} = 0,\ \vartheta = 0\ at\ r = a,\ b$. Imposing these boundary conditions in equation (5.13) gives four homogeneous equations for the determination of A, B, C and D. These equations have a nontrivial solution only if

$$\frac{kaJ_1(ka)\ln a/b + k_{26}^2[J_0(ka) - J_0(kb)]}{kbJ_1(kb)\ln a/b + k_{26}^2[J_0(ka) - J_0(kb)]} = \frac{ka\,Y_1(ka)\ln a/bk_{26}^2[Y_0(ka) - Y_0(kb)]}{kb\,Y_1(kb)\ln a/bk_{26}^2[Y_0(ka) - Y_0(kb)]} \tag{5.19}$$

where $J_0' = -J_1$ and $Y_0' = -Y_1$.

5.4.2 Numerical results and discussion for a piezoelectric cylindrical panel

The frequency equation (5.19) is numerically solved for PZT-5A material. For the purpose of numerical computation we consider a closed circular cylindrical shell with the centre angle $\alpha = 2\pi$ and the integer n must be even since the shell vibrates in circumferential full wave. The material properties of PZT-5A are taken as in Ashida and Tauchert (2001)

$$c_{11} = 13.9 \times 10^{10}\,\mathrm{N\,m^{-2}},\ c_{12} = 7.78 \times 10^{10}\,\mathrm{N\,m^{-2}},\ c_{13} = 7.78 \times 10^{10}\,\mathrm{N\,m^{-2}}$$
$$c_{33} = 11.3 \times 10^{10}\,\mathrm{N\,m^{-2}},\ c_{44} = 11.3 \times 10^{10}\,\mathrm{N\,m^{-2}},\ e_{15} = 13.4\,\mathrm{C\,m^{-2}},$$
$$e_{33} = 13.8\,\mathrm{C\,m^{-2}},\ \epsilon_{11} = 60 \times 10^{-10}\,\mathrm{C^2\,N^{-1}\,m^{-2}},\ \rho = 7750\,\mathrm{Kg\,m^{-3}}.$$

Figures 5.15 and 5.16 represent the variations of the lowest frequency of a simply supported cylindrical shell of PZT-5A shown for two values of the parameter $t_L = m\pi R/L$ with respect to the thickness-to-mean radius ratio (h/R) for different values of the length-to-inner radius ratio of the shell ($L/b = 0.1$ 0.25 and 0.5). From these figures, it is observed that the lowest frequency of the shell is non-dispersive

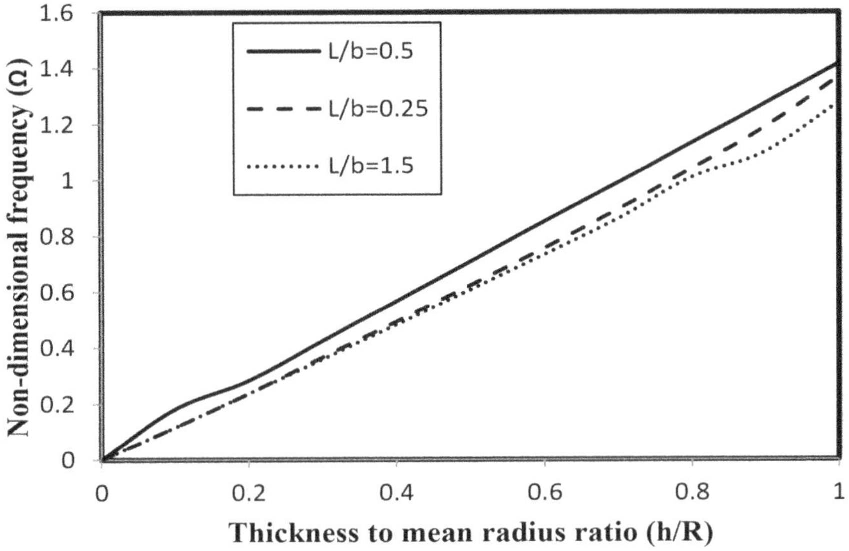

Figure 5.15. Variation of non-dimensional frequency Ω with thickness-to-mean radius ratio h/R for $t_L = 1$.

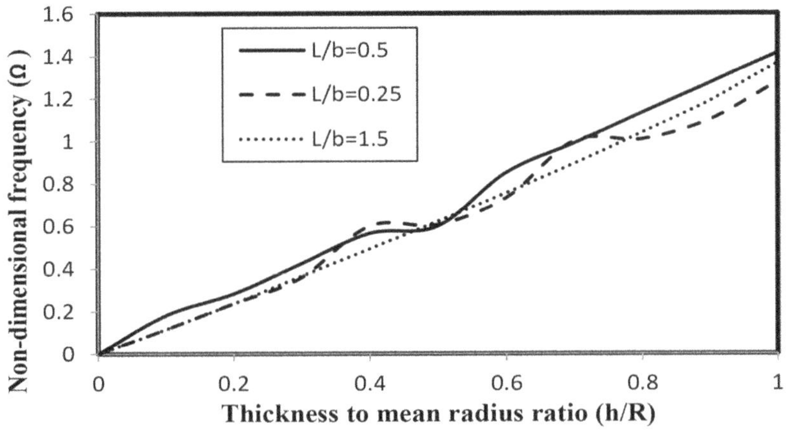

Figure 5.16. Variation of non-dimensional frequency Ω with thickness-to-mean radius ratio h/R for $t_L = 2$.

only for small values of thickness-to-mean radius ratio in the range $0 \leqslant h/R \leqslant 1$ and for higher values of h/R these become dispersive. For $h/R \geqslant 1$ in the case of $L/b = 0.5$ and 0.25 for the two values $t_L = 1$ and 2, increase in lowest frequency is quite dispersive and asymptotically increasing in nature. These results may also provide engineers and designers with useful reference solutions for modelling and design.

References

Ashida F and Tauchert T R 2001 A general plane-stress solution in cylindrical coordinates for a piezoelectric plate *J. Solids Struct.* **30** 4969–85

Buchanan G R 2003 Free vibration of an infinite magneto-electro-elastic cylinder *J. Sound Vib.* **268** 413–26

Ponnusamy P and Selvamani R 2013 Wave propagation in a transversely isotropic magneto thermo elastic cylindrical panel *Eur. J. Mech.* A*39 76–85*

Sharma J N, Pal M and Chand D 2004 Three dimensional vibration analysis of a piezo-thermo elastic cylindrical panel *Int. J. Eng. Sci.* **42** 1655–73

Chapter 6

Modelling of elastic waves in a fluid-loaded and immersed piezoelectric hollow cylinder

6.1 Introduction

Piezoelectric materials have strong physical properties and lack of chemical strength, which means high expenses for manufacturing. Lead zirconate titanate-4 material is used in polymer fibre materials as piezoelectric ceramics which produce greater sensitivity and higher processing temperatures, due to its flexible characteristics, and it plays a vital role in the applications of polymers which include audio-microphones, high- and low-frequency speakers, acoustic modems, pressure switches, position switches, accelerometers, impact detectors, flow meters and load cells.

Meeker and Meitzler (1964) developed the scientific wave phenomenon in cylindrical structures. The analysis of elastic waves along with an anisotropic circular cylinder with respect to hexagonal symmetry has been analysed by Morse (1954). The behaviour of a smart cylindrical shell with imperfect bonding with the multiphysics conditions has been investigated by Saadatfar (2015). Later, Saadatfar and Aghaie-Khafri (2015) studied the behaviour of a functionally graded rotating hybrid cylindrical shell with imperfect bonding in a humid thermal environment. The developments of piezoelectric materials have been investigated by Tiersten (1969). Parton and Kudryavtsev (1988) discussed the governing equations of motion for piezoelectric materials. Shulga (2002) reported axisymmetric and non-axisymmetric vibration analysis in anisotropic cylinders with piezoceramic material using different boundary conditions and polarization directions. The thermomagneto-electroelastic behaviour of an imperfect hybrid functionally graded rotating hollow cylinder. Saadatfar and Aghaie-Khafri (2016) later studied the magnetic effect and thermoelastic analysis of a rotating cylindrical shell on an elastic foundation.

Based on Nagaya (1981a, 1981b) Fourier expansion collocation method, the wave propagation in infinite piezoelectric solid cylinders of arbitrary cross-section

doi:10.1088/978-0-7503-5638-1ch6

was studied by Paul and Venkatesan (1987, 1989). By using Fourier integral transforms, the coupled electro-elastic equations for a long piezoceramic cylinder were solved analytically by Rajapakse and Zhou (1997). Bending analysis of cylindrical shells with a coated piezoelectric material was constructed by Wang (2002). By using the Bessel series, on the flat surfaces, polarized piezoelectric cylinders were developed by Ebenezer and Ramesh (2003). Berg *et al* (2004) reported that the electric field energy is not constant over the thickness of piezoceramic cylindrical shells. Later, this concept was extended and compared with computational results (Botta and Cerri 2007). Piezoelectric cylindrical transducers with radial polarization were developed by Kim and Lee (2007) and their results were compared with the results of finite element method. Graff (1991) and Achenbach (1973) discussed the wave propagation and their characteristics in elastic solid. Free vibration analysis of fibre-reinforced (CGFR) plates resting on the elastic foundations was developed by Yas and Aragh (2010). The quasistatic and impact property of a novel fibre/metal laminate system based on tough glass-fibre-reinforced polypropylene has been reported by Reyes and Cantwell (2000).

The vibration analysis of an isotropic cylinder with inner and outer fluid was studied by Berliner and Solecki (1996). The wave propagation characteristics of an isotropic rod introducing scalar and vector potentials were analysed by Dayal (1993). The propagation of longitudinal guided waves in a fluid-loaded transversely isotropic rod based on the superposition of partial waves was constructed by Nagy (1995). Shanker *et al* (2013) studied the vibrations in a fluid-loaded poroelastic hollow cylinder surrounded by a fluid in plane-strain form. The wave propagation in a magneto-thermoelastic cylindrical panel immersed in an inviscid fluid was discussed by Ponnusamy and Selvamani (2013, 2016).

In this chapter, the modelling of elastic waves in a hollow fibre of circular cross-section immersed in fluid composed of piezoelectric material is considered.

6.2 Model of the solid medium

The equations of motion without body force in cylindrical coordinates (r, θ, z) are given from Paul and Raju (1982).

$$\sigma_{rr,\, r} + \frac{1}{r}\sigma_{r\theta,\, \theta} + \sigma_{rz,\, z} + \frac{1}{r}(\sigma_{rr} - \sigma_{\theta\theta}) = \rho u_{r,\, tt}$$

$$\sigma_{r\theta,\, r} + \frac{1}{r}\sigma_{\theta\theta,\, \theta} + \sigma_{\theta z,\, z} + \frac{2}{r}\sigma_{r\theta} = \rho u_{\theta,\, tt}$$

$$\sigma_{rz,\, r} + \frac{1}{r}\sigma_{\theta z,\, \theta} + \sigma_{zz,\, z} + \frac{1}{r}\sigma_{rz} = \rho u_{z,\, tt}. \tag{6.1}$$

The components D_r, D_θ and D_z is given as

$$\frac{1}{r}\frac{\partial}{\partial r}(rD_r) + \frac{1}{r}\frac{\partial D_\theta}{\partial \theta} + \frac{\partial D_z}{\partial r} = 0. \tag{6.2}$$

The elastic and piezoelectric relations of transversely isotropic fibre can be given as

$$\sigma_{rr} = c_{11}e_{rr} + c_{12}e_{\theta\theta} + c_{13}e_{zz} - e_{31}E_z$$

$$\sigma_{\theta\theta} = c_{12}e_{rr} + c_{11}e_{\theta\theta} + c_{13}e_{zz} - e_{31}E_z$$

$$\sigma_{zz} = c_{13}e_{rr} + c_{13}e_{\theta\theta} + c_{33}e_{zz} - e_{33}E_z$$

$$\sigma_{r\theta} = c_{66}e_{r\theta}$$

$$\sigma_{\theta z} = c_{44}e_{\theta z} - e_{15}E_\theta$$

$$\sigma_{rz} = 2c_{44}e_{rz} - e_{15}E_r \tag{6.3}$$

and

$$D_r = e_{15}e_{rz} + \varepsilon_{11}E_r$$

$$D_\theta = e_{15}e_{\theta z} + \varepsilon_{11}E_\theta$$

$$D_z = e_{31}(e_{rr} + e_{\theta\theta}) + e_{33}e_{zz} + \varepsilon_{33}E_z \tag{6.4}$$

in which σ_{rr}, $\sigma_{\theta\theta}$, σ_{zz}, $\sigma_{r\theta}$, $\sigma_{\theta z}$, σ_{rz} and e_{rr}, $e_{\theta\theta}$, e_{zz}, $e_{r\theta}$, $e_{\theta z}$, e_{rz} denotes the stress and strain component, respectively. c_{11}, c_{12}, c_{13}, c_{33}, c_{44}, $c_{66} = (c_{11} - c_{12})/2$ stands for five elastic constants. $e_{31,}e_{15,}e_{33}$ denotes piezoelectric constant, ε_{11}, ε_{33} denotes dielectric constants, ρ is mass density of the transversely isotropic fiber.

The strain-displacement relations can be expressed as

$$e_{rr} = u_{r,\,r},\;\; e_{\theta\theta} = r^{-1}(u_r + u_{\theta,\,\theta}),\;\; e_{zz} = u_{z,\,z} \tag{6.5a}$$

$$e_{r\theta} = u_{\theta,\,r} + r^{-1}(u_{r,\,\theta} - u_\theta),\;\; e_{z\theta} = (u_{\theta,\,z} + r^{-1}u_{z,\,\theta}),\;\; e_{rz} = u_{z,\,r} + u_{r,\,z}. \tag{6.5b}$$

Substituting equations (6.3), (6.4) and (6.5) in equations (6.1) and (6.2), the following equations of motion are obtained:

$$\begin{aligned} &c_{11}(u_{rr,\,r} + r^{-1}u_{r,\,r} - r^{-2}u_r) - r^{-2}(c_{11} + c_{66})u_{\theta,\,\theta} \\ &+ r^{-2}c_{66}u_{r,\,\theta\theta} \\ &+ c_{44}u_{r,\,zz} + (c_{44} + c_{13})u_{z,\,rz} + r^{-1}(c_{66} + c_{12})u_{\theta,\,r\theta} \\ &+ (e_{31} + e_{15})V_{,rz} = \rho u_{r,\,tt} \end{aligned} \tag{6.6a}$$

$$\begin{aligned} &r^{-1}(c_{12} + c_{66})u_{r,\,r\theta} + r^{-2}(c_{66} + c_{11})u_{r,\,\theta} \\ &+ c_{66}(u_{\theta,\,rr} + r^{-1}u_{\theta,\,r} - r^{-2}u_\theta) + r^{-2}c_{11}u_{\theta,\,\theta\theta} + c_{44}u_{\theta,\,zz} \\ &+ r^{-1}(c_{44} + c_{13})u_{z,\,\theta z} + (e_{31} + e_{15})V_{,\theta z} = \rho u_{\theta,\,tt} \end{aligned} \tag{6.6b}$$

$$\begin{aligned} &c_{44}(u_{z,\,rr} + r^{-1}u_{z,\,r} + r^{-2}u_{z,\,\theta\theta}) + r^{-1}(c_{44} + c_{13})(u_{r,\,z} + u_{\theta,\,\theta z}) \\ &+ (c_{44} + c_{13})u_{r,\,rz} + c_{33}u_{z,\,zz} + e_{33}V_{,zz} \\ &+ e_{15}(V_{,rr} + r^{-1}V_{,r} + r^{-2}V_{,\theta\theta}) = \rho u_{z,\,tt} \end{aligned} \tag{6.6c}$$

$$e_{15}(u_{z,\,rr} + r^{-1}u_{z,\,r} + r^{-2}u_{z,\,\theta\theta}) + (e_{31} + e_{15})$$
$$(u_{r,\,zr} + r^{-1}u_{r,\,z} + r^{-1}u_{\theta,\,z\theta}) + e_{33}u_{z,\,zz} - \varepsilon_{33}V_{,zz} \tag{6.6d}$$
$$- \varepsilon_{11}(V_{,rr} + r^{-1}V_{,r} + r^{-2}V_{,\theta\theta}) = 0.$$

6.3 Solutions of the field equation

The following displacement components are used to uncouple equation (6.6) from Paul and Raju (1982)

$$u_r(r,\,\theta,\,z,\,t) = \left(\phi_{,r} + r^{-1}\psi_{,\theta}\right)\exp(\iota(kz + \omega t))$$

$$u_\theta(r,\,\theta,\,z,\,t) = \left(r^{-1}\phi_{,\theta} - \psi_{,r}\right)\exp(\iota(kz + \omega t))$$

$$u_z(r,\,\theta,\,z,\,t) = \left(\frac{i}{a}\right)W\,\exp(\iota(kz + \omega t))$$

$$V(r,\,\theta,\,z,\,t) = iV\,\exp(\iota(kz + \omega t))$$

$$E_r(r,\,\theta,\,z,\,t) = -E_{,r}\,\exp(\iota(kz + \omega t))$$

$$E_\theta(r,\,\theta,\,z,\,t) = -r^{-1}E_{,\theta}\,\exp(\iota(kz + \omega t))$$

$$E_z(r,\,\theta,\,z,\,t) = E_{,z}\,\exp(\iota(kz + \omega t)) \tag{6.7}$$

where $i = \sqrt{-1}$, k is the wave number, ω is the angular frequency, $\phi(r,\,\theta)$, $W(r,\,\theta)$, $\psi(r,\,\theta)$ and $E(r,\,\theta)$ stands for displacement potentials and $V(r,\,\theta)$ denote electric potentials and a denote geometrical parameter of the fibre.

The dimensionless quantities are introduced as $x = r/a$, $\zeta = ka$, $\Omega^2 = \rho\omega^2a^2/c_{44}$, $\bar{c}_{11} = c_{11}/c_{44}$, $\bar{c}_{13} = c_{13}/c_{44}$, $\bar{c}_{33} = c_{33}/c_{44}$, $\bar{e}_{13} = e_{13}/e_{33}$, $\bar{e}_{15} = e_{15}/e_{33}$, $\bar{\varepsilon}_{11} = \varepsilon_{11}/\varepsilon_{33}$, $\bar{c}_{66} = c_{66}/c_{44}$ and using equation (6.7) in equation (6.6), we can get the following non-dimensional form of equations

$$(\bar{c}_{11}\,\nabla^2 + (\Omega^2 - \zeta^2))\phi + \zeta(1 + \bar{c}_{13})W + \zeta(\bar{e}_{31} + \bar{e}_{15})V = 0$$

$$\zeta(1 + \bar{c}_{13})\,\nabla^2\,\phi + (\nabla^2 + (\Omega^2 - \zeta^2\bar{c}_{33}))W + (\bar{e}_{15}\,\nabla^2 - \zeta^2)V = 0$$

$$\zeta(\bar{e}_{31} + \bar{e}_{15})\,\nabla^2\,\phi + (\bar{e}_{15}\,\nabla^2 - \zeta^2)W + (\zeta^2\bar{\varepsilon}_{33} - \bar{\varepsilon}_{11}\nabla^2)V = 0 \tag{6.8}$$

and

$$(\bar{c}_{66}\,\nabla^2 + (\Omega^2 - \zeta^2))\psi = 0 \tag{6.9}$$

where $\nabla^2 = \dfrac{\partial^2}{\partial x^2} + \dfrac{1}{x}\dfrac{\partial}{\partial x} + \dfrac{1}{x^2}\dfrac{\partial^2}{\partial\theta^2}$.

The equation (6.8) is rewritten as follows

$$\begin{vmatrix} (\bar{c}_{11}\nabla^2 + (\Omega^2 - \zeta^2)) & -\zeta(1+\bar{c}_{13}) & -\zeta(\bar{e}_{31}+\bar{e}_{15}) \\ \zeta(1+\bar{c}_{13})\nabla^2 & (\nabla^2 + (\Omega^2 - \zeta^2\bar{c}_{33})) & (\bar{e}_{15}\nabla^2 - \zeta^2) \\ \zeta(\bar{e}_{31}+\bar{e}_{15})\nabla^2 & (\bar{e}_{15}\nabla^2 - \zeta^2) & (\zeta^2\bar{\varepsilon}_{33} - \bar{\varepsilon}_{11}\nabla^2) \end{vmatrix} (\phi, W, V) = 0. \tag{6.10}$$

Solving the above determinant, the following partial differential equation can be obtained as

$$(P\,\nabla^6 + Q\,\nabla^4 + R\,\nabla^2 + S)(\phi, W, V) = 0 \tag{6.11}$$

where

$$P = c_{11}(\bar{e}_{15}^2 + \varepsilon_{11})$$

$$\begin{aligned} Q = {} & [(1 + \bar{c}_{11})\bar{\varepsilon}_{11} + \bar{e}_{15}^2]\Omega^2 \\ & + \{2(\bar{e}_{31} + \bar{e}_{15})\bar{c}_{13}\bar{e}_{15} - (1 + \bar{\varepsilon}_{11}\bar{c}_{33})\bar{c}_{11} + \bar{c}_{13}^2\bar{\varepsilon}_{11} + 2\bar{c}_{13}\bar{\varepsilon}_{11} - 2\bar{e}_{15}\bar{c}_{11} + 2\bar{e}_{13}^2\}\varsigma^2 \end{aligned}$$

$$\begin{aligned} R = {} & \bar{\varepsilon}_{11}\Omega^4 - [(1 + \bar{c}_{13})\bar{\varepsilon}_{11} + (1 + \bar{c}_{11}) + (\bar{e}_{31} + \bar{e}_{15}) + 2\bar{e}_{15}]\varsigma^2\Omega^2 \\ & + \{\bar{c}_{11}(1 + \bar{c}_{33}\bar{\varepsilon}_{33}) - [(\bar{e}_{31} + \bar{e}_{15})^2 + \bar{\varepsilon}_{11}] \\ & - 2\bar{e}_{31}(1 + \bar{c}_{13}) - \bar{c}_{13}\bar{\varepsilon}_{33}(\bar{c}_{33} + \bar{c}_{13}) + 2\bar{e}_{15}\}\varsigma^4 \end{aligned}$$

$$S = -\{(1 + \bar{c}_{33})\varsigma^6 - [2(1 + \bar{c}_{33})\bar{\varepsilon}_{33} + 1]\varsigma^4\Omega^2 + \bar{\varepsilon}_{33}\varsigma^2\Omega^4\}.$$

Solving equation (6.11), we get solutions for a circular fibre as (Selvamani 2016a)

$$\phi = \sum_{i=1}^{3}[A_i J_n(\alpha_i ax) + B_i Y_n(\alpha_i ax)]\cos n\theta$$

$$W = \sum_{i=1}^{3} a_i[A_i J_n(\alpha_i ax) + B_i Y_n(\alpha_i ax)]\cos n\theta$$

$$V = \sum_{i=1}^{3} b_i[A_i J_n(\alpha_i ax) + B_i Y_n(\alpha_i ax)]\cos n\theta. \tag{6.12}$$

The constants a_i, b_i defined in equation (6.12) is computed from the following equations

$$(1 + \bar{c}_{13})\varsigma a_i + (\bar{e}_{31} + \bar{e}_{15})\varsigma b_i = -(\bar{c}_{11}(\alpha_i a)^2 - \Omega^2 + \varsigma^2)$$

$$((\alpha_i a)^2 - \Omega^2 + \varsigma^2\bar{c}_{33})a_i + (\bar{e}_{15}(\alpha_i a)^2 + \varsigma^2)b_i = -(\bar{c}_{13} + 1)\varsigma(\alpha_i a)^2. \tag{6.14}$$

Solving equation (6.9), we get

$$\psi = [A_4 J_n(\alpha_4 ax) + B_4 Y_n(\alpha_4 ax)]\sin n\theta \tag{6.15}$$

where $(\alpha_4 a)^2 = \Omega^2 - \zeta^2$. The Bessel functions J_n and Y_n are replaced by the modified Bessel functions I_n and K_n when the root $(\alpha_4 a)^2 < 0$.

6.4 Model of the fluid medium

The inviscid fluid acoustic pressure and radial displacement are (Selvamani 2016b, 2016c)

$$p^f = -B^f\left(u^f_{r,\,r} + r^{-1}(u^f_r + u^f_{\theta,\,\theta}) + u^f_{z,\,z}\right) \tag{6.16}$$

and

$$c_f^{-2} u^f_{r,\,tt} = \Delta_{,r}. \tag{6.17}$$

Here, B^f and ρ^f stands for adiabatic bulk modulus and density of the fluid, $c^f = \sqrt{B^f/\rho^f}$ is the acoustic phase velocity in the fluid, and $(u^f_r,\ u^f_\theta,\ u^f_z)$ is the displacement vector.

$$\Delta = \left(u^f_{r,\,r} + r^{-1}(u^f_r + u^f_{\theta,\,\theta}) + u^f_{z,\,z}\right). \tag{6.18}$$

Substituting the following fluid displacement component

$$u^f_r = \phi^f_{,r},\quad u^f_\theta = r^{-1}\phi^f_{,\theta}\, u^f_z = \phi^f_{,z} \tag{6.19}$$

and seeking the solution of equation (6.17) in the form

$$\phi^f = (r,\,\theta,\,z,\,t) = \sum_{n=0}^{\infty} \phi^f\,(r)\cos n\theta e^{i(kz+\omega t)}. \tag{6.20}$$

The oscillating waves for annular region in the inner fluid can be defined as

$$\phi^f = A_5 J_n(\alpha_5 ax) \tag{6.21}$$

where $(\alpha_5 a)^2 = \Omega^2/\overline{\rho}^f_1 \overline{B}^f_1 - \varsigma^2$, in which $\overline{\rho}^f_1 = \rho_1/\rho^f$, $\overline{B}^f_1 = B^f_1/c_{44}$. The Bessel function J_n in (6.21) is to be replaced by modified Bessel function I_n for $(\alpha_5 a)^2 < 0$, Similarly, for the outer fluid that represents the oscillatory waves propagating away it is given as

$$\phi^f = B_5 H^{(2)}_n(\alpha_6 ax) \tag{6.22}$$

where $(\alpha_6 a)^2 = \Omega^2/\overline{\rho}^f_2 \overline{B}^f_2 - \varsigma^2$, in which $\overline{\rho}^f_2 = \rho_2/\rho^f$, $\overline{B}^f_2 = B^f_2/c_{44}$, $H^{(2)}_n$ is the Hankel function of the second kind. It is replaced by K_n, where K_n is the modified Bessel function of the second kind for $(\alpha_6 a)^2 < 0$. By inserting equation (6.19) in (6.16) along with (6.21) and (6.22), the acoustic pressure for the inner and outer fluid is given as

$$p^f_1 = A_5 \Omega^2 \overline{\rho}_1 J_n(\alpha_5 ax)\cos n\theta e^{i(\varsigma\overline{z}+\Omega T_a)} \tag{6.23}$$

$$p^f_2 = B_5 \Omega^2 \overline{\rho}_2 H^{(2)}_n(\alpha_6 ax)\cos n\theta e^{i(\varsigma\overline{z}+\Omega T_a)}. \tag{6.24}$$

6.5 Solid–fluid boundary conditions and frequency equations

The inner boundary conditions are expressed as

$$[\sigma_{rr}, \sigma_{r\theta}, \sigma_{rz}, V, u_r] = \left[-p_1^f, 0, 0, 0, u_r^f\right], r = a \tag{6.25}$$

the outer boundary conditions are defined as

$$[\sigma_{rr}, \sigma_{r\theta}, \sigma_{rz}, V, u_r] = \left[-p_2^f, 0, 0, 0, u_r^f\right], r = b. \tag{6.26}$$

Inserting the solution in equations (6.12), (6.15), (6.23) and (6.24) in equations (6.25) and (6.26), we obtain the following form

$$[M]\{X\} = \{0\}. \tag{6.27}$$

The elements of the matrix $[M]$ are given in appendix A. The solution of equation (6.27) is nontrivial when the determinant of the coefficient of $\{X\}$ vanishes, that is

$$|M| = 0. \tag{6.28}$$

Equation (6.28) represents the frequency equation consisting of a transversely isotropic piezoelectric fluid-loaded circular fibre immersed in inviscid fluid.

6.5.1 Electro-mechanical coupling

The electro-mechanical coupling (E^2) for the cylindrical fibre can be expressed as

$$E^2 = \left|\frac{P_e - P_f}{P_e}\right| \tag{6.29}$$

where P_e and P_f denotes phase velocities of the wave under electrically shorted and charge free boundary conditions at the surface of the fibre.

6.6 Numerical results and discussion

The secular equation given in equation (6.28) is a general complex transcendental equation with unknown frequency and wave number. The material properties are taken as Berlincourt *et al* (1964): $c_{11} = 13.9 \times 10^{10}\,\mathrm{N\,m^{-2}}$, $c_{12} = 7.78 \times 10^{10}\,\mathrm{N\,m^{-2}}$, $c_{13} = 7.43 \times 10^{10}\,\mathrm{N\,m^{-2}}$, $c_{33} = 11.5 \times 10^{10}\,\mathrm{N\,m^{-2}}$, $c_{44} = 2.56 \times 10^{10}\,\mathrm{N\,m^{-2}}$, $c_{66} = 3.06 \times 10^{10}\,\mathrm{N\,m^{-2}}$, $e_{31} = -5.2\,\mathrm{C\,m^{-2}}$, $e_{33} = 15.1\,\mathrm{C\,m^{-2}}$, $e_{15} = 12.7\,\mathrm{C\,m^{-2}}$, $\varepsilon_{11} = 6.46 \times 10^{-9}\,\mathrm{C^2\,N^{-1}\,m^{-2}}$, $\varepsilon_{33} = 5.62 \times 10^{-9}\,\mathrm{C^2\,N^{-1}\,m^{-2}}$, $\rho = 7500\,\mathrm{Kg\,m^{-2}}$ and the density fluid $\rho^f = 1000\,\mathrm{Kg\,m^{-3}}$, phase velocity $c = 1500\,\mathrm{m\,s^{-1}}$ and used for the numerical calculations.

6.6.1 Comparison table

The secular equation is obtained and compared with the secular equations of Paul and Raju (1982), which gives very good agreement with the author. The secant method is used to calculate the dimensionless frequencies for the wave number range $0 < \varsigma \leqslant 0.1$ and is given in table 6.1. From table 6.1, it is observed that, by

Table 6.1. The dimensionless frequencies (Ω) of longitudinal and flexural mode for lead zirconate titanate-4 material frequencies (Ω) without fluid medium.

	$n = 0$		$n = 1$		$n = 2$	
Σ	Author	Saadatfar and Aghaie-Khafri (2016)	Author	Saadatfar and Aghaie-Khafri (2016)	Author	Saadatfar and Aghaie-Khafri (2016)
0.01	4.6654	4.6655	1.8911	1.8911	3.1815	3.1816
	8.5421	8.5424	6.4147	6.4143	8.0351	8.0351
	12.3876	12.3875	10.3462	10.3469	12.0568	12.0561
	16.2235	16.2234	14.2201	14.2195	15.9751	15.9750
0.04	4.6667	4.6641	1.8913	1.8923	3.1824	3.1825
	8.5423	8.5425	6.4145	6.4142	8.0351	8.0351
	12.3844	12.3873	10.3471	10.3479	12.0564	12.0563
	16.2237	16.2234	14.2195	14.2195	15.9754	15.9751
0.1	4.6566	4.6566	1.8991	1.8992	3.1870	3.1869
	8.5426	8.5427	6.4135	6.4137	8.0351	8.0348
	12.3859	12.3858	10.3472	10.3475	12.0573	12.0578
	16.2234	16.2235	14.2193	14.2195	15.9756	15.9757

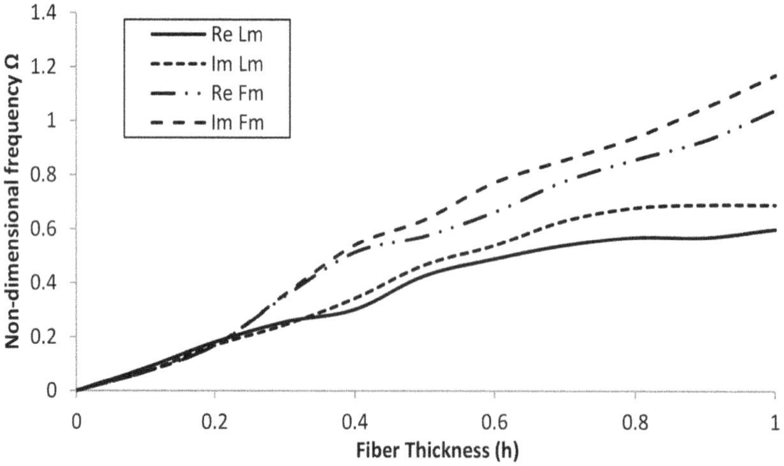

Figure 6.1. The fibre thickness (h) versus dimensionless frequency (**Ω**) for a piezoelectric fluid-loaded fibre.

increasing the wave number (ς) increases the dimensionless frequencies (Ω) and also this nature clearly displays the structure of the frequency spectrum near the cut-off frequencies of the fibre without fluid environment.

6.6.2 Dispersion analysis

The dispersion curves are drawn for longitudinal and flexural modes of the fibres. The notations ReLm, ImLm and ReFm, ImFm stand for the real and imaginary

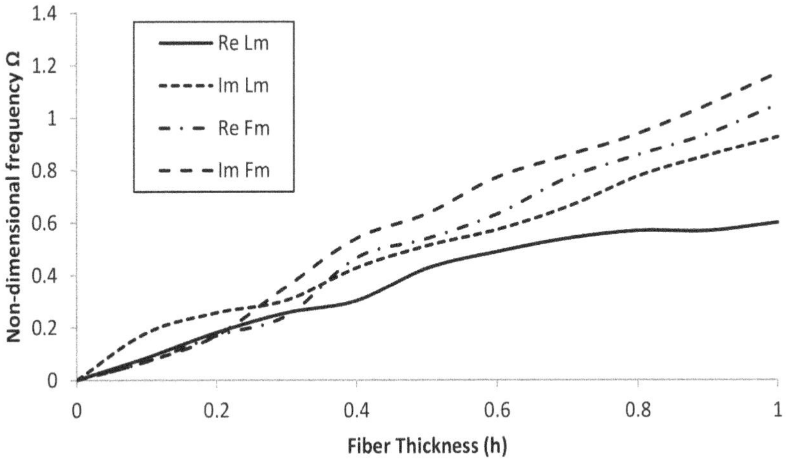

Figure 6.2. The effect fibre thickness (h) along with dimensionless frequency (Ω) for a fluid-loaded and immersed piezoelectric fibre.

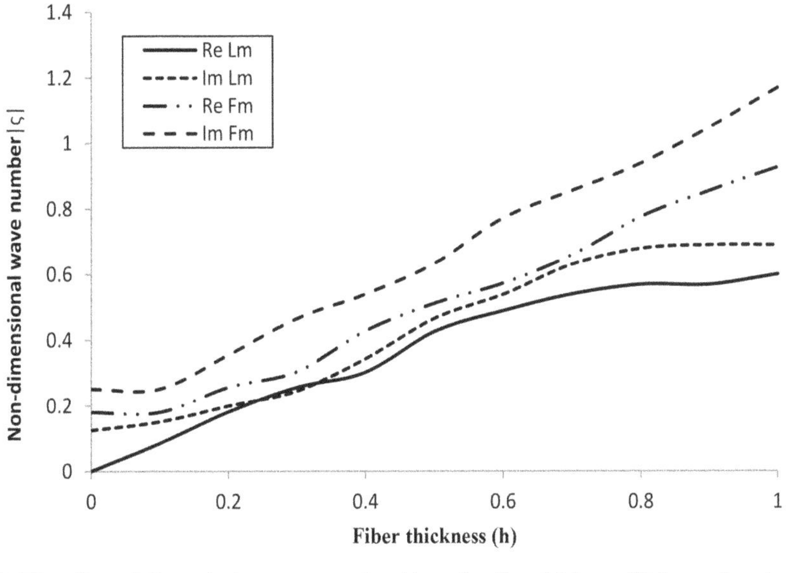

Figure 6.3. The effect of dimensionless wave number $|\varsigma|$ on the fibre thickness (h) for a piezoelectric fluid-loaded fibre.

part of longitudinal mode, real and imaginary part of flexural mode. The graphs are drawn for dimensionless frequency $\Omega(=\sqrt{\frac{\rho\omega^2a^2}{c_{44}}})$ with fibre (h) thickness for the real and imaginary part of longitudinal and flexural modes of piezoelectric fluid-loaded and immersed cylindrical fibre are shown in figures 6.1 and 6.2. From these figures, it is noticed that increasing the fibre thickness increases the dimensionless frequencies. Due to the surrounding fluid, the influence of dissipation energy helps to dominant the dimensionless frequency in the flexural and dispersive at higher thickness range. Figures 6.3 and 6.4 shows the dimensionless wave number versus thickness of the

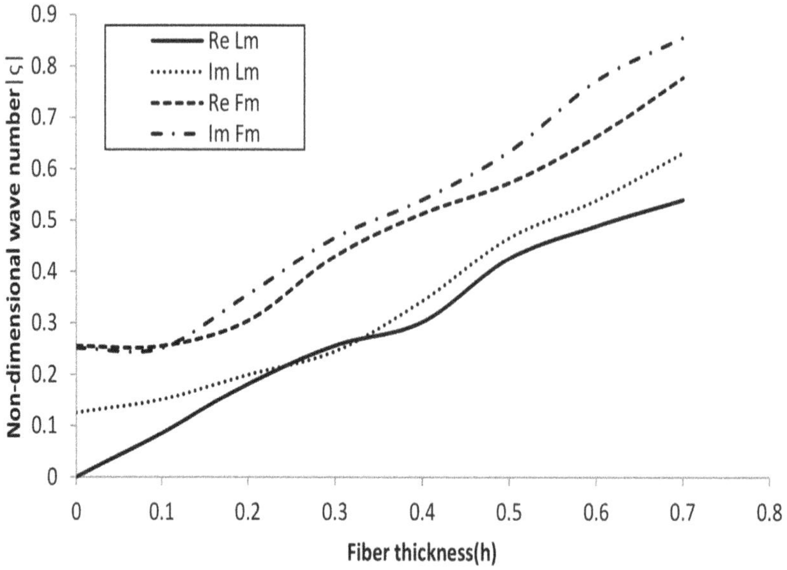

Figure 6.4. The effect of dimensionless wave number $|\varsigma|$ on the fibre thickness (h) for a fluid-loaded and immersed piezoelectric fibre.

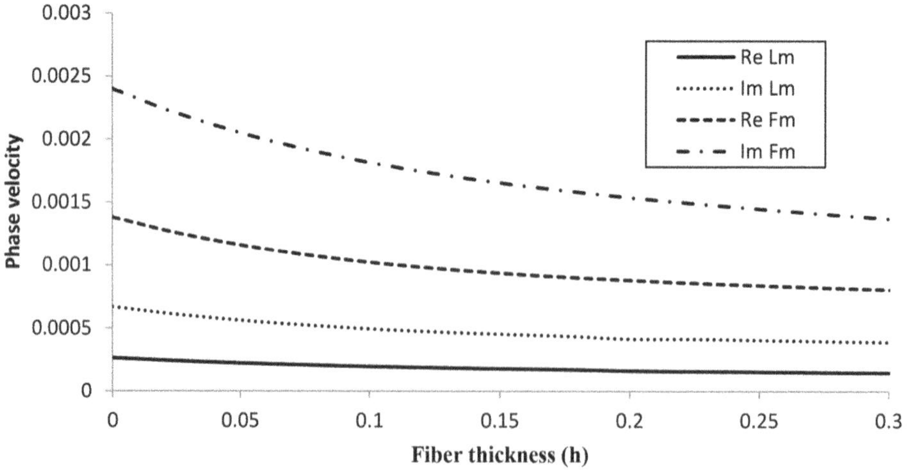

Figure 6.5. The influence of phase velocity along with fibre thickness (h) for a fluid-loaded piezoelectric fibre.

fibre with respect to fluid-loaded and immersed piezoelectric fibres. From these, both the real and imaginary parts of wave numbers merge for initial range thickness and increase in the remaining range. The crossing of points between the non-dimensional wave number and thickness of the fibre denotes the travelling of energy modes along the solid–fluid interface.

The dispersion waves are drawn between the thickness of the fibre (h) and phase velocity ($=\frac{\omega}{\varsigma}$) in figures 6.5 and 6.6. From this, it can be seen that increasing the

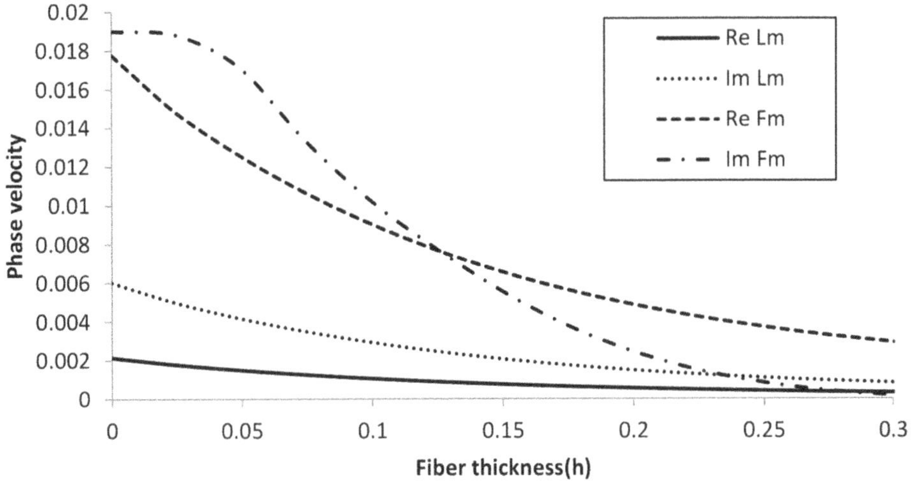

Figure 6.6. The influence of fibre thickness (h) along with dimensionless wave number $|\varsigma|$ for a fluid-loaded and immersed piezoelectric fibre.

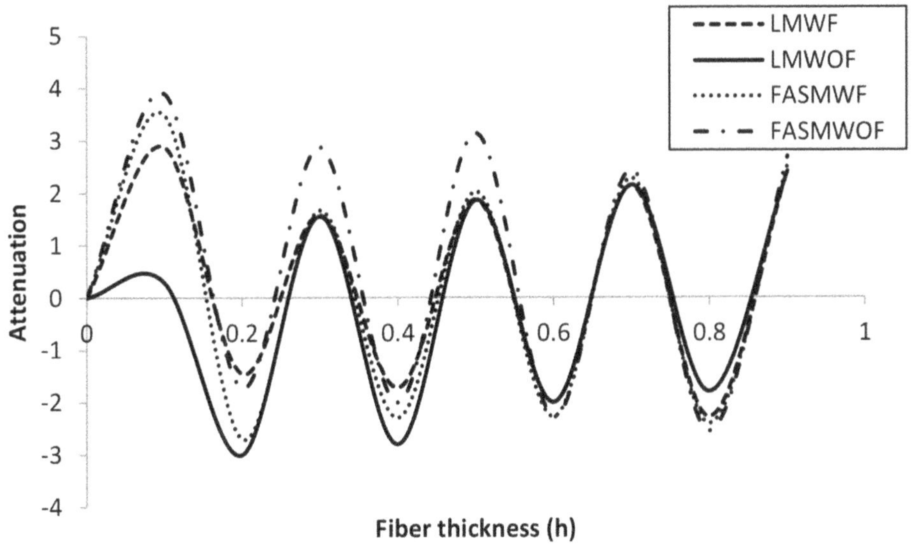

Figure 6.7. The effect of attenuation along with fibre thickness (h) for a fluid-loaded and immersed piezoelectric fibre.

thickness of the fibre decreases the phase velocity. The dispersion curves become smoother in this case than those in the absence of surrounding liquid because of the shock absorption nature of the liquid. Figure 6.7 shows a graph between the thickness of fibre (h) and the attenuation ($c^{-1} = v^{-1} + i\omega^{-1}q$) versus thickness of the fibre (h) for longitudinal and flexural antisymmetric modes of fluid-loaded and immersed piezoelectric fibre. From figure 6.7, it is observed that, by the variation of fibre thickness, the attenuation profile becomes oscillatory.

In figures 6.8 and 6.9, graphs are drawn between the piezoelectric fibre of wave number $|\varsigma|$ and electro-mechanical coupling (E^2) for circular cross-section for the longitudinal and flexural (symmetric and antisymmetric) modes. From this, it is observed that, both the longitudinal and flexural modes merge for $|\varsigma| > 0.2$ and start increasing monotonically. Also, in the wave number range $0.6 \leqslant |\varsigma| \leqslant 0.8$, the coupling effect of various interacting media gets maximum value in symmetric modes. The merging of vibration modes and oscillating nature of curves between the different modes shows that there is a transfer of energy between the modes of vibrations by the effect of the inner and outer fluid in the system. The electro-mechanical coupling is high in magnitude in flexural modes and dispersive at higher thickness range, which might happen because the fluid medium acts as an added mass for the fibre.

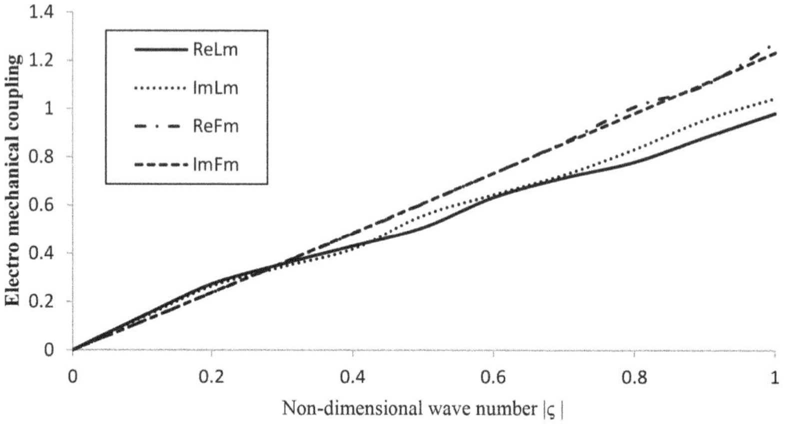

Figure 6.8. The effect of electro-mechanical coupling (E^2) with dimensionless wave number $|\varsigma|$ for a fluid-loaded piezoelectric fibre.

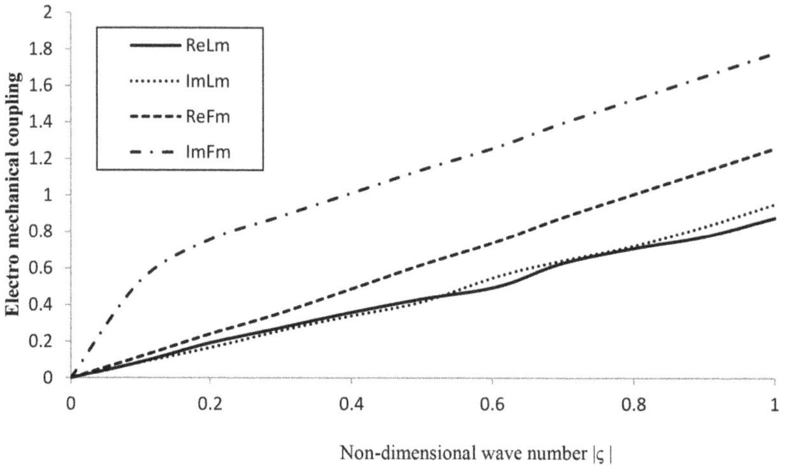

Figure 6.9. The effect of electro-mechanical coupling (E^2) on dimensionless wave number $|\varsigma|$ for a fluid-loaded and immersed piezoelectric fibre.

Appendix A

$$m_{1i} = 2\bar{c}_{66}\{n(n-1) - \bar{c}_{11}(\alpha_i a)^2 - \varsigma(\bar{c}_{13}a_i + \bar{e}_{31}b_i)\} \\ J_n(\alpha_i a) + 2\bar{c}_{66}(\alpha_i a)J_{n+1}(\alpha_i a),\ \ i = 1, 2, 3 \tag{A.1}$$

$$m_{14} = 2\bar{c}_{66}n\{(n-1)J_n(\alpha_4 a) - (\alpha_4 a)J_{n+1}(\alpha_4 a)\} \tag{A.2}$$

$$m_{15} = 2\bar{c}_{66}n\{(n-1)Y_n(\alpha_4 a) - (\alpha_4 a)Y_{n+1}(\alpha_4 a)\} \tag{A.3}$$

$$m_{1i} = 2\bar{c}_{66}\{n(n-1) - \bar{c}_{11}(\alpha_i a)^2 - \varsigma(\bar{c}_{13}a_i + \bar{e}_{31}b_i)\} \\ Y_n(\alpha_i a) + 2\bar{c}_{66}(\alpha_i a)Y_{n+1}(\alpha_i a),\ \ i = 6, 7, 8 \tag{A.4}$$

$$m_{19} = \overline{\rho}_1\Omega^2 J_n(\alpha_5 a) \tag{A.5}$$

$$m_{10} = \overline{\rho}_2\Omega^2 H_n^{(2)}(\alpha_6 a) \tag{A.6}$$

$$m_{2i} = 2n\{(n-1)J_n(\alpha_i a) + (\alpha_i a)J_{n+1}(\alpha_i a)\},\ i = 1, 2, 3 \tag{A.7}$$

$$m_{24} = \{[(\alpha_4 a)^2 - 2n(n-1)]J_n(\alpha_4 a) - 2(\alpha_4 a)J_{n+1}(\alpha_4 a)\} \tag{A.8}$$

$$m_{25} = \{[(\alpha_4 a)^2 - 2n(n-1)]Y_n(\alpha_4 a) - 2(\alpha_4 a)Y_{n+1}(\alpha_4 a)\} \tag{A.9}$$

$$m_{2i} = 2n\{(n-1)Y_n(\alpha_i a) + (\alpha_i a)Y_{n+1}(\alpha_i a)\},\ i = 6, 7, 8 \tag{A.10}$$

$$m_{29} = 0,\ \ m_{210} = 0 \tag{A.11}$$

$$m_{3i} = ((\varsigma + a_i) + \bar{e}_{15}b_i)\{nJ_n(\alpha_i a) - (\alpha_i a)J_{n+1}(\alpha_i a)\},\ i = 1, 2, 3 \tag{A.12}$$

$$m_{34} = n\varsigma J_n(\alpha_4 a) \tag{A.13}$$

$$m_{35} = n\varsigma Y_n(\alpha_4 a) \tag{A.14}$$

$$m_{3i} = ((\varsigma + a_i) + \bar{e}_{15}b_i)\{nJ_n(\alpha_i a) - (\alpha_i a)J_{n+1}(\alpha_i a)\},\ i = 6, 7, 8 \tag{A.15}$$

$$m_{39} = 0,\ \ m_{310} = 0, \tag{A.16}$$

$$m_{4i} = b_i J_n(\alpha_i a),\ i = 1, 2, 3 \tag{A.17}$$

$$m_{44} = 0,\ m_{45} = 0 \tag{A.18}$$

$$m_{4i} = b_i Y_n(\alpha_i a),\ i = 6, 7, 8 \tag{A.19}$$

$$m_{49} = 0,\ m_{410} = 0 \tag{A.20}$$

$$m_{5i} = \{nJ_n(\alpha_i a) - (\alpha_i a)J_{n+1}(\alpha_i a)\},\ i = 1, 2, 3 \tag{A.21}$$

$$m_{54} = nJ_n(\alpha_4 a) \tag{A.22}$$

$$m_{55} = nY_n(\alpha_4 a) \tag{A.23}$$

$$m_{5i} = \{nY_n(\alpha_i a) - (\alpha_i a)Y_{n+1}(\alpha_i a)\},\ i = 6,\ 7,\ 8 \tag{A.24}$$

$$m_{59} = \Omega^2\overline{\rho}_1\{nJ_n(\alpha_5 a) - (\alpha_5 a)J_{n+1}(\alpha_5 a)\} \tag{A.25}$$

$$m_{510} = \Omega^2\overline{\rho}_2\left\{nH_n^{(2)}(\alpha_6 a) - (\alpha_6 a)H_{n+1}^{(2)}(\alpha_6 a)\right\}. \tag{A.26}$$

This chapter was reprinted from Selvamani (2016c) by permission from Springer Nature Customer Service Centre GmbH.

References

Achenbach J D 1973 *Wave Motion in Elastic Solids* (New York: Dover)

Berg M, Hagedorn P and Gutschmidt S 2004 On the dynamics of piezoelectric cylindrical shell *J. Sound Vib.* **274** 91–109

Berlincourt D A, Curran D R and Jaffe H 1964 *Piezoelectric and Piezomagnetic Materials and their Function in Transducers* (New York: Academic)

Botta F and Cerri G 2007 Wave propagation in Reissner–Mindlin piezoelectric coupled cylinder with non-constant electric field through the thickness *Int. J. Solids Struct.* **44** 6201–19

Berliner J and Solecki R 1996 Wave propagation in a fluid-loaded transversely isotropic cylinder. Part I. Analytical formulation; Part II numerical results *J. Acoust. Soc. Am.* **99** 1841–53

Dayal V 1993 Longitudinal waves in homogeneous anisotropic cylindrical bars immersed in fluid *J. Acoust. Soc. Am.* **93** 1249–55

Ebenezer D D and Ramesh R 2003 Analysis of axially polarized piezoelectric cylinders with arbitrary boundary conditions on the flat surfaces *J. Acoust. Soc. Am.* **113** 1900–8

Graff K F 1991 *Wave Motion in Elastic Solids* (New York: Dover)

Kim J O and Lee J G 2007 Dynamic characteristics of piezoelectric cylindrical transducers with radial polarization *J. Sound Vibr.* **300** 241–9

Morse R W 1954 Compressional waves along an anisotropic circular cylinder having hexagonal symmetry *J. Acoust. Soc. Am.* **26** 1018–21

Meeker T R and Meitzler A H 1964 Guided wave propagation in elonged cylinders and plates *Physical Acoustics* **vol 1** ed W P Mason (New York: Academic)

Nagaya K 1981a Vibration of thick walled pipe or ring of arbitrary shape in its plane *J. Appl. Mech.* **50** 757–64

Nagaya K 1981b Dispersion of elastic waves in bars with polygonal cross-section *J. Acoust. Soc. Am.* **70** 763–70

Nagy B 1995 Longitudinal guided wave propagation in a transversely isotropic rod immersed in fluid *J. Acoust. Soc. Am.* **98** 454–7

Parton V Z and Kudryavtsev B A 1988 *Electromagnetoelasticity* (New York: Gordon and Breach)

Paul H S and Raju D P 1982 Asymptotic analysis of the modes of wave propagation in a piezoelectric solid cylinder *J. Acoust. Soc. Am.* **71** 255–63

Paul H S and Venkatesan M 1987 Wave propagation in a piezoelectric ceramic cylinder of arbitrary cross section *J. Acoust. Soc. Am.* **82** 2010–3

Paul H S and Venkatesan M 1989 Wave propagation in a piezoelectric ceramic cylinder of arbitrary cross section with a circular cylindrical cavity *J. Acoust. Soc. Am.* **85** 163–70

Ponnusamy P and Selvamani R 2013 Wave propagation in a transversely magneto thermo elastic cylindrical panel *Eur. J. Mech. A. Solids* **39** 76–85

Ponnusamy P and Selvamani R 2016 Wave propagation in a transversely isotropic magneto-electro-elastic solid bar immersed in an inviscid fluid *J. Egypt. Math. Soc.* **24** 92–9

Rajapakse R K N D and Zhou Y 1997 Stress analysis of piezoceramic cylinders *Smart Mater. Struct.* **6** 169–77

Reyes V G and Cantwell W J 2000 The mechanical properties of fiber-metal laminates based on glass fiber reinforced polypropylene *Compos. Sci. Technol.* **60** 1085–94

Saadatfar M 2015 Effect of multiphysics conditions on the behavior of an exponentially graded smart cylindrical shell with imperfect bonding *Meccanica* **50** 2135–52

Saadatfar M and Aghaie-Khafri M 2015 On the behavior of a rotating functionally graded hybrid cylindrical shell with imperfect bonding subjected to hygrothermal condition *J. Therm. Stress.* **38** 854–88

Saadatfar M and Aghaie-Khafri M 2016 Thermoelastic analysis of a rotating functionally graded cylindrical shell with functionally graded sensor and actuator layers on an elastic foundation placed in a constant magnetic field *J. Intell. Mater. Syst. Struct.* **27** 512–27

Shanker B, Nath C N, Shah S A and Reddy P M 2013 Vibrations in a fluid-loaded poroelastic hollow cylinder surrounded by a fluid in plane-strain form *Int. J. Appl. Mech. Eng.* **18** 189–216

Selvamani R 2016a Influence of thermo-piezoelectric field in a circular bar subjected to thermal loading due to laser pulse *Mater. Phys. Mech.* **27** 1–8

Selvamani R 2016b Dispresion analysis in a fluid-filled and immersed transversely isotropic thermo-electro-elastic hollow cylinder *Mech. Mech. Eng.* **20** 209–31

Selvamani R 2016c Modeling of elastic waves in a fluid loaded and immersed piezoelectric hollow fiber *Int. J. Appl. Comput. Math.* **3** 3263–77

Shulga N A 2002 Propagation of harmonic waves in anisotropic piezoelectric cylinders. Homogeneous piezoceramic wave guides *Int. Appl. Mech.* **38** 933–53

Tiersten H F 1969 *Linear Piezoelectric Plate Vibrations* (New York: Plenum)

Wang Q 2002 Axi-symmetric wave propagation in cylinder coated with a piezoelectric layer *Int. J. Solids Struct.* **39** 3023–37

Yas M H and Aragh B S 2010 Free vibration analysis of continuous grading fiber reinforced plates on elastic foundation *Int. J. Eng. Sci.* **48** 1881–95

IOP Publishing

Mathematical Modelling and Characterization of Cylindrical Structures

Farzad Ebrahimi and Rajendran Selvamani

Chapter 7

Wave propagation in a generalized piezothermoelastic rotating bar of circular cross-section

A mathematical model is developed to study the wave propagation in a generalized piezothermelastic rotating bar of circular cross-section by using the Lord–Shulman (L–S) and Green–Lindsay (G–L) theories of thermoelasticity. The frequency equations are obtained by using the thermally insulated/isothermal and electrically shorted/charge-free boundary conditions. The secant method is used to find the roots of the frequency equation, applicable for complex roots. The analytical terms have been derived to include the time requirement for the acceleration of the heat flow and the coupling between the temperature and strain fields, for the non-classical thermoelastic theories, L–S and G–L theories. The dispersion curves are presented for the physical quantities such as thermomechanical coupling, electromechanical coupling, specific loss, frequency shift and frequency. The effect of thermal relaxation time and rotational speed on wave number for the case of generalized piezothermoelastic material of circular cross-section have not been reported in the literature. These results are new and original.

7.1 Introduction

This section has been reproduced with permission from Selvamani and Ponnusamy (2015). Copyright 2015, Emerald Group Publishing Limited.

A rotating piezothermoelastic bar of circular cross-section has gained considerable importance in construction of gyroscopes to measure the angular velocity of a rotating body. The coupled behaviour thermoelectro-elastic materials is useful in space shuttles, supersonic airplanes, rockets and missiles, plasma physics and smart structure applications for self-monitoring and self-controlling processes.

doi:10.1088/978-0-7503-5638-1ch7

The composite consisting of piezoelectric and piezomagnetic components has found increasing application in engineering structures, particularly in smart/intelligent structure systems. A polymer piezoelectric material plays an important role for the flexible characteristics of these polymers.

Morse (1954) reported the vibration analysis in an anisotropic circular cylinder with hexagonal symmetry. Mindlin (1961) studied the thermopiezoelectric theory and later developed the governing equations for a thermopiezoelectric plate (1979). Nowacki (1978, 1979) proposed the physical laws for thermopiezoelectric materials. The generalized theory of thermopiezoelectricity considering the finite speed of propagation of thermal signals was analysed by Chandrasekhariah (1984, 1988) and he also estimated the thermal relaxation times in 1986. The effect of heat conduction on a linear piezoelectric body with the help of perturbation method was estimated by Yang and Batra (1995). The free vibration analysis of a homogeneous, transversely isotropic, piezothermoelastic cylindrical panel based on three-dimensional piezoelectric thermoelasticity was discussed by Sharma *et al* (2004). The propagation of straight and circular crested waves in generalized piezothermoelastic materials was presented by Sharma and Walia (2006).

Tang and Xu (1995) derived the general dynamic equations based on the anisotropic composite laminated plate theory. Also, analytical dynamical solutions for the case of general force acting on simply supported piezothermoelastic laminated plates have been examined as a special case. The thermopiezoelectricity theory for a composite plate was reported by Tauchert (1992). These corresponding equations for an isotropic case were obtained by Dhaliwal and Sherief (1980). Green and Laws (1972) proposed a generalization of this inequality. Also, they developed an explicit version of the constitutive equations. An analytical solution of piezolaminated rectangular plate with arbitrary clamped/simply supported boundary conditions was analysed by Peyman *et al* (2013). Sabzikar Boroujerdy and Eslami (2014) developed the axisymmetric snap-through behaviour of Piezo-FGM shallow clamped spherical shells under thermoelectromechanical loading. By using Love's first approximation theory, the vibration of a rotating thin cylindrical panel was discussed by Loy and Lam (1995). The effect of rotation and relaxation times on plane waves in generalized thermo-visco-elasticity was proposed by Roychoudhuri and Mukhopadhyay (2000). The energy dissipation and critical speed of granular flow in a rotating cylinder was studied by Sergiu *et al* (2014) and they obtained that the coefficient of friction has the greatest significance on the centrifuging speed. Bayat *et al* (2014) presented the one-dimensional analysis for magneto-thermomechanical response in a functionally graded annular variable-thickness rotating disk.

The wave propagation in a thermoelastic plate of arbitrary cross-sections considering Fourier expansion collocation method was developed by Ponnusamy (2011). The wave propagation in a piezoelectric solid bar of circular cross-section immersed in fluid was studied by Ponnusamy (2013). Ponnusamy and Selvamani (2012) proposed the wave propagation in a magneto-thermoelastic wave in a transversely isotropic cylindrical panel. Later, Selvamani and Ponnusamy (2013) analysed the wave propagation in a generalized thermoelastic plate immersed in fluid. Also, the dynamic response of a solid bar of cardioidal cross-sections immersed in an inviscid fluid using Fourier expansion collocation method has been studied by Selvamani and Ponnusamy (2014).

In this chapter, wave propagation in a transversely isotropic piezothermoelastic bar of circular cross-section is studied. The displacement functions to represent three displacement components on the basis of three-dimensional generalized piezothermoelasticity for transversely isotropic media are considered. The frequency equations are obtained for L–S and G–L theories of thermoelasticity. The computed thermomechanical coupling, electromechanical coupling, frequency shift, specific loss and frequency have been presented in the form of dispersion curves.

7.2 Formulation of a piezothermoelastic rotating bar

This section has been reproduced with permission from Selvamani and Ponnusamy (2015). Copyright 2015, Emerald Group Publishing Limited.

The initially uniform temperature T_0 is considered in the undistributed state, assuming that the medium is rotating with uniform angular velocity $\vec{\Omega}$. The equations of motion in the absence of body force and including the Coriolis effect and centripetal forces are

$$\frac{\partial}{\partial r}\sigma_{rr} + \frac{1}{r}\frac{\partial}{\partial \theta}\sigma_{r\theta} + \frac{\partial}{\partial z}\sigma_{rz} + \frac{(\sigma_{rr} - \sigma_{\theta\theta})}{r} + \rho(\vec{\Omega} \times (\vec{\Omega} \times \vec{u}) + 2(\vec{\Omega} \times \vec{u}_{,t})) = \rho\frac{\partial^2 u_r}{\partial t^2}$$

$$\frac{\partial}{\partial r}\sigma_{r\theta} + \frac{1}{r}\frac{\partial}{\partial \theta}\sigma_{\theta\theta} + \frac{\partial}{\partial z}\sigma_{\theta z} + \frac{2\sigma_{r\theta}}{r} = \rho\frac{\partial^2 u_\theta}{\partial t^2}$$

$$\frac{\partial}{\partial r}\sigma_{rz} + \frac{1}{r}\frac{\partial}{\partial \theta}\sigma_{\theta z} + \frac{\partial}{\partial z}\sigma_{zz} + \frac{\sigma_{rz}}{r} + \rho(\vec{\Omega} \times (\vec{\Omega} \times \vec{u}) + 2(\vec{\Omega} \times \vec{u}_{,t})) = \rho\frac{\partial^2 u_z}{\partial t^2}. \quad (7.1)$$

The heat conduction equation is

$$K_1\left(\frac{\partial^2 T}{\partial r^2} + \frac{1}{r}\frac{\partial T}{\partial r} + \frac{1}{r^2}\frac{\partial^2 T}{\partial \theta^2}\right) + K_3\frac{\partial^2 T}{\partial z^2} - \rho C_v\left(\frac{\partial T}{\partial t} + t_0\frac{\partial^2 T}{\partial t^2}\right)$$
$$= T_0\left(\frac{\partial}{\partial t} + \delta_{1k} t_0\frac{\partial^2}{\partial t^2}\right)[\beta_1(S_{rr} + S_{\theta\theta}) + \beta_3 S_{zz} - p_3 E_z]. \quad (7.2)$$

The electric displacements D_r, D_θ and D_z satisfy the Gaussian equation

$$\frac{1}{r}\frac{\partial}{\partial r}(rD_r) + \frac{1}{r}\frac{\partial D_\theta}{\partial \theta} + \frac{\partial D_z}{\partial r} = 0. \quad (7.3)$$

The elastic, piezoelectric, thermal conduction and dielectric matrices of the 6 mm crystal class for the piezothermoelastic relations are

$$\sigma_{rr} = c_{11}S_{rr} + c_{12}S_{\theta\theta} + c_{13}S_{zz} - \beta_1(T + t_1\delta_{2k}\dot{T}) + e_{31}E_z$$

$$\sigma_{\theta\theta} = c_{12}S_{rr} + c_{11}S_{\theta\theta} + c_{13}S_{zz} - \beta_1(T + t_1\delta_{2k}\dot{T}) + e_{31}E_z$$

$$\sigma_{zz} = c_{13}S_{rr} + c_{13}S_{\theta\theta} + c_{33}S_{zz} - \beta_3(T + t_1\delta_{2k}\dot{T}) + e_{33}E_z \tag{7.4}$$

$$\begin{aligned}\sigma_{r\theta} &= 2c_{66}S_{r\theta}\\ \sigma_{\theta z} &= 2c_{44}S_{\theta z} - e_{15}E_\theta\\ \sigma_{rz} &= 2c_{44}S_{rz} - e_{15}E_r\end{aligned} \tag{7.5}$$

and

$$\begin{aligned}D_r &= 2e_{15}S_{rz} + \varepsilon_{11}E_r\\ D_\theta &= 2e_{15}S_{\theta z} + \varepsilon_{11}E_\theta\\ D_z &= e_{31}(S_{rr} + S_{\theta\theta}) + e_{33}S_{zz} + \varepsilon_{33}E_z + p_3T\end{aligned} \tag{7.6}$$

where σ_{rr}, $\sigma_{\theta\theta}$, σ_{zz}, $\sigma_{r\theta}$, $\sigma_{\theta z}$, σ_{rz} and S_{rr}, $S_{\theta\theta}$, S_{zz}, $S_{r\theta}$, $S_{\theta z}$, S_{rz} are the stress and strain components, c_{11}, c_{12}, c_{13}, c_{33}, c_{44} and $c_{66} = \frac{(c_{11} - c_{12})}{2}$ denote five elastic constants, e_{31}, e_{15}, e_{33} are piezoelectric constants, ε_{11}, ε_{33} are the dielectric constants, T is the temperature change about the equilibrium temperature T_0, C_v stands for heat capacity, β_1, β_3 are the coefficients of thermal expansion, K_1, K_3 denotes thermal conductivity, t_0 and t_1 are the two times for thermal relaxation, and ρ is the mass density. The displacement equation of motion has the additional terms with a time-dependent centripetal acceleration $\vec{\Omega} \times (\vec{\Omega} \times \vec{u})$ and $2(\vec{\Omega} \times \vec{u}_{,t})$, where $\vec{u} = (u, 0, w)$ is the displacement vector and $\vec{\Omega} = (0, \Omega, 0)$ is the angular velocity. The comma notation is used for spatial derivatives; the superposed dot represents time differentiation and δ_{ik} is the Kronecker delta. In addition, $k = 1$ for L–S theory and $k = 2$ for G–L theory. The thermal relation times t_0 and t_1 satisfy the inequalities $t_0 \geqslant t_1 \geqslant 0$ for G–L theory only, and we assume that $\rho > 0$, $T_0 > 0$ and $C_v > 0$.

The strain S_{ij} are related to the displacements and can be expressed as

$$\begin{aligned}&S_{rr} = \frac{\partial u_r}{\partial r},\ S_{\theta\theta} = \left(\frac{1}{r}\frac{\partial u_\theta}{\partial \theta} + \frac{u_r}{r}\right),\ S_{zz} = \frac{\partial u_z}{\partial z},\\ &S_{\theta z} = \frac{1}{2}\left(\frac{\partial u_\theta}{\partial z} + \frac{1}{r}\frac{\partial u_z}{\partial \theta}\right),\ S_{r\theta} = \frac{1}{2}\left(\frac{1}{r}\frac{\partial u_r}{\partial \theta} + \frac{\partial u_\theta}{\partial r} - \frac{u_\theta}{r}\right),\\ &S_{rz} = \frac{1}{2}\left(\frac{\partial u_r}{\partial z} + \frac{\partial u_z}{\partial r}\right),\ E_r = -\frac{\partial E}{\partial r},\ E_\theta = -\frac{1}{r}\frac{\partial E}{\partial \theta},\ E_z = -\frac{\partial E}{\partial z}.\end{aligned} \tag{7.7}$$

7.2.1 Lord–Shulman theory

The L–S theory of heat conduction equation and normal stresses for a three-dimensional piezothermoelasticity is obtained by substituting $K = 1$ in equations (7.2) and (7.4), and we arrive at

$$\begin{aligned}&K_1\left(\frac{\partial^2 T}{\partial r^2} + \frac{1}{r}\frac{\partial T}{\partial r} + \frac{1}{r^2}\frac{\partial^2 T}{\partial \theta^2}\right) + K_3\frac{\partial^2 T}{\partial z^2} - \rho C_v\left(\frac{\partial T}{\partial t} + t_0\frac{\partial^2 T}{\partial t^2}\right)\\ &= T_0\left(\frac{\partial}{\partial t} + t_0\frac{\partial^2}{\partial t^2}\right)[\beta_1(S_{rr} + S_{\theta\theta}) + \beta_3 S_{zz} - p_3E_z]\end{aligned} \tag{7.8}$$

and

$$\sigma_{rr} = c_{11}S_{rr} + c_{12}S_{\theta\theta} + c_{13}S_{zz} - \beta_1 T + e_{31}E_z$$

$$\sigma_{\theta\theta} = c_{12}S_{rr} + c_{11}S_{\theta\theta} + c_{13}S_{zz} - \beta_1 T + e_{31}E_z$$

$$\sigma_{zz} = c_{13}S_{rr} + c_{13}S_{\theta\theta} + c_{33}S_{zz} - \beta_3 T + e_{33}E_z. \tag{7.9}$$

Inserting equations (7.5), (7.7) and (7.9) in equations (7.1) and (7.8), we get

$$c_{11}\left(\frac{\partial^2 u_r}{\partial r^2} + \frac{1}{r}\frac{\partial u_r}{\partial r} - \frac{u_r}{r^2}\right) + \frac{(c_{12}+c_{66})}{r}\frac{\partial^2 u_\theta}{\partial r\partial\theta} + (c_{13}+c_{44})\frac{\partial^2 u_z}{\partial r\partial z} - \frac{(c_{11}+c_{66})}{r^2}\frac{\partial u_\theta}{\partial\theta} + \frac{c_{66}}{r^2}\frac{\partial^2 u_r}{\partial\theta^2}$$
$$+ c_{44}\frac{\partial^2 u_r}{\partial z^2} - \beta_1\frac{\partial T}{\partial r} + (e_{15}+e_{31})E_{,rz} + \rho(\Omega^2 u + 2\Omega w_{,t}) = \rho\frac{\partial^2 u_r}{\partial t^2}$$

$$c_{66}\left(\frac{\partial^2 u_\theta}{\partial r^2} + \frac{1}{r}\frac{\partial u_\theta}{\partial r} - \frac{u_\theta}{r^2}\right) + \frac{c_{11}}{r^2}\frac{\partial^2 u_\theta}{\partial\theta^2} + c_{44}\frac{\partial^2 u_\theta}{\partial z^2} + \frac{(c_{66}+c_{12})}{r}\frac{\partial^2 u_r}{\partial r\partial\theta} + \frac{(c_{66}+c_{11})}{r^2}\frac{\partial u_r}{\partial\theta}$$
$$+ \frac{(c_{44}+c_{13})}{r}\frac{\partial^2 u_z}{\partial\theta\partial z} - \frac{\beta_1}{r}\frac{\partial T}{\partial\theta} + \frac{(e_{31}+e_{15})}{r}E_{,\theta z} = \rho\frac{\partial^2 u_\theta}{\partial t^2}$$

$$c_{44}\left(\frac{\partial^2 u_z}{\partial r^2} + \frac{1}{r}\frac{\partial u_z}{\partial r} + \frac{1}{r^2}\frac{\partial^2 u_z}{\partial\theta^2}\right) + c_{33}\frac{\partial^2 u_z}{\partial z^2} + \frac{(c_{13}+c_{44})}{r}\left(\frac{\partial u_r}{\partial z} + \frac{\partial^2 u_\theta}{\partial\theta\partial z}\right) + (c_{13}+c_{44})\frac{\partial^2 u_r}{\partial r\partial z} + e_{33}E_{,zz}$$
$$+ e_{15}\left(\frac{\partial^2 E}{\partial r^2} + \frac{1}{r}\frac{\partial E}{\partial r} + \frac{1}{r^2}\frac{\partial^2 E}{\partial\theta^2}\right) - \beta_3 T + \rho(\Omega^2 w + 2\Omega u_{,t}) = \rho\frac{\partial^2 u_z}{\partial t^2}$$

$$K_1\,\nabla^2\,T + K_3\frac{\partial^2 T}{\partial z^2} = T_0\left(\frac{\partial}{\partial t} + t_0\frac{\partial^2}{\partial t^2}\right)$$
$$\left[\beta_1\left(\frac{\partial u_r}{\partial r} + \frac{1}{r}\frac{\partial u_\theta}{\partial\theta} + \frac{u_r}{r}\right) + \beta_3\frac{\partial u_z}{\partial z} - p_3\frac{\partial E}{\partial z}\right] + \rho C_v\left(\frac{\partial T}{\partial t} + t_0\frac{\partial^2 T}{\partial t^2}\right)$$

$$e_{15}\left(\frac{\partial^2 u_z}{\partial r^2} + \frac{1}{r}\frac{\partial u_z}{\partial r} + \frac{1}{r^2}\frac{\partial^2 u_z}{\partial\theta^2}\right) + (e_{15}+e_{31})\left(\frac{\partial^2 u_r}{\partial r\partial z} + \frac{1}{r}\frac{\partial^2 u_\theta}{\partial\theta\partial z} + \frac{1}{r}\frac{\partial u_r}{\partial z}\right)$$
$$- \varepsilon_{11}\left(\frac{\partial^2 E}{\partial r^2} + \frac{1}{r}\frac{\partial E}{\partial r} + \frac{1}{r^2}\frac{\partial^2 E}{\partial\theta^2}\right) + e_{33}\frac{\partial^2 u_z}{\partial z^2} - \varepsilon_{33}\frac{\partial^2 E}{\partial z^2} + p_3\frac{\partial T}{\partial z} = 0. \tag{7.10}$$

7.3 Analytical solution

The following displacement components are used to obtain the propagation of harmonic waves in a piezothermoelastic circular bar, which are taken from Paul (1966). Thus, we seek the solution of equation (7.10) in the form

$$u_r(r,\,\theta,\,z,\,t) = \left(\varphi_{,r} + \frac{1}{r}\psi_{,\theta}\right)e^{i(kz+\omega t)}$$

$$u_\theta(r,\theta,z,t) = \left(\frac{1}{r}\varphi_{,\theta} - \psi_{,r}\right)e^{i(kz+\omega t)}$$

$$u_z(r,\theta,z,t) = \left(\frac{i}{a}\right)We^{i(kz+\omega t)}$$

$$E(r,\theta,z,t) = \left(\frac{ic_{44}}{ae_{33}}\right)Ee^{i(kz+\omega t)}$$

$$T(r,\theta,z,t) = \frac{c_{44}}{\beta_3 a^2}Te^{i(kz+\omega t)}$$

$$E_r(r,\theta,z,t) = -E_{,r}e^{i(kz+\omega t)}$$

$$E_\theta(r,\theta,z,t) = -\frac{1}{r}E_{,\theta}e^{i(kz+\omega t)}$$

$$E_z(r,\theta,z,t) = E_{,z}e^{i(kz+\omega t)} \tag{7.11}$$

where $i=\sqrt{-1}$, k and ω are the wave number and angular frequency, $\varphi(r,\theta)$, $\psi(r,\theta)$, $E(r,\theta)$ and $T(r,\theta)$ are the displacement potentials and a is the geometrical parameter of the bar.

The non-dimensional form of material quantities are

$$x=\frac{r}{a},\ \zeta=ka,\ \varpi^2=\frac{\rho\omega^2a^2}{c_{44}},\ \bar{c}_{11}=\frac{c_{11}}{c_{44}},\ \bar{c}_{13}=\frac{c_{13}}{c_{44}},\ \bar{c}_{33}=\frac{c_{33}}{c_{44}},$$

$$\bar{c}_{66}=\frac{c_{66}}{c_{44}},\ \bar{\beta}=\frac{\beta_1}{\beta_3},\ \bar{K}_i=\frac{K_ic_{44}}{\omega\beta_3^2T_0a^2},\ \bar{d}=\frac{\rho C_vc_{44}}{\beta_3^2T_0},\ \bar{p}_3=\frac{p_3c_{44}}{\beta_3e_{33}},$$

$$\bar{\varepsilon}_{11}=\frac{\varepsilon_{11}c_{44}}{e_{33}^2},\ \bar{e}_{31}=\frac{e_{31}}{e_{33}},\ \bar{e}_{15}=\frac{e_{15}}{e_{33}}\Gamma=\frac{\rho\Omega^2R^2}{2+\bar{\lambda}}$$

and inserting equation (7.11) in equation (7.10), we obtain

$$[\bar{c}_{11}\nabla^2+(\varpi^2+\Gamma-\zeta^2)]\varphi-\zeta(1+\bar{c}_{13})W-\zeta(\bar{e}_{15}+\bar{e}_{31})E-\bar{\beta}T=0$$

$$\zeta(1+\bar{c}_{13})\nabla^2\varphi+[\nabla^2+(\varpi^2+\Gamma-\zeta^2\bar{c}_{33})]W+(\bar{e}_{15}\nabla^2-\zeta^2)E-\zeta T=0$$

$$\tau_0\bar{\beta}\nabla^2\varphi-\tau_0\zeta W+\tau_0\zeta\bar{p}_3E+[\tau_0\bar{d}+i\bar{K}_1\nabla^2-i\bar{K}_3\zeta^2]T=0$$

$$\zeta(\bar{e}_{15}+\bar{e}_{31})\nabla^2\varphi+(\bar{e}_{15}\nabla^2-\zeta^2)W+(\zeta^2\bar{\varepsilon}_{33}-\bar{\varepsilon}_{11}\nabla^2)E+\bar{p}_3\zeta T=0 \tag{7.12}$$

and

$$(\bar{c}_{66}\nabla^2+(\varpi^2+\Gamma-\zeta^2))\psi=0. \tag{7.13}$$

Equation (7.12) can be written as

$$\begin{vmatrix} \bar{c}_{11}\nabla^2 + g_3 & -\zeta g_6 & -\zeta g_5 & -\bar{\beta} \\ \zeta g_6 \nabla^2 & \nabla^2 + g_1 & (\bar{e}_{15}\nabla^2 - \zeta^2) & -\zeta \\ \zeta g_5 \nabla^2 & (\bar{e}_{15}\nabla^2 - \zeta^2) & (\zeta^2 \bar{\varepsilon}_{33} - \bar{\varepsilon}_{11}\nabla^2) & \bar{p}_3 \zeta \\ \tau_0 \bar{\beta}\nabla^2 & -\tau_0 \zeta & \tau_0 \zeta \bar{p}_3 & (i\bar{K}_1 \nabla^2 + g_2) \end{vmatrix} (\varphi, W, E, T) = 0 \tag{7.14}$$

where

$$g_1 = \varpi^2 + \Gamma - \zeta^2 \bar{c}_{33},\ g_2 = \tau_0 \bar{d} - i\bar{K}_3\zeta^2,$$
$$g_3 = \varpi^2 + \Gamma - \zeta^2,\ g_4 = \bar{\varepsilon}_{11}^2 + \bar{e}_{15}^2,\ g_5 = \bar{e}_{31} + \bar{e}_{15} \text{ and } g_6 = 1 + \bar{c}_{13}.$$

By solving the above determinant, we get the following form

$$(A\nabla^8 + B\nabla^6 + C\nabla^4 + D\nabla^2 + E)(\varphi, W, E, T) = 0. \tag{7.15}$$

where

$$A = -i\bar{K}_1 \bar{c}_{11} g_4$$

$$B = i\bar{K}_1 \bar{c}_{11}(\zeta^2(\bar{\varepsilon}_{33} + \bar{e}_{15}) - \bar{\varepsilon}_{11} g_1) - \bar{c}_{11}\bar{\varepsilon}_{11} g_2 - i\bar{K}_1 g_3 g_4 - \zeta^2 g_5 g_6 i\bar{K}_1 \bar{e}_{15}$$
$$- \zeta^2 g_5 i\bar{K}_1(\bar{e}_{15} g_6 - g_5) - \bar{e}_{15}^2 \tau_0 \bar{\beta}^2$$

$$\begin{aligned} C = {} & \bar{c}_{11} g_1 (i\bar{K}_1 \zeta^2 \bar{\varepsilon}_{33} - \bar{\varepsilon}_{11} g_2) - \bar{c}_{11}\bar{e}_{15}[\bar{e}_{15} g_2 + \zeta^2(-i\bar{K}_1 + i\bar{K}_3\zeta^2 + \tau_0(\bar{p}_3 \bar{e}_{15} - \bar{d} - \bar{\beta}_3 \bar{p}_3 \zeta^2))] \\ & + \bar{c}_{11}\left[g_2(\zeta^2 \bar{\varepsilon}_{33} + \bar{e}_{15}) + \zeta^2(\tau_0 \bar{p}_3 \bar{e}_{15} - i\bar{K}_1) + \tau_0\left(-\bar{p}_3^2 - \bar{p}_3 \bar{e}_{15} + \bar{\varepsilon}_{11}\right)\right] \\ & + i\bar{K}_3 g_3[\zeta^2(\bar{\varepsilon}_{33} + \bar{e}_{15}) - \bar{\varepsilon}_{11} g_1] - \bar{\varepsilon}_{11} g_2 g_3 - \zeta^2 g_6 \bar{e}_{15}(g_2 g_5 - \tau_0 \bar{\beta}\bar{p}_3) \\ & - \zeta^2 g_5 g_6 \bar{e}_{15}(g_2 - i\bar{K}_1\zeta^2) + \zeta^2(i\bar{K}_1(\zeta^2 g_5 g_6 + g_1 g_5^2) - g_2 g_5^2) \\ & + \tau_0 \bar{\beta}\zeta^2(-\bar{\varepsilon}_{11} g_6 + g_5(-\bar{p}_3 + g_1 \bar{e}_{15})) + \zeta^2 g_6 \tau_0 \bar{\beta}(\bar{e}_{15}\bar{p}_3 - \bar{\varepsilon}_{11}) \\ & + \tau_0 \bar{\beta}(-g_1 \bar{\varepsilon}_{11}\bar{\beta} + \bar{e}_{15}\zeta^2(\bar{\beta} - g_5)) + \tau_0 \bar{\beta}^2(\zeta^2 \bar{e}_{15} - \bar{\varepsilon}_{11}) \end{aligned}$$

$$\begin{aligned} D = {} & \bar{c}_{11}\zeta^2\left[g_1\left(\bar{\varepsilon}_{33} g_2 - \bar{p}_3^2 \tau_0\right) + \zeta^2(\tau_0(\bar{d} + \bar{p}_3 - \bar{\varepsilon}_{33}) + \zeta^2(\tau_0 \bar{p}_3 - i\bar{K}_3))\right] + g_1 g_3 (i\bar{K}_1 \zeta^2 \bar{\varepsilon}_{33} - \bar{\varepsilon}_{11} g_2) \\ & - \bar{e}_{15} g_3[\bar{e}_{15} g_2 + \zeta^2(-i\bar{K}_1 - g_2 + \tau_0 \bar{p}_3(\bar{e}_{15} - \zeta^2)) \\ & + \zeta^2 g_3[g_2(\bar{\varepsilon}_{33} + \bar{e}_{15}) - i\bar{K}_1\zeta^2 + \tau_0(\bar{p}_3(\zeta^2 \bar{e}_{15} - \bar{p}_3 - \bar{e}_{15}) + \bar{\varepsilon}_{11})] + \zeta^4 g_6^2\left(\bar{\varepsilon}_{33} g_2 - \bar{p}_3^2 g_5\right) \\ & + \zeta^2 g_1 g_5 (g_2 g_5 - \tau_0 \bar{p}_3 \bar{\beta}) - \tau_0 \zeta^4 (g_5(\bar{\beta} - g_5) - g_6 \bar{\beta}\bar{\varepsilon}_{33}\bar{p}_3) - \zeta^2 \tau_0 \bar{\beta}(g_5 \bar{p}_3 - \bar{\varepsilon}_{33}\bar{\beta}) \\ & - \zeta^4 \tau_0 \bar{\beta}(\bar{\beta} - g_5) \end{aligned}$$

$$\begin{aligned} E = {} & g_3 \zeta^2\left[g_1\left(\bar{\varepsilon}_{33} g_2 - \bar{p}_3^2 \tau_0\right) + \zeta^2(\tau_0(\bar{d} + \bar{p}_3 - \bar{\varepsilon}_{33}) + \zeta^2(\tau_0 \bar{p}_3 - i\bar{K}_3))\right] \\ & + \zeta^4 g_6 (g_5(g_2 - \tau_0 \bar{p}_3) - \tau_0 \bar{p}_3 \bar{\beta}) - g_1 \bar{\beta}\zeta^2 \tau_0 (\bar{p}_3 g_5 - \bar{\varepsilon}_{33}\bar{\beta}). \end{aligned} \tag{7.16}$$

Solving equation (7.15), we get the following solution as

$$\varphi = \sum_{i=1}^{4} A_i J_n(\alpha_i a x) \cos n\theta$$

$$W = \sum_{i=1}^{4} a_i A_i J_n(\alpha_i ax) \cos n\theta$$

$$E = \sum_{i=1}^{4} b_i A_i J_n(\alpha_i ax) \cos n\theta$$

$$T = \sum_{i=1}^{4} c_i A_i J_n(\alpha_i ax) \cos n\theta \tag{7.17}$$

where $(\alpha_i a)^2 > 0$, $(i = 1,2,3,4)$ are the roots of the algebraic equation

$$A(\alpha a)^8 - B(\alpha a)^6 + C(\alpha a)^4 - D(\alpha a)^2 + E = 0. \tag{7.18}$$

The solutions corresponding to the root $(\alpha_i a)^2 = 0$ are not considered here, since $J_n(0)$ is zero, except for $n = 0$. The Bessel function J_n is used when the roots $(\alpha_i a)^2$, $(i = 1,2,3,4)$ are real or complex and the modified Bessel function I_n is used when the roots $(\alpha_i a)^2$, $(i = 1,2,3,4)$ are imaginary.

The constants a_i, b_i and c_i defined in equation (7.17) can be calculated from the following equations

$$a_i = \frac{\zeta g_6(\bar{e}_{15}\,\nabla^2 - \zeta^2) - \zeta g_5(\nabla^2 + g_1)}{H}$$

$$b_i = \frac{-\zeta^2 g_5 - \bar{\beta}(\bar{e}_{15}\,\nabla^2 - \zeta^2)}{H}$$

$$c_i = \frac{-\bar{\beta}\zeta g_6 + \zeta(\bar{c}_{11}\,\nabla^2 + g_3)}{H}$$

where

$$H = \bar{c}_{11}\,\nabla^2 + g_3(\nabla^2 + g_1) + \zeta^2 g_6^2\,\nabla^2. \tag{7.19}$$

Solving the equation (7.13), we obtain

$$\psi = A_5 J_n(\alpha_5 ax) \sin n\theta \tag{7.20}$$

where $(\alpha_5 a)^2 = \frac{(\varpi^2 + \Gamma - \zeta^2)}{\bar{c}_{66}}$. The modified Bessel function I_n is used instead of the Bessel function J_n, when $(\alpha_5 a)^2 < 0$.

7.4 Boundary conditions and frequency equations

The free vibration of a transversely isotropic piezothermoelastic solid bar of circular cross-section is considered in this problem.

(i) **Mechanical boundary conditions**

$$\sigma_{rr} = \sigma_{r\theta} = \sigma_{rz} = 0 \text{ at } r = a. \tag{7.21a}$$

(ii) **Thermal boundary conditions**

$$\text{Isothermal surfaces } T = 0 \tag{7.21b}$$

$$\text{Thermally insulated surfaces } T_{,r} = 0.$$

(iii) **Electrical boundary conditions**

$$\text{Charge free (Open circuits) surfaces } E = 0 \tag{7.21c}$$

$$\text{Electrically shorted (Closed circuits) surfaces } D_r = 0.$$

By inserting s the solution in equations (7.17) and (7.20) in the boundary condition in equation (7.21), we arrive at the following form

$$[A]\{X\} = \{0\} \tag{7.22}$$

where $[A]$ is a 5×5 matrix of unknown wave amplitudes and $\{X\}$ is an 5×1 column vector of the unknown amplitude coefficients A_1, A_2, A_3, A_4, A_5. The elements of $[A]$ are given in appendix A of chapter 6. The solution of equation (7.22) is nontrivial when the determinant of the coefficient of the wave amplitudes $\{X\}$ vanishes, that is

$$|A| = 0. \tag{7.23}$$

Equation (7.23) is said to be the frequency equation of the coupled system and the elements in the determinants can be expressed as

$$a_{1i} = 2\bar{c}_{66}\{n(n-1)J_n(\alpha_i ax) + (\alpha_i ax)J_{n+1}(\alpha_i ax)\}$$
$$- x^2[(\alpha_i a)^2\bar{c}_{11} + \zeta\bar{c}_{13}a_i + \zeta b_i + \bar{\beta}c_i]J_n(\alpha_i ax),\ i = 1, 2, 3, 4$$

$$a_{15} = 2\bar{c}_{66}\{n(n-1)J_n(\alpha_5 ax) - (\alpha_5 ax)J_{n+1}(\alpha_5 ax)\}$$

$$a_{2i} = 2n\{(\alpha_i ax)J_{n+1}(\alpha_i ax) - (n-1)J_n(\alpha_i ax)\},\ i = 1, 2, 3, 4$$

$$a_{25} = \{[(\alpha_5 ax)^2 - 2n(n-1)]J_n(\alpha_5 ax) - 2(\alpha_5 ax)J_{n+1}(\alpha_5 ax)\}$$

$$a_{3i} = (\zeta + a_i + \bar{e}_{15}b_i)[nJ_n(\alpha_i ax) - (\alpha_i ax)J_{n+1}(\alpha_i ax)],\ i = 1, 2, 3, 4$$

$$a_{35} = n\zeta J_n(\alpha_5 ax)$$

$$a_{4i} = (\bar{e}_{15}\zeta a_i - \bar{\varepsilon}_{11}b_i)\{nJ_n(\alpha_i ax) - (\alpha_i ax)J_{n+1}(\alpha_i ax)\},\ i = 1, 2, 3, 4$$

$$a_{45} = \bar{e}_{15}\zeta nJ_n(\alpha_5 ax)$$

$$a_{5i} = c_i\{nJ_n(\alpha_i ax) - (\alpha_i ax)J_{n+1}(\alpha_i ax)\},\ i = 1, 2, 3, 4.$$

$$a_{55} = 0$$

7.4.1 Green–Lindsay theory

The governing equations of heat conduction and motion for a piezothermoelastic cylinder of circular bar, in the context of G–L theory is obtained by substituting $K = 2$ in the heat conduction equation (7.2) and in the stress–strain relation equation (7.4), thus we get

$$K_1\left(\frac{\partial^2 T}{\partial r^2} + \frac{1}{r}\frac{\partial T}{\partial r} + \frac{1}{r^2}\frac{\partial^2 T}{\partial \theta^2}\right) + K_3\frac{\partial^2 T}{\partial z^2} - \rho C_v\left(\frac{\partial T}{\partial t} + t_0\frac{\partial^2 T}{\partial t^2}\right) = T_0\frac{\partial}{\partial t}[\beta_1(S_{rr} + S_{\theta\theta}) + \beta_3 S_{zz} - p_3 E_z] \tag{7.24}$$

and

$$\begin{aligned}\sigma_{rr} &= c_{11}S_{rr} + c_{12}S_{\theta\theta} + c_{13}S_{zz} - \beta_1(T + t_1\dot{T}) - e_{31}E_z \\ \sigma_{\theta\theta} &= c_{12}S_{rr} + c_{11}S_{\theta\theta} + c_{13}S_{zz} - \beta_3(T + t_1\dot{T}) - e_{31}E_z \\ \sigma_{zz} &= c_{13}S_{rr} + c_{13}S_{\theta\theta} + c_{33}S_{zz} - \beta_3(T + t_1\dot{T}) - e_{33}E_z.\end{aligned} \tag{7.25}$$

Substituting equations (7.5), (7.6), (7.25) in equation (7.1) along with equation (7.24), we obtain the following equation as:

$$c_{11}\left(\frac{\partial^2 u_r}{\partial r^2} + \frac{1}{r}\frac{\partial u_r}{\partial r} - \frac{u_r}{r^2}\right) + \frac{(c_{12}+c_{66})}{r}\frac{\partial^2 u_\theta}{\partial r\partial\theta} + (c_{13} + c_{44})\frac{\partial^2 u_z}{\partial r\partial z} - \frac{(c_{11}+c_{66})}{r^2}\frac{\partial u_\theta}{\partial\theta}$$
$$+ \frac{c_{66}}{r^2}\frac{\partial^2 u_r}{\partial\theta^2} + c_{44}\frac{\partial^2 u_z}{\partial z^2} - \beta_1(C) + (e_{15} + e_{31})E_{,rz} + \rho(\Omega^2 u + 2\Omega w_{,t}) = \rho\frac{\partial^2 u_r}{\partial t^2}$$

$$c_{66}\left(\frac{\partial^2 u_\theta}{\partial r^2} + \frac{1}{r}\frac{\partial u_\theta}{\partial r} - \frac{u_\theta}{r^2}\right) + \frac{(c_{11})}{r^2}\frac{\partial^2 u_\theta}{\partial r^2} + c_{44}\frac{\partial^2 u_\theta}{\partial z^2} + \frac{(c_{66}+c_{12})}{r}\frac{\partial^2 u_r}{\partial r\partial\theta}$$
$$+ \frac{(c_{66}+c_{11})}{r}\frac{\partial u_r}{\partial\theta} + \frac{(c_{44}+c_{13})}{r}\frac{\partial^2 u_z}{\partial\theta\partial z} - \frac{\beta_1}{r}(T + t_1\dot{T}) + \frac{(e_{31}+e_{15})}{r}E_{,\theta z} = \rho\frac{\partial^2 u_\theta}{\partial t^2}$$

$$c_{44}\left(\frac{\partial^2 u_z}{\partial r^2} + \frac{1}{r}\frac{\partial u_z}{\partial r} + \frac{1}{r^2}\frac{\partial^2 u_z}{\partial\theta^2}\right) + c_{33}\frac{\partial^2 u_z}{\partial z^2} + \frac{(c_{13}+c_{44})}{r}\left(\frac{\partial u_r}{\partial z} + \frac{\partial^2 u_\theta}{\partial\theta\partial z}\right) + (c_{13} + c_{44})\frac{\partial^2 u_r}{\partial r\partial z}$$
$$+ e_{33}E_{,zz} + e_{15}\left(E_{,rr} + \frac{1}{r}E_{,r} + \frac{1}{r^2}E_{\theta\theta}\right) - \beta_3(T + t_1\dot{T})_{,z} + \rho(\Omega^2 w + 2\Omega u_{,t}) = \rho\frac{\partial^2 u_z}{\partial t^2}$$

$$e_{15}\left(\frac{\partial^2 u_z}{\partial r^2} + \frac{1}{r}\frac{\partial u_z}{\partial r} + \frac{1}{r^2}\frac{\partial^2 u_z}{\partial\theta^2}\right) + (e_{15} + e_{31})\left(\frac{\partial^2 u_z}{\partial r\partial z} + \frac{1}{r^2}\frac{\partial^2 u_\theta}{\partial z\partial\theta} + \frac{1}{r}\frac{\partial u_z}{\partial z}\right)$$
$$- \varepsilon_{11}\left(E_{,rr} + \frac{1}{r}E_{,r} + \frac{1}{r^2}E_{,\theta\theta}\right) + e_{33}\frac{\partial^2 u_z}{\partial z^2} - \varepsilon_{33}E_{,zz} + p_3 T_{,z} = 0$$

$$K_1\nabla^2 T + K_3 T_{,zz} = T_0\frac{\partial}{\partial t}\left[\beta_1\left(\frac{\partial u_r}{\partial r} + \frac{1}{r}\frac{\partial u_\theta}{\partial\theta} + \frac{u_r}{r}\right) + \beta_3\frac{\partial u_z}{\partial z} - p_3\frac{\partial E}{\partial z}\right] + \rho C_v\left(\frac{\partial T}{\partial t} + t_0\frac{\partial^2 T}{\partial t^2}\right). \tag{7.26}$$

Inserting the solution in equations (7.11) and (7.26), we get the following form

$$\begin{vmatrix} \left(\bar{c}_{11}\,\nabla^2 + g_3\right) & -\zeta g_6 & -\zeta g_5 & -\overline{\beta}\tau_1 \\ \zeta g_6 \nabla^2 & \nabla^2 + g_1 & \left(\bar{e}_{15}\,\nabla^2 - \zeta^2\right) & -\zeta \\ \zeta g_5 & \left(\bar{e}_{15}\,\nabla^2 - \zeta\right) & \zeta^2 \bar{\varepsilon}_{33} - \bar{\varepsilon}_{11}\nabla^2 & \overline{p}_3\zeta \\ \tau_0{}'\overline{\beta}\nabla^2 & -\tau_0{}'\zeta & \tau_0{}'\zeta\overline{p}_3 & i\overline{K}_1\,\nabla^2 + g_2 \end{vmatrix} \left(\phi,\, W,\, E,\, T\right) = 0. \quad (7.27)$$

Solving the above determinant, we reach the following equation

$$(P\,\nabla^8 + Q\,\nabla^6 + R\,\nabla^4 + S\,\nabla^2 + U)(\phi,\, W,\, E,\, T) = 0 \quad (7.28)$$

where $P = -i\overline{K}_1\bar{c}_{11}g_4$

$$Q = i\overline{K}_1\bar{c}_{11}(\zeta^2(\bar{\varepsilon}_{11} + \bar{e}_{15}) - \bar{\varepsilon}_{11}g_1) - \bar{c}_{11}\bar{\varepsilon}_{11}g_2 - i\overline{K}_1 g_3 g_4 - \zeta^2 g_5 g_6 i\overline{K}_1\bar{e}_{15} - \zeta^2 g_5 i\overline{K}_1(\bar{e}_{15}g_6 - g_5) - \bar{e}_{15}{}^2\tau_0{}'\overline{\beta}^2$$

$$\begin{aligned} R = {} & \bar{c}_{11}g_1(i\overline{K}_1\zeta^2\bar{\varepsilon}_{33} - \bar{\varepsilon}_{11}g_2) - \bar{c}_{11}\bar{e}_{15}[\bar{e}_{15}g_2 + \zeta^2(-i\overline{K}_1 + i\overline{K}_3\zeta^2 - \tau_0{}'\overline{d} + \tau_0{}'(\overline{p}_3\bar{e}_{15} - \overline{\beta}_3\overline{p}_3\zeta^2))] \\ & + \bar{c}_{11}[g_2(\zeta^2\bar{\varepsilon}_{33} + \bar{e}_{15}) + \zeta^2(\tau_0{}'\overline{p}_3\bar{e}_{15} - i\overline{K}_1) + \tau_0{}'(-\overline{p}_3{}^2 - \overline{p}_3\bar{e}_{15} + \bar{\varepsilon}_{11})] \\ & + i\overline{K}_3 g_3[\zeta^2(\bar{\varepsilon}_{33} + \bar{e}_{15}) - \bar{\varepsilon}_{11}g_1] - \bar{\varepsilon}_{11}g_2g_3 - \zeta^2 g_6\bar{e}_{15}(g_2g_5 - \tau_0{}'\overline{\beta}\overline{p}_3) \\ & - \zeta^2 g_5g_6\bar{e}_{15}(g_2 - i\overline{K}_1\zeta^2) + \zeta^2(i\overline{K}_1(\zeta^2 g_5g_6 + g_1g_5{}^2) - g_2g_5{}^2) \\ & + \tau_0{}'\overline{\beta}\zeta^2(-\bar{\varepsilon}_{11}g_6 + g_5(-\overline{p}_3 + g_1\bar{e}_{15})) + \zeta^2 g_6\tau_0{}'\tau_1\overline{\beta}(\overline{p}_3\bar{e}_{15} - \bar{\varepsilon}_{11}) + \tau_0{}'\tau_1\overline{\beta}(-g_1\bar{\varepsilon}_{11}\overline{\beta}) \\ & + \bar{e}_{15}\zeta^2(\overline{\beta} - g_5)) + \tau_0{}'\tau_1\overline{\beta}^2(\zeta^2\bar{e}_{15} - \bar{\varepsilon}_{11}) \end{aligned}$$

$$\begin{aligned} S = {} & \bar{c}_{11}\zeta^2[g_1(\bar{\varepsilon}_{33}g_2 - \overline{p}_3{}^2\tau_0) + \zeta^2(\tau_0\overline{d} + \tau_0{}'(\overline{p}_3 - \bar{\varepsilon}_{33}) + \zeta^2(\tau_0{}'\overline{p}_3 - i\overline{K}_3))] + g_1g_3(i\overline{K}_1\zeta^2\bar{\varepsilon}_{33} - \bar{\varepsilon}_{11}g_2) \\ & - \bar{e}_{15}g_3[\bar{e}_{15}g_2 + \zeta^2(-i\overline{K}_1 - g_2) + \tau_0{}'\overline{p}_3(\bar{e}_{15} - \zeta^2)] \\ & + \zeta^2 g_3[g_2(\bar{\varepsilon}_{33} + \bar{e}_{15}) - -i\overline{K}_1\zeta^2 + \tau_0{}'\overline{p}_3(\zeta^2\bar{e}_{15} - \overline{p}_3 - \zeta^2\bar{e}_{15}) +{}^2\bar{\varepsilon}_{33})] \\ & + \zeta^4 g_6^2(\bar{\varepsilon}_{33}g_2 - \overline{p}_3{}^2 g_5) + \zeta^2 g_1g_5(g_2g_5 - \tau_0{}'\tau_1\overline{p}_3\overline{\beta}) \\ & - \tau_0{}'\zeta^4(g_5(\overline{\beta} - g_5) - g_6\overline{\beta}(\bar{\varepsilon}_{33} - \overline{p}_3)) - \zeta^2\tau_0{}'\tau_1\overline{\beta}(g_5\overline{p}_3 - \bar{\varepsilon}_{33}\overline{\beta}) - \tau_0{}'\zeta^4\tau_1\overline{\beta}((\overline{\beta} - g_5)) \end{aligned}$$

$$\begin{aligned} U = {} & g_3\zeta^2[g_2(\overline{\varepsilon}_{33}g_2 - \overline{p}_3{}^2\tau_0{}') + \zeta^2(\tau_0\overline{d} + \tau_0{}'(\overline{p}_3 - \bar{\varepsilon}_{33}) + \zeta^2(\tau_0{}'\overline{p}_3 - -i\overline{K}_3)] \\ & + \zeta^4 g_6(g_5(g_2 - \tau_0{}'\overline{\beta}\overline{p}_3) - \zeta^2\tau_0{}'\tau_1\overline{\beta}g_1 - (\overline{p}_3 g_5 - \bar{\varepsilon}_{33}\overline{\beta}) \end{aligned}$$

and

$$d_i = \zeta g_6(\bar{e}_{15}\,\nabla^2 - \zeta^2) - \zeta g_5(\,\nabla^2 + g_1)/H$$

$$e_i = -\zeta^2 g_5 - \overline{\beta}\tau_1(\bar{e}_{15}\,\nabla^2 - \zeta^2)/H$$

$$f_i = -\overline{\beta}\tau_1\zeta g_6 + \zeta(\bar{c}_{11}\,\nabla^2 + g_3)/H. \quad (7.29)$$

7.4.2 Specific loss

This subsection has been reproduced with permission from Selvamani and Ponnusamy (2015). Copyright 2015, Emerald Group Publishing Limited.

Specific loss is used to determine the energy of internal frictional of a material. It is the ratio of the amount of energy (ΔE) dissipated in a specimen through a stress cycle to the elastic energy (E) stored in that specimen at maximum strain. According to Kolsky (1963), specific loss ($\Delta E/E$) is equal to 4π times of the absolute value of the imaginary part of α_i α_i to the real part α_i, so that we obtain

$$|\Delta E/E| = 4\pi\ |\mathrm{Im}(\zeta)/\mathrm{Re}(\zeta)| = |v_q/\omega|. \tag{7.30}$$

7.4.3 Relative frequency shift

This subsection has been reproduced with permission from Selvamani and Ponnusamy (2015). Copyright 2015, Emerald Group Publishing Limited.

Relative frequency shift plays an important role in construction of rotating gyroscopes, acoustic sensors and actuators. Due to rotation, the frequency shift is defined as $\Delta\omega = \omega(\Omega) - \omega(0)$. Ω denotes angular rotation; the relative frequency shift is expressed as

$$\mathrm{R.\ F.\ S} = \left|\frac{\Delta\omega}{\omega}\right| = \left|\frac{\omega(\Omega) - \omega(0)}{\omega(0)}\right| \tag{7.31}$$

where $\omega(0)$ is the frequency of the waves in the absence of rotation.

7.4.4 Thermomechanical coupling

This subsection has been reproduced with permission from Selvamani and Ponnusamy (2015). Copyright 2015, Emerald Group Publishing Limited.

Thermomechanical coupling can be expressed as

$$\kappa^2 = \left|\frac{V_t - V_i}{V_t}\right| \tag{7.32}$$

where V_t and V_i are the phase velocities of the wave under thermally insulated and isothermal boundary, respectively.

7.4.5 Electromechanical coupling

The electromechanical coupling (ε^2) for a cylindrical bar is important for alteration of structural responses through applied electric fields in the design of sensors and surface acoustic damping wave filters. Electromechanical coupling is defined by Teston *et al* (2002) as

$$\varepsilon^2 = \left|\frac{V_e - V_f}{V_e}\right| \tag{7.33}$$

where V_e and V_f are the phase velocities of the wave under electrically shorted and charge-free boundary conditions at the surface of the bar.

7.5 Numerical results and discussion

This section has been reproduced with permission from Selvamani and Ponnusamy (2015). Copyright 2015, Emerald Group Publishing Limited.

PZT-5A material is chosen for the numerical calculation. The material properties of PZT-5A are considered from Sharma *et al* (2004)

$$c_{11} = 13.9 \times 10^{10}\,\mathrm{N\,m^{-2}},\ c_{12} = 7.78 \times 10^{10}\,\mathrm{N\,m^{-2}},$$
$$c_{13} = 7.43 \times 10^{10}\,\mathrm{N\,m^{-2}},\ c_{33} = 11.5 \times 10^{10}\,\mathrm{N\,m^{-2}},$$

$$c_{44} = 2.56 \times 10^{10}\,\mathrm{N\,m^{-2}},\ c_{66} = 3.06 \times 10^{10}\,\mathrm{N\,m^{-2}},\ \beta_1 = 1.52 \times 10^{6}\,\mathrm{N\,K^{-1}m^{-2}},$$

$$\beta_3 = 1.53 \times 10^{6}\,\mathrm{N\,K^{-1}\,m^{-2}},\ T_0 = 298\,\mathrm{K},\ c_v = 420\,\mathrm{J\,kg^{-1}\,K^{-1}},$$
$$p_3 = -452 \times 10^{-6}\,\mathrm{C\,K^{-1}\,m^{-2}},\ K_1 = K_3 = 1.5\,\mathrm{W\,m^{-1}\,K^{-1}},$$
$$e_{13} = -6.98\,\mathrm{C\,m^{-2}},\ e_{33} = 13.8\,\mathrm{C\,m^{-2}},\ e_{15} = 13.4\,\mathrm{C\,m^{-2}},$$
$$\varepsilon_{11} = 60.0 \times 10^{-10}\,\mathrm{C^2\,N^{-1}m^{-2}},\ \varepsilon_{33} = 54.7 \times 10^{-10}\mathrm{C^2\,N^{-1}\,m^{-2}},\ \rho = 7750\,\mathrm{kg\,m^{-2}}.$$

The relaxation times of thermal values are considered as $t_0 = 0.75 \times 10^{-13}$ and $t_0 = 0.5 \times 10^{-13}$ s from Chandrasekhariah (1986). The longitudinal and flexural modes are considered in this problem.

7.5.1 Dispersion curves

The dispersion curves are drawn for electromechanical coupling, thermomechanical coupling, frequency shift, specific loss and frequency. The longitudinal, flexural symmetric and flexural antisymmetric modes notations are used in the figures, namely L (0, 1), F_S (1, 3) and F_A (1, 3), respectively.

The influence of wave number $|\varsigma|$ versus thermomechanical coupling for longitudinal and flexural (symmetric and antisymmetric) modes of a piezothermoelastic rotating bar of circular cross-section is shown in figures 7.1–7.3 with rotational speed $\Omega = 0.2$ and $\Omega = 0.4$. From these figures, it is noticed that increasing the wave number increases the thermomechanical coupling in all the three modes of vibration. In flexural antisymmetric modes, the thermomechanical coupling becomes dispersive due to thermal relaxation times and rotational speed.

In figures 7.4–7.6, the graphs are drawn between the wave number $|\varsigma|$ and electromechanical coupling of a piezothermoelastic rotating bar of circular cross-section for the longitudinal and flexural (symmetric and antisymmetric) modes with L–S and G–L theories. From this, it is cleared that, both the L–S and G–L modes merge for $|\varsigma| > 0.2$ and begin increasing monotonically. Also, in the wave number range $0.6 \leqslant |\varsigma| \leqslant 0.8$, the coupling effect gets higher value in symmetric modes. The effect of thermal relaxation times and rotation gives the energy transportation between the modes.

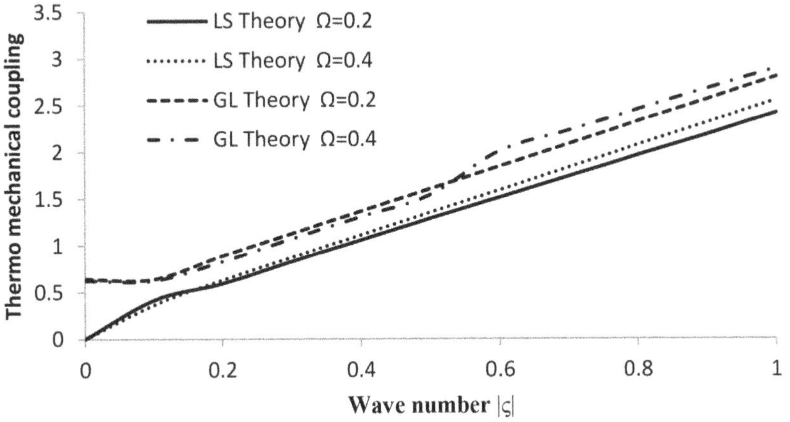

Figure 7.1. The effect of thermomechanical coupling along with wave number $|\varsigma|$ of L (0, 1) for a piezothermoelastic rotating cylindrical bar.

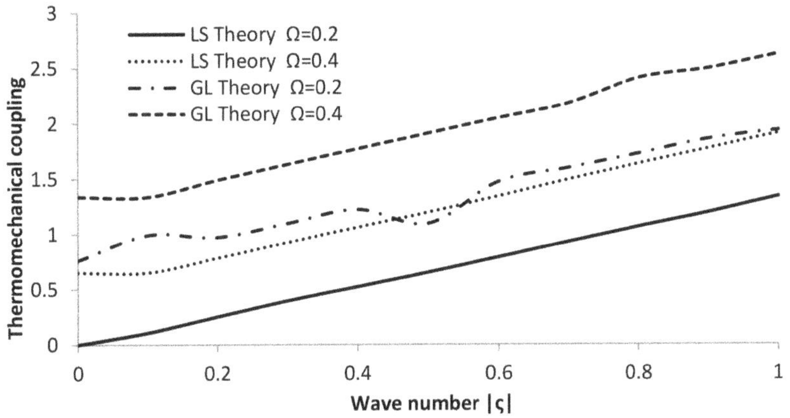

Figure 7.2. The effect of thermomechanical coupling along with wave number $|\varsigma|$ of F_S (1,3) for a piezothermoelastic rotating cylindrical bar.

The effect of wave number $|\varsigma|$ along with frequency shift for the longitudinal and flexural (symmetric and antisymmetric) modes are shown in figures 7.7–7.9 with the non-conventional thermal modes L–S, G–L and rotational speed. At lower wave number, the relative frequency shift is quite high and becomes steady by increasing the wave number. The relative frequency shift profiles are more dispersive in trend for symmetrical modes than in longitudinal modes and experience oscillation in the wave number range $0.2 \leqslant |\varsigma| \leqslant 0.6$ for flexural antisymmetric modes of vibration. The cross-over points between the vibration modes represent the transfer of energy between the thermal modes.

The comparison between the wave number $|\varsigma|$ and the specific loss are observed in figures 7.10–7.12 for the longitudinal and flexural (symmetric and antisymmetric)

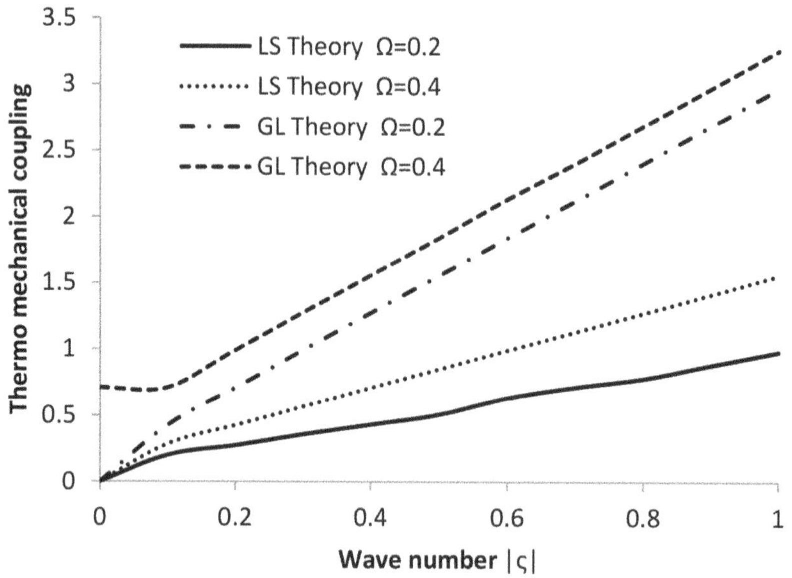

Figure 7.3. The effect of thermomechanical coupling along with wave number $|\varsigma|$ of $F_A(1,3)$ for a piezothermoelastic rotating cylindrical bar.

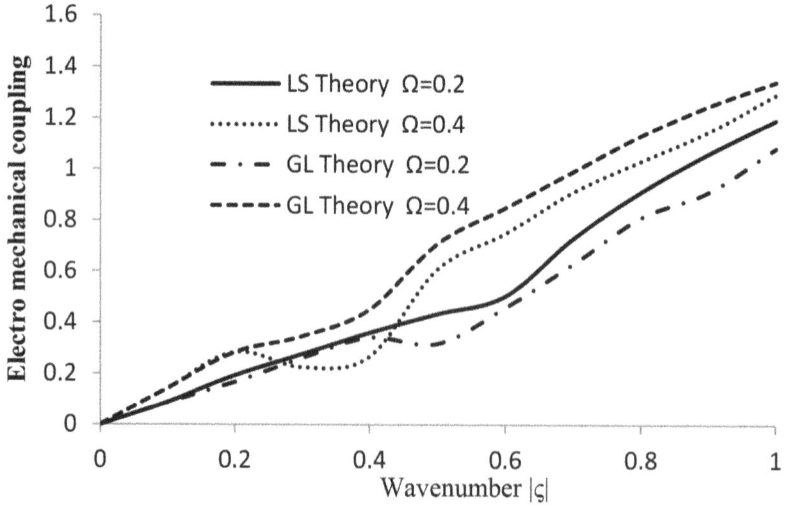

Figure 7.4. The effect of electromechanical coupling along with wave number $|\varsigma|$ of L(0,1) for a piezothermoelastic rotating cylindrical bar.

modes with L–S, G–L theories and rotational speed. In the large wavelength, the specific loss gets the maximum magnitude value and becomes steady and linear in character in the smaller wavelength. It is noticed from the dispersion curves that the specific loss at longer wavelength is more sensitive to finite thermal signal and rotation than at shorter wavelength (higher wave number) due to the anisotropy of the material and the effect of thermal relaxation times. From figures 7.11 and 7.12, it

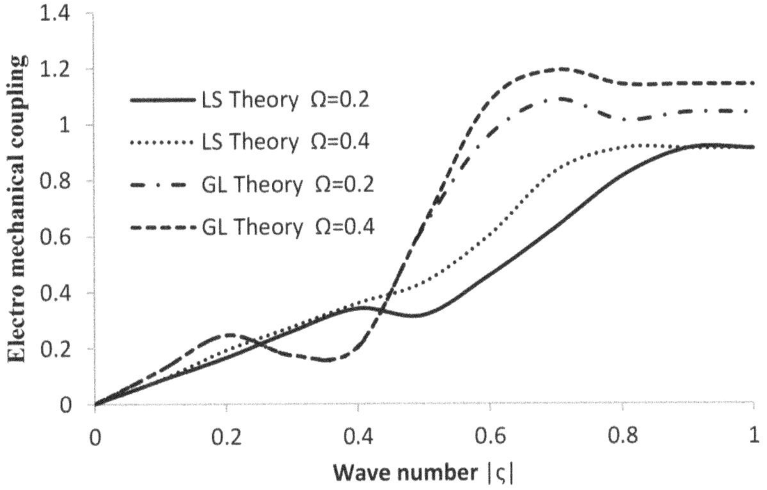

Figure 7.5. The effect of electromechanical coupling along with wave number $|\varsigma|$ of $F_S(1,3)$ for a piezothermoelastic rotating cylindrical bar.

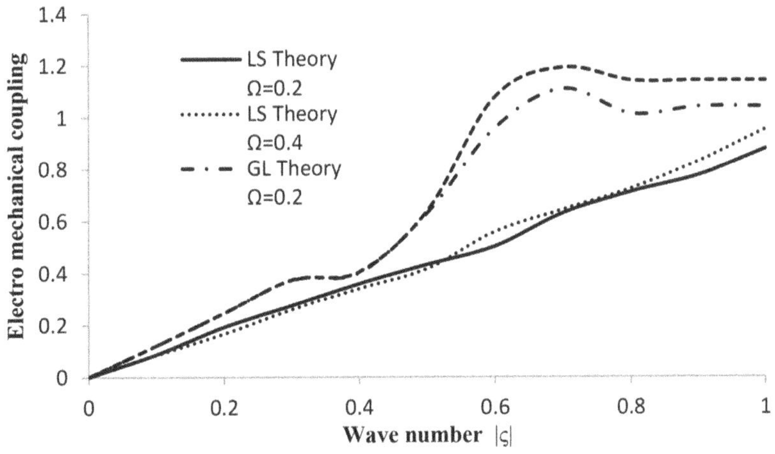

Figure 7.6. The influence of electromechanical coupling on wave number $|\varsigma|$ of $F_A(0,1)$ for a piezothermoelastic rotating cylindrical bar.

is clear that the specific loss is less in L–S theory compared with G–L theory due to relaxation times. This cross-over point represents the transfer of energy between modes of vibration due to the combined effect of thermal relaxation times and rotational speed.

In figures 7.13 and 7.14, the effect of the frequency with respect to the wave number of the thermoelastic and piezothermoelastic cylindrical bar are shown for the longitudinal and flexural modes with L–S and G–L theories for the angular velocity $\Omega = 0$. From this, it is observed that increasing the wave number increases

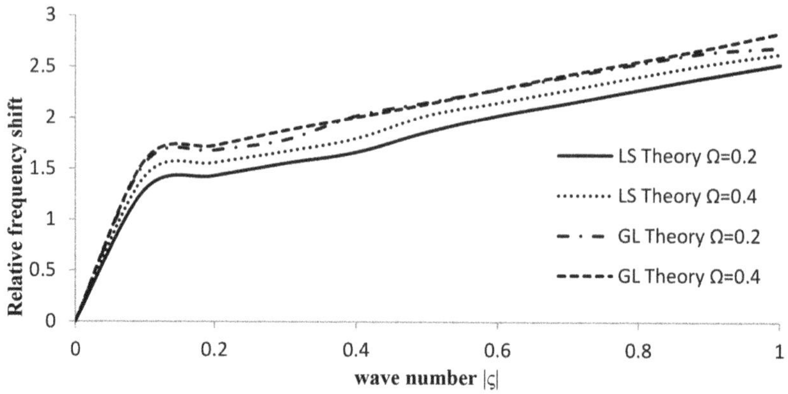

Figure 7.7. The influence of relative frequency shift on wave number |ς|of L(0,1) for a piezothermoelastic rotating cylindrical bar.

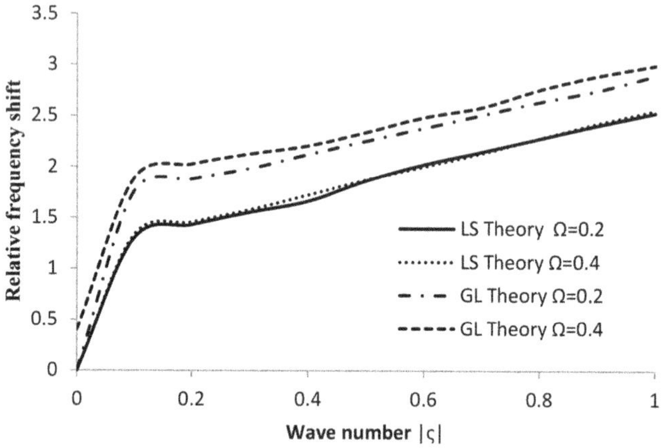

Figure 7.8. The influence of relative frequency shift on wave number |ς| of $F_S(1,3)$ for a piezothermoelastic rotating cylindrical bar.

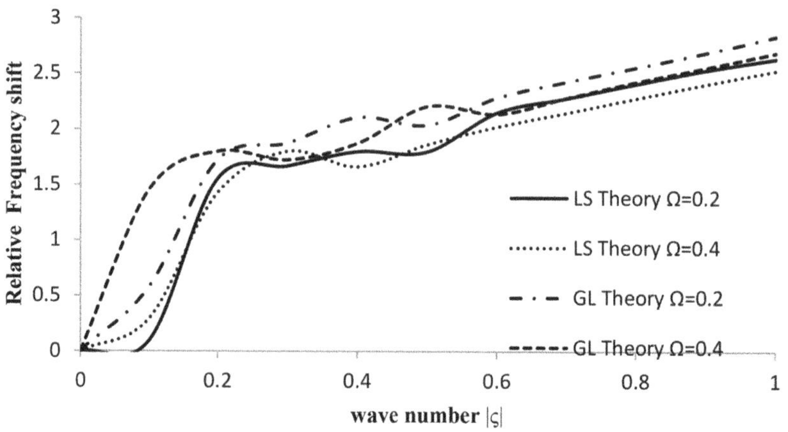

Figure 7.9. The influence of relative frequency shift on wave number |ς| of $F_A(1,3)$ for a piezothermoelastic rotating cylindrical bar.

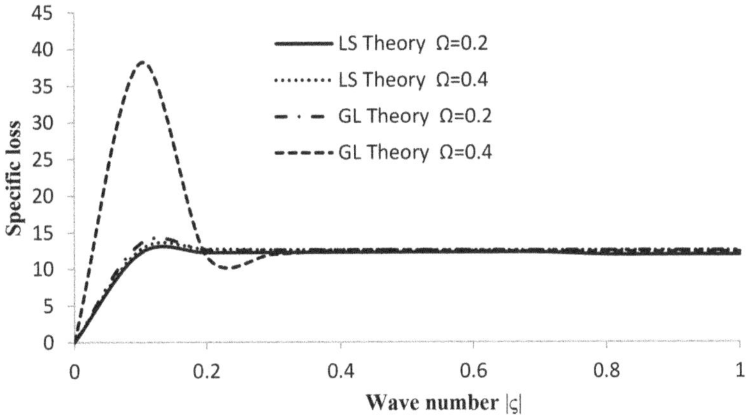

Figure 7.10. The influence of specific loss along with wave number |ς| of L(0,1) for a piezothermoelastic rotating cylindrical bar.

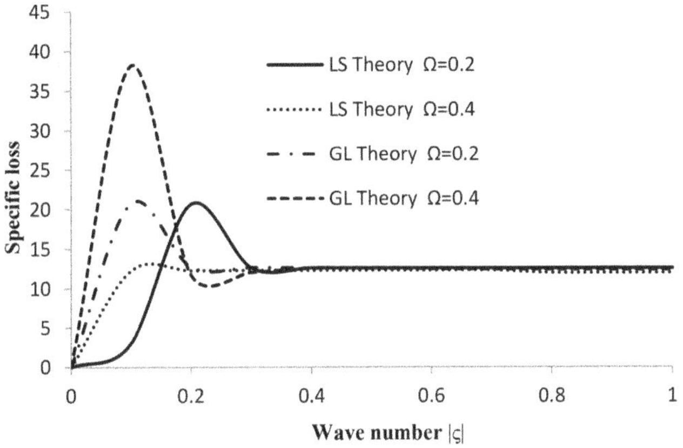

Figure 7.11. The influence of specific loss along with wave number |ς| of $F_S(1,3)$ for a piezothermoelastic rotating cylindrical bar.

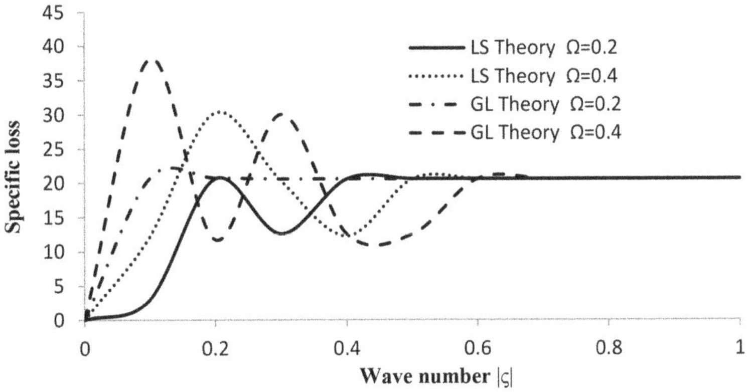

Figure 7.12. The influence of specific loss along with wave number |ς| of $F_A(1,3)$ for a piezothermoelastic rotating cylindrical bar.

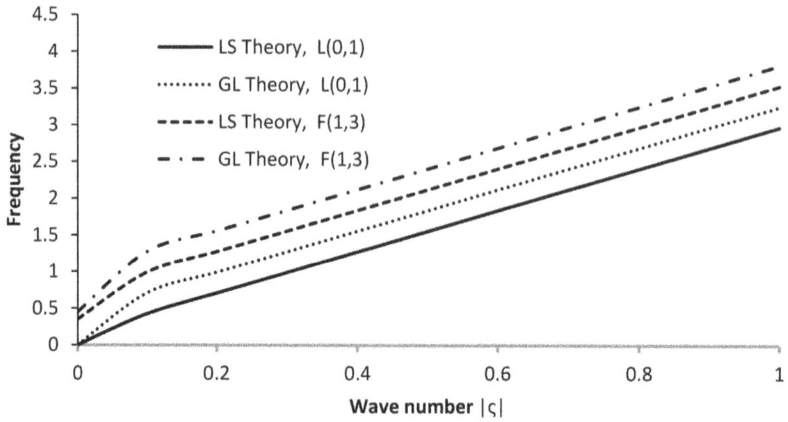

Figure 7.13. The effect of wave number $|\varsigma|$ versus frequency for a thermoelastic cylindrical bar with $\Omega = 0$.

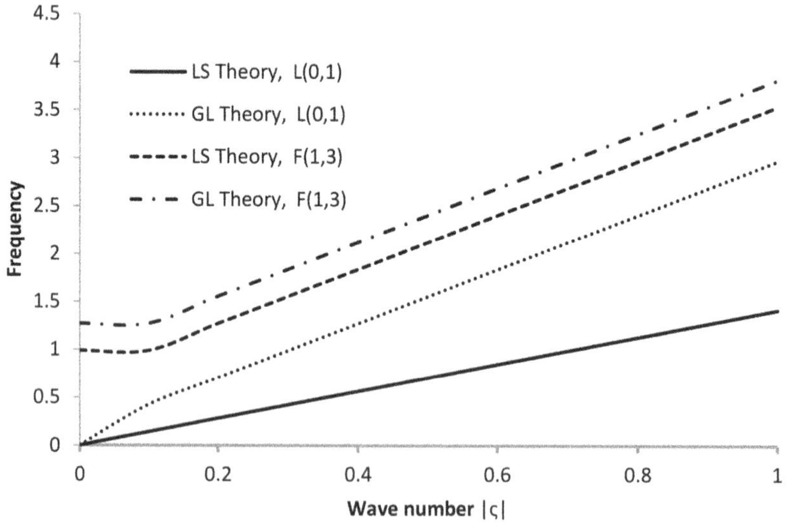

Figure 7.14. The effect of wave number $|\varsigma|$ versus frequency for a piezothermoelastic cylindrical bar with $\Omega = 0$.

the dimensionless frequency for both a thermoelastic and a piezothermoelastic bar. But although the behaviour of the variation of the frequency of a piezothermoelastic bar with increasing wave number in figure 7.13 is linear, it becomes dispersive in flexural mode of the vibration beyond $|\varsigma| = 0.2$ due to the significant effect of piezoelectricity in the vibrational modes.

References

Bayat M, Rahimi M, Saleem M, Mohazzab A H M, Wudtke I and Talebi H 2014 One-dimensional analysis for magneto-thermo-mechanical response in a functionally graded annular variable-thickness rotating disk *Appl. Math. Modell.* **38** 4625–39

Chandrasekhariah D S 1984 A temperature rate dependent theory of piezoelectricity *J. Therm. Stresses* **7** 293–306

Chandrasekharaiah M N 1986 Geometrical analysis of field-ion images *J. Phys.Coll.* **47** 432–7

Chandrasekhariah D S 1988 A generalized linear thermoelasticity theory of piezoelectric media *Acta Mech.* **71** 39–49

Dhaliwal R S and Sherief H H 1980 Generalized thermo elasticity for anisotropic media *Q. J. Appl. Math. Mech.* **38** 1–8

Green A E and Laws N 1972 On the entropy production inequality *Arch. Ration. Mech. Anal.* **45** 47–53

Kolsky H 1963 *Stress Waves in Solids* (New York, London: Clarendon Press, Oxford Dover Press) **1935**

Loy C T and Lam K Y 1995 Vibration of rotating thin cylindrical panels *J. Appl. Acoust.* **46** 327–43

Mindlin R D 1961 On the equations of motion of piezoelectric crystals *Problems of Continuum Mechanics* **vol 70** (Philadelphia, PA: SIAM) pp 282–90 (Philadelphia, N. I. Muskelishvili's Birthday)

Morse R W 1954 Compressional waves along an anisotropic circular cylinder having hexagonal symmetry *J. Acoust. Soc. Am.* **26** 1018–21

Nowacki W 1978 Some general theorems of thermo-piezoelectricity *J. Therm. Stresses* **1** 171–82

Nowacki W 1979 Foundations of linear piezoelectricity *Electromagnetic Interactions in Elastic Solids* ed H Parkus (Wien: Springer) ch 1

Paul H S 1966 Vibrations of circular cylindrical shells of piezo-electric silver iodide crystals *J. Acoust. Soc. Am.* **40** 1077–80

Peyman Y M, Tahani M and Naserian-Nik A M 2013 Analytical solution of piezolaminated rectangular plates with arbitrary clamped/simply-supported boundary conditions under thermo-electro-mechanical loadings *Appl. Math. Model.* **37** 3228–41

Ponnusamy P 2011 Wave propagation in thermo-elastic plate of arbitrary cross-sections *Multidiscipl. Model. Mater. Struct.* **7** 1573–605

Ponnusamy P 2011 Dispersion analysis of generalized thermo elastic plate of polygonal cross-section *Appl. Math. Model.* **36** 3343–58

Ponnusamy P 2013 Wave propagation in a piezoelectric solid bar of circular cross section immersed in fluid *Int. J. Press. Vessel Pip.* **105** 12–8

Ponnusamy P and Selvamani R 2012 Dispersion analysis of generalized magneto-thermoelastic waves in a transversely isotropic cylindrical panel *J. Therm. Stresses* **35** 1119–42

Roychoudhuri R S and Mukhopadhyay S 2000 Effect of rotation and relaxation times on plane waves in generalized thermo visco elasticity *Int. J. Math. Math. Sci.* **23** 497–505

Sabzikar Boroujerdy M and Eslami M R 2014 Axisymmetric snap-through behavior of piezo-FGM shallow clamped spherical shells under thermo-electro-mechanical loading *Int. J. Press. Vessels Pip.* **120** 19–26

Selvamani R and Ponnusamy P 2013 Extensional waves in a transversely isotropic solid bar immersed in an inviscid fluid calculated using chebyshev polynomials *Mater. Phys. Mech.* **16** 82–91

Selvamani R and Ponnusamy P 2013 Generalized thermo elastic waves in a rotating plate subjected to ramp type increase in boundary temperature *J. Mech. Eng.* **1** 57–65

Selvamani R and Ponnusamy P 2013 Flexural vibration in a heat conducting cylindrical panel resting on Winkler elastic foundation *Mater. Phys. Mech.* **17** 121–34

Selvamani R and Ponnusamy P 2013 Wave propagation in a generalized thermo elastic circular plate immersed in fluid *Struct. Eng. Mech.* **46** 827–42

Selvamani R and Ponnusamy P 2013 Elasto dynamic wave propagation in a transversely isotropic piezoelectric circular plate immersed in fluid *Mater. Phys. Mech.* **17** 164–77

Selvamani R and Ponnusamy P 2013 Generalized thermoelastic waves in a rotating ring shaped circular plate immersed in an inviscid fluid *Mater. Phys. Mech.* **18** 77–92

Selvamani R and Ponnusamy P 2014 Modeling and analysis of waves in a heat conducting thermo-elastic plate of elliptical shape *Latin Am. J. Solids Struct.* **11** 2589–606

Selvamani R and Ponnusamy P 2014 Dynamic response of a solid bar of cardioidal cross-sections immersed in an inviscid fluid *Appl. Math. Inform. Sci.* **8** 2909

Selvamani R and Ponnusamy P 2015 Wave propagation in a generalized piezothermoelastic rotating bar of circular cross section *Multidisc. Model. Mater. Struct.* **11** 216–37

Sergiu C, Dragomir , Sinnott M D, Semercigil S E and Turan Ö F 2014 A study of energy dissipation and critical speed of granular flow in a rotating cylinder *J. Sound Vib.* **333** 6815–27

Sharma J N, Pal M and Chand D 2004 Three dimensional vibration analysis of a piezo-thermo elastic cylindrical panel *Int. J. Eng. Sci.* **42** 1655–73

Sharma J N, Pal M and Chand D 2004 Three dimensional vibrational analysis of a piezothermoelastic cylindrical panel *Int. J. Eng. Sci.* **42** 1655–73

Sharma J N and Walia V 2006 Straight and circular crested waves in generalized piezothermoelastic materials *J. Therm. Stresses* **29** 529–51

Tang Y X and Xu K 1995 Dynamic analysis of a piezothermoelastic laminated plate *J. Therm. Stresses* **18** 87–104

Tauchert T R 1992 Piezothermoelastic behavior of a laminated plate *J. Therm. Stresses* **15** 25–37

Teston F, Chenu C, Felix N and Lethiecq M 2002 Acoustoelectric effect in piezocomposites sensors *J. Mater. Sci. Eng.* **21** 177–81

Yang J S and Batra R C 1995 Free vibrations of a linear thermo-piezoelectric body *J. Therm. Stresses* **18** 247–62

IOP Publishing

Farzad Ebrahimi and Rajendran Selvamani

Chapter 8

Dispersion analysis of magneto-electroelastic plate of arbitrary cross-sections immersed in fluid

The three-dimensional linear elasticity theories are used to design the analytical formulation of the problem. Since the inner and outer boundaries of the arbitrary cross-sectional plate is irregular, Fourier expansion collocation method (FECM) is considered to form the frequency equations for arbitrary cross-sectional boundary conditions. The secant method is used to find the roots of the frequency equation for complex solutions.

The dispersion curves are plotted for radial stress, hoop strain, non-dimensional frequency, magnetic potential, and electric potential and their characteristics are discussed. The FECM and finite element method (FEM) are used to study the wave numbers of longitudinal modes of arbitrary (elliptic, cardioid) cross-sectional plates which are presented in tables. This result can be applied for optimum design of composite plates with arbitrary cross-sections.

8.1 Introduction

Arbitrary cross-sectional plates are the widely used structural components in constructing high strength and reliable structures. They are acted upon by non-uniform loads, in such fields as engineering and mining industries as well as in other branches where modern techniques are used.

Nagaya (1981a, 1981b, 1983a, 1983c, 1983d) reported on the wave propagation in polygonal and arbitrary cross-sectional plates. He formulated the Fourier expansion collocation method for this purpose and the same method is used in this problem. Ech-Cherif El-Kettani *et al* (2004) developed a method to study the down slope or upslope of waves propagating on trapezoidal sections whose thickness varies linearly

doi:10.1088/978-0-7503-5638-1ch8

with small angles and continuously adapts to the varying thickness. Guo *et al* (2016) investigated the analytical three-dimensional solutions of anisotropic multilayered composite plates with modified couple-stress effect. Finite element analysis integrated with numerical computation is an effective tool in most studies on ultrasonic wave propagation (Moser *et al* 1999, Jeong and Park 2005). The electro-magneto-thermoelasticity of semi-infinite transversely isotropic materials are analysed by Hon *et al* (2008). Guo (2011) developed a new analytical model for thermopiezoelectric materials and Kuang (2009) derived the governing equations for thermopiezoelectricity materials. The wave propagation in a generalized thermoelastic cylinder was studied by Ponnusamy (2007) for arbitrary cross-section and later, by using three-dimensional linear theory, the wave propagation in a generalized piezothermoelastic rotating bar of circular cross-section was discussed by Selvamani and Ponnusamy (2015).

The free vibration of clamped-clamped magneto-electroelastic cylindrical shells, free vibration behaviour of multiphase and layered magneto-electro-elastic beam, free vibrations of simply supported layered and multiphase magneto-electroelastic cylindrical shells were constructed by Annigeri *et al* (2006, 2007a, 2007b). Pan and Heyliger (2003) investigated the three-dimensional behaviour of magneto-electroelastic laminates under simple supported condtions. The dynamic fracture behaviour of an internal interfacial crack between two dissimilar magneto-electroelastic plates were derived by Feng and Pan (2008).

The effect of wall compliance on lowest order mode propagation in fluid-filled or submerged impedance tubes were contributed by Easwaran and Munjal (1995). Eigen equations are derived using a closed form of the coupled wave equations and applying the boundary conditions at the fluid–solid interface. Also, they investigated axial attenuation characteristics of plane waves along water-filled tubes submerged in water or air. Dispersion analysis of generalized magneto-thermoelastic waves in a transversely isotropic cylindrical panel was devised by Ponnusamy and Selvamani (2012). Selvamani (2017) analysed the modelling of elastic waves in a fluid-loaded and immersed piezoelectric hollow fibre.

The vibration analysis of a magneto-electroelastic plate of arbitrary cross-sections immersed in fluid is investigated in this problem using FECM on the irregular boundaries. The displacement potential functions are used to uncouple the equations of motion. The boundary conditions are used to obtain the frequency equations. The dispersion curves are drawn for radial stress, hoop strain, non-dimensional frequency, magnetic potential, and electric potential, and their characteristics are discussed.

8.2 Formulation of the problem

The displacement vectors and stress components are defined by the cylindrical coordinates r, θ and z.

By the absence of body force, the governing differential equations of motion can be expressed as

$$\sigma_{rr,r} + r^{-1}\sigma_{r\theta,\theta} + \sigma_{rz,z} + r^{-1}(\sigma_{rr} - \sigma_{\theta\theta}) = \rho u_{tt}$$

$$\sigma_{r\theta,r} + r^{-1}\sigma_{\theta\theta,\theta} + \sigma_{\theta z,z} + 2r^{-1}\sigma_{r\theta} = \rho v_{tt}$$

$$\sigma_{rz,r} + r^{-1}\sigma_{\theta z,\theta} + r^{-1}\sigma_{zz,z} + r^{-1}\sigma_{rz} = \rho w_{tt}. \tag{8.1}$$

The Gaussian electric conduction equation can be defined as

$$D_{r,r} + r^{-1}D_r + r^{-1}D_{\theta,\theta} + D_{z,z} = 0. \tag{8.2}$$

The Maxwell magnetic conduction equation is defined as

$$B_{r,r} + r^{-1}B_r + r^{-1}B_{\theta,\theta} + B_{z,z} = 0. \tag{8.3}$$

The stress–strain relation of isotropic material is given as

$$\sigma_{rr} = c_{11}e_{rr} + c_{12}e_{\theta\theta} + c_{13}e_{zz} + c_{31}E_z - q_{31}H_z$$

$$\sigma_{\theta\theta} = c_{12}e_{rr} + c_{11}e_{\theta\theta} + c_{13}e_{zz} - e_{31}E_z - q_{31}H_z$$

$$\sigma_{zz} = c_{13}e_{rr} + c_{13}e_{\theta\theta} + c_{33}e_{zz} - e_{33}E_z - q_{33}H_z$$

$$\sigma_{r\theta} = 2c_{66}e_{r\theta}$$

$$\sigma_{\theta z} = 2c_{44}e_{\theta z} - e_{15}E_\theta - q_{15}H_\theta$$

$$\sigma_{rz} = 2c_{44}e_{rz} - e_{15}E_r - q_{15}H_r \tag{8.4}$$

$$D_r = 2e_{15}e_{rz} + \varepsilon_{11}E_r + m_{11}H_r$$

$$D_\theta = 2e_{15}e_{\theta z} + \varepsilon_{11}E_\theta + m_{11}H_\theta$$

$$D_z = e_{31}(e_{rr} + e_{\theta\theta}) + \varepsilon_{33}e_{zz} + m_{33}H_z \tag{8.5}$$

and

$$B_r = 2q_{15}e_{rz} + m_{11}E_r + \mu_{11}H_r$$

$$B_\theta = 2q_{15}e_{\theta z} + m_{11}E_\theta + \mu_{11}H_\theta$$

$$B_z = q_{31}(e_{rr} + e_{\theta\theta}) + q_{33}e_{zz} + m_{33}E_z + \mu_{33}H_z \tag{8.6}$$

where σ_{rr}, $\sigma_{\theta\theta}$, $\sigma_{r\theta}$, σ_{rz}, $\sigma_{\theta z}$ are the stress components, c_{11}, c_{12}, c_{13}, c_{33}, c_{44} and c_{66} are elastic constants and $c_{66} = (c_{11} - c_{12})/2$, ε_{11}, ε_{33} stands for dielectric constants, μ_{11},μ_{33} denotes the coefficient of magnetic permeability, e_{31}, e_{33}, e_{15} are the coefficients of piezoelectric material, q_{31}, q_{33}, q_{15} denotes the coefficients of piezomagnetic material, m_{11}, m_{33} stands for coefficient of magneto-electric material, ρ is the mass density of the material, D_r, D_θ and D_z and are the electric displacements, B_r, B_θ and $B_z$are the magnetic displacements components.

The strain e_{ij} is related to the displacements corresponding to the cylindrical coordinates (r, θ, z) given by

$$e_{rr} = u_r,\ e_{\theta\theta} = r^{-1}(v_\theta + u),\ e_{zz} = w_z$$

$$e_{\theta z} = \frac{1}{2}(v_z + r^{-1}w_\theta),\ e_{r\theta} = \frac{1}{2}(r^{-1}u_\theta + v_r - r^{-1}v),\ e_{rz} = \frac{1}{2}(u_z + w_r) \tag{8.7}$$

where u, v and w denote the mechanical displacements along the radial, circumferential and axial directions, respectively.

The relation between the electric potential E and the electric field vector E_i, $(i = (r, \theta, z)$ can be given as

$$E_r = -\frac{\partial E}{\partial r},\ E_\theta = -\frac{1}{r}\frac{\partial E}{\partial \theta}\ \text{and}\ E_z = -\frac{\partial E}{\partial z}. \tag{8.8}$$

Similarly, the magnetic potential H and the magnetic field vector H_i, $(i = (r, \theta, z)$ can be expressed as

$$H_r = -\frac{\partial H}{\partial r},\ H_\theta = -\frac{1}{r}\frac{\partial H}{\partial \theta}\ \text{and}\ H_z = -\frac{\partial H}{\partial z}. \tag{8.9}$$

Substituting equations (8.4)–(8.9) in equations (8.1)–(8.3), we get the following form

$$\begin{aligned} &c_{11}(u_{rr} + r^{-1}u_r - r^{-2}u) + c_{66}r^{-2}u_{\theta\theta} \\ &+ c_{44}u_{zz} + (c_{66} + c_{12})r^{-1}v_{r\theta} - (c_{11} + c_{66})r^{-2}v_\theta \\ &+ (c_{44} + c_{13})w_{rz} + (e_{13} + e_{15})E_{rz} + (q_{31} + q_{15})H_{rz} = \rho u_{tt} \end{aligned} \tag{8.10a}$$

$$\begin{aligned} &(c_{66} + c_{12})r^{-1}u_{r\theta} + (c_{11} + c_{66})r^{-2}u_\theta \\ &+ c_{66}(v_{rr} + r^{-1}v_r - r^{-2}v) + c_{44}v_{zz} + c_{11}r^{-2}v_{\theta\theta} \\ &+ (c_{44} + c_{13})r^{-1}w_{\theta z} + (e_{13} + e_{15})r^{-1}E_{\theta z} + (q_{31} + q_{15})r^{-1}H_{\theta z} = \rho v_{tt} \end{aligned} \tag{8.10b}$$

$$\begin{aligned} &c_{44} + c_{13}(u_{rz} + r^{-1}u_z + r^{-1}v_{\theta z}) + c_{44}(w_{rr} + r^{-1}w_r + w_{\theta\theta}) \\ &+ c_{33}w_{zz} + e_{33}E_{zz} + q_{33}H_{zz} + e_{15}(E_{rr} + r^{-1}E_r + r^{-2}E_{\theta\theta}) \\ &+ q_{15}(H_{rr} + r^{-1}H_r + r^{-2}H_{\theta\theta}) = \rho w_{tt} \end{aligned} \tag{8.10c}$$

$$\begin{aligned} &e_{15}(w_{rr} + r^{-1}w_r + r^{-2}w_{\theta\theta}) + (e_{31} + e_{15})(u_{rz} + r^{-1}u_z + r^{-1}v_{\theta z}) \\ &+ e_{33}w_{zz} - \varepsilon_{33}E_{zz} - m_{33}H_{zz} - \varepsilon_{11}(E_{rr} + r^{-1}E_r + r^{-2}E_{\theta\theta}) \\ &- m_{11}(H_{rr} + r^{-1}H_r + r^{-2}H_{\theta\theta}) = 0 \end{aligned} \tag{8.10d}$$

$$\begin{aligned} &q_{15}(w_{rr} + r^{-1}w_r + r^{-2}w_{\theta\theta}) + (q_{31} + q_{15})(u_{rz} + r^{-1}u_z + r^{-1}v_{\theta z}) \\ &+ q_{33}w_{zz} - \mu_{33}H_{zz} - m_{11}H_{zz} - m_{11}(E_{rr} + r^{-1}E_r + r^{-2}E_{\theta\theta}) \\ &- \mu_{11}(H_{rr} + r^{-1}H_r + r^{-2}H_{\theta\theta}) = 0. \end{aligned} \tag{8.10e}$$

8.3 Equations of motion for a solid medium

Equation (8.10) is a coupled partial differential equation with three displacements, magnetic, thermal and electric conduction components. The following solutions are used to uncouple equation (8.10),

$$u = \varepsilon_n[(r^{-1}\psi_{n,\theta} - \phi_{n,r}) + (r^{-1}\overline{\psi}_{n,\theta} - \overline{\phi}_{n,r})]$$

$$v = \sum\varepsilon_n[(-r^{-1}\phi_{n,\theta} - \psi_{n,r}) + (-r^{-1}\overline{\phi}_{n,\theta} - \overline{\psi}_{n,r})]$$

$$W = \sum\varepsilon_n[W_{n,z} + \overline{W}_{n,z}]$$

$$E = \sum\varepsilon_n[E_{n,z} + \overline{E}_{n,z}]$$

$$H = \sum\varepsilon_n[H_{n,z} + \overline{H}_{n,z}] \tag{8.11}$$

where $\varepsilon_n = \frac{1}{2}$ for $n = 0$, $\varepsilon_n = 1$ for $n \geqslant 1$, $\phi_n(r, \theta)$, $\psi_n(r, \theta)$, $W_n(r, \theta)$, $E_n(r, \theta)$, $H_n(r, \theta)$ stands for the symmetric mode of displacement potentials and $\overline{\phi}_n(r, \theta)$, $\overline{\psi}_n(r, \theta)$, $\overline{W}_n(r, \theta)$, $\overline{E}_n(r, \theta)$, and $\overline{H}_n(r, \theta)$ are the antisymmetric mode of displacement potentials.

Inserting equation (8.11) in (8.10), we obtained the following transformed equation

$$\left(c_{11}\nabla_1^2 + c_{44}\frac{\partial^2}{\partial z^2} - \rho\frac{\partial^2}{\partial t^2}\right)\phi_n - (c_{13} + c_{44})\frac{\partial W_n}{\partial z} - (e_{31} + e_{15})\frac{\partial E_n}{\partial z} - (q_{31} + q_{15})\frac{\partial H_n}{\partial z} = 0 \tag{8.12a}$$

$$\left(c_{11}\nabla_1^2 + c_{44}\frac{\partial^2}{\partial z^2} - \rho\frac{\partial^2}{\partial t^2}\right)\phi_n - (c_{13} + c_{44})\frac{\partial W_n}{\partial z} - (e_{31} + e_{15})\frac{\partial E_n}{\partial z} - (q_{31} + q_{15})\frac{\partial H_n}{\partial z} = 0 \tag{8.12b}$$

$$\left(c_{44}\nabla_1^2 + c_{33}\frac{\partial^2}{\partial z^2} - \rho\frac{\partial^2}{\partial t^2}\right)W_n - (c_{13} + c_{44})\frac{\partial}{\partial z}\nabla_1^2\phi_n + \left(e_{33}\frac{\partial^2}{\partial z^2} + e_{15}\nabla_1^2\right)E_n + \left(q_{33}\frac{\partial^2}{\partial z^2} + q_{15}\nabla_1^2\right)H_n = 0 \tag{8.12c}$$

$$\left(e_{33}\frac{\partial^2}{\partial z^2} + e_{15}\nabla_1^2\right)W_n - (e_{31} + e_{15})\frac{\partial}{\partial z}\nabla_1^2\phi_n - \left(\varepsilon_{33}\frac{\partial^2}{\partial z^2} + \varepsilon_{11}\nabla_1^2\right)E_n + \left(m_{33}\frac{\partial^2}{\partial z^2} + m_{11}\nabla_1^2\right)H_n = 0 \tag{8.12d}$$

$$\left(q_{33}\frac{\partial^2}{\partial z^2}+q_{15}\nabla_1^2\right)W_n-(q_{31}+q_{15})\,\nabla_1^2\ \phi_n$$
$$-\left(m_{33}\frac{\partial^2}{\partial z^2}+m_{11}\nabla_1^2\right)E_n-\left(\mu_{33}\frac{\partial^2}{\partial z^2}+\mu_{11}\nabla_1^2\right)H_n=0 \tag{8.12e}$$

and

$$\left(c_{66}\ \nabla_1^2+c_{44}\frac{\partial^2}{\partial z^2}-\rho\frac{\partial^2}{\partial t^2}\right)\psi_n=0 \tag{8.13}$$

where

$$\nabla_1^2=\frac{\partial^2}{\partial r^2}+\frac{1}{r}\frac{\partial}{\partial r}+\frac{1}{r^2}\frac{\partial^2}{\partial\theta^2}.$$

Equation (8.13) gives a purely transverse wave, which is not affected by the transmission of energy through magnetic, thermal and electric field. This is said to be simple harmonic wave and polarized in perpendicular to the z-axis. The disturbance was assumed as time harmonic factor $e^{i\omega t}$ and the angular velocity ω. Hence, equations (8.12a)–(8.12e) become

$$\left(c_{11}\ \nabla_1^2+c_{44}\frac{\partial^2}{\partial z^2}+\rho\omega^2\right)\phi_n-(c_{13}+c_{44})\frac{\partial W_n}{\partial z}$$
$$-(e_{31}+e_{15})\frac{\partial E_n}{\partial z}-(q_{31}+q_{15})\frac{\partial H_n}{\partial z}=0 \tag{8.14a}$$

$$\left(c_{11}\ \nabla_1^2+c_{44}\frac{\partial^2}{\partial z^2}+\rho\omega^2\right)\phi_n-(c_{13}+c_{44})\frac{\partial W_n}{\partial z}$$
$$-(e_{31}+e_{15})\frac{\partial E_n}{\partial z}-(q_{31}+q_{15})\frac{\partial H_n}{\partial z}=0 \tag{8.14b}$$

$$\left(c_{44}\ \nabla_1^2+c_{33}\frac{\partial^2}{\partial z^2}+\rho\omega^2\right)W_n-(c_{13}+c_{44})\frac{\partial}{\partial z}\ \nabla_1\ \phi_n$$
$$+\left(e_{33}\frac{\partial^2}{\partial z^2}+e_{15}\nabla_1^2\right)E_n+\left(q_{33}\frac{\partial^2}{\partial z^2}+q_{15}\nabla_1^2\right)H_n=0 \tag{8.14c}$$

$$\left(e_{33}\frac{\partial^2}{\partial z^2}+e_{15}\nabla_1^2\right)W_n-(e_{31}+e_{15})\frac{\partial}{\partial z}\ \nabla_1^2\phi_n$$
$$-\left(\varepsilon_{33}\frac{\partial^2}{\partial z^2}+\varepsilon_{11}\nabla_1^2\right)E_n-\left(m_{33}\frac{\partial^2}{\partial z^2}+m_{11}\nabla_1^2\right)H_n=0. \tag{8.14d}$$

The distributing function for the free vibration of an arbitrary cross-sectional plate is assumed as

$$\phi_n(r,\theta,z,t) = \phi_n^*(r)\cos(m\pi\varsigma)\cos n\theta$$

$$W_n(r,\theta,z,t) = W_n^*(r)\cos(m\pi\varsigma)\cos n\theta$$

$$E_n(r,\theta,z,t) = \left(\frac{c_{44}}{q_{33}}\right)E_n^*(r)\sin(m\pi\varsigma)\cos n\theta$$

$$H_n(r,\theta,z,t) = \left(\frac{c_{44}}{q_{33}}\right)H_n^*(r)\sin(m\pi\varsigma)\cos n\theta \quad (8.15)$$

and

$$\psi_n(r,\theta,z,t) = \psi_n^*(r)\cos(m\pi\varsigma)\sin n\theta \quad (8.16)$$

where

$$\zeta = \frac{z}{L}.$$

The non-dimensional quantities are given as

$$x = \frac{r}{a},\ t_L = \frac{m\pi a}{L},\ \bar{c}_{ij} = \frac{c_{ij}}{c_{44}},\ \bar{e}_{ij} = \frac{e_{ij}}{e_{33}},\ \bar{q}_{ij} = \frac{q_{ij}}{q_{33}},$$
$$\Omega^2 = \frac{\rho\omega^2a^2}{c_{44}},\ \overline{m}_{ij} = \frac{m_{ij}c_{44}}{q_{33}e_{33}},\ P_v = \frac{\rho\omega^2a^2}{c_{44}\zeta}$$

$$\overline{\mu}_{ij} = \frac{\mu_{ij}c_{44}}{q_{33}^2},\ \bar{\varepsilon}_{ij} = \frac{\varepsilon_{ij}c_{44}}{e_{33}^2}$$

and using equations (8.15) and (8.16) in equations (8.14) and (8.13), we get

$$(\bar{c}_{11}\nabla_2^2 - t_L^2 + \Omega^2)\phi_n - (\bar{c}_{13}+1)t_L W_n$$
$$-\left(-t_L^2 + \bar{e}_{15}\right)t_L E_n - (\bar{q}_{31}+\bar{q}_{15})t_L H_n = 0$$

$$\left(\nabla_2^2 - \bar{c}_{33}t_L^2 + \Omega^2\right)W_n + (\bar{c}_{13}+1)t_L\ \nabla_2^2\ \phi_n$$
$$+\left(-t_L^2 + \bar{e}_{15}\nabla_2^2\right)E_n + \left(-t_L^2 + \bar{q}_{15}\nabla_2^2\right)H_n = 0$$

$$\left(\bar{e}_{15}\nabla_2^2 - t_L^2\right)W_n + (\bar{e}_{31}+\bar{e}_{15})t_L\ \nabla_2^2\ \phi_n$$
$$+\left(\bar{\varepsilon}_{33}t_L^2 - \bar{\varepsilon}_{11}\nabla_2^2\right)E_n + \left(\overline{m}_{33}t_L^2 + \overline{m}_{11}\nabla_2^2\right)H_n = 0$$

$$\left(\bar{q}_{15}\nabla_2^2 - t_L^2\right)W_n + (\bar{q}_{31} + \bar{q}_{15})t_L\ \nabla_2^2\ \phi_n \\ + \left(\overline{m}_{33}t_L^2 - \overline{m}_{11}\nabla_2^2\right)E_n + \left(\bar{\mu}_{33}t_L^2 - \bar{\mu}_{11}\nabla_2^2\right)H_n = 0 \tag{8.17}$$

and

$$(\bar{c}_{66}\ \nabla_2^2 - t_L^2 + \Omega^2)\psi_n = 0 \tag{8.18}$$

where

$$\nabla_2^2 = \frac{\partial^2}{\partial x^2} + \frac{1}{x}\frac{\partial}{\partial x} + \frac{n^2}{x^2}.$$

Equation (8.17) has a trivial solution, by taking the coefficient of matrix as equal to zero, we obtain the non-trivial solution

$$\begin{vmatrix} (\bar{c}_{11}\ \nabla_2^2 + g_1) & -g_2 t_L & -g_3 t_L & -g_4 t_L \\ g_2 t_L \nabla_2^2 & \left(\nabla_2^2 + g_5\right) & \left(\bar{e}_{15}\ \nabla_2^2 - t_L^2\right) & \left(\bar{q}_{15}\ \nabla_2^2 - t_L^2\right) \\ g_3 t_L \nabla_2^2 & \left(\bar{e}_{15}\ \nabla_2^2 - t_L^2\right) & (-\bar{\varepsilon}_{11}\ \nabla_2^2 + g_6) & (-\overline{m}_{11}\ \nabla_2^2 + g_7) \\ g_4 t_L \nabla_2^2 & \left(\bar{q}_{15}\ \nabla_2^2 - t_L^2\right) & (-\overline{m}_{11}\ \nabla_2^2 + g_7) & (-\bar{\mu}_{11}\ \nabla_2^2 + g_8) \end{vmatrix} (\phi_n, W_n, E_n, H_n) = 0. \tag{8.19}$$

Here

$$g_1 = \Omega^2 - t_L^2,\ g_2 = 1 + \bar{c}_{13},\ g_3 = \bar{e}_{31} + \bar{e}_{15},\ g_4 = \bar{q}_{31} + \bar{q}_{15}, \\ g_5 = -\bar{c}_{33}t_L^2 + \Omega^2,\ g_6 = \bar{\varepsilon}_{33}t_L^2,\ g_7 = \overline{m}_{33}t_L^2,\ g_8 = \bar{\mu}_{33}t_L^2.$$

By solving the above determinant, the following differential equation is obtained in the form

$$(A\nabla_2^{10} + \mathrm{B}\nabla_2^8 + \mathrm{C}\nabla_2^6 + \mathrm{D}\nabla_2^4 + E\nabla_2^2 + F)\phi_n = 0 \tag{8.20}$$

where

$$A = \mathrm{i}\,\bar{k}_1\bar{c}_{11}\left\{(\bar{\varepsilon}_{11}\left[\bar{\mu}_{11} + \bar{q}_{15}^2\right] + (\bar{\mu}_{11} + \bar{e}_{15}^2 - \overline{m}_{11}[\overline{m}_{11} + 2\bar{e}_{15}\bar{q}_{15}]\right\}$$

$$B = \bar{c}_{11}\{-g_6\bar{\mu}_{11} - g_8\bar{\varepsilon}_{11} + 2g_7\overline{m}_{11} + g_5 i\bar{k}_1[\bar{\varepsilon}_{11}\bar{\mu}_{11} - \overline{m}_{11}] - \bar{e}_{15}[g_8\bar{e}_{15} + 2t_L^2(\bar{\mu}_{11} - \overline{m}_{11})] \\ + \bar{q}_{15}[-g_6\bar{q}_{15} + 2t_L^2(\overline{m}_{11} - \bar{\varepsilon}_{11})] + 2g_7\bar{e}_{15}\bar{q}_{15}\} \\ + g_1 i\bar{k}_1\left\{\bar{\varepsilon}_{11}\left[\bar{\mu}_{11} + \bar{q}_{15}^2\right] + \bar{\mu}_{11}\bar{e}_{15}^2 - \overline{m}_{11}[\overline{m}_{11} + 2\bar{e}_{15}\bar{q}_{15}]\right\} \\ + g_2 t_L^2\{g_2 i\bar{k}_1[\bar{\varepsilon}_{11}\bar{\mu}_{11} - \overline{m}_{11}^2] - \bar{e}_{15}[-g_3\bar{\mu}_{11} + g_4\overline{m}_{11}] + \bar{q}_{15}[g_4\bar{\varepsilon}_{11} - g_3\overline{m}_{11}]\} \\ - g_3 t_L^2\{g_2[-\bar{e}_{15}\bar{\mu}_{15} + \overline{m}_{11}\bar{q}_{15}] + g_3\bar{\mu}_{11} + g_4\overline{m}_{11} + \bar{q}_{15}[g_3\bar{q}_{15} - g_4\bar{e}_{15}]\} \\ + g_4 t_L^2\{g_2[-\bar{e}_{15}\overline{m}_{11} + \bar{\varepsilon}_{11}\bar{q}_{15}] + g_3\overline{m}_{11} - g_4\bar{\varepsilon}_{11} + \bar{e}_{15}[g_3\bar{q}_{15} - g_4\bar{e}_{15}]\}$$

$$
\begin{aligned}
C = {} & \bar{c}_{11}\{g_6 g_8 - g_7^2 + g_5 i\bar{k}_1[-g_6\bar{\mu}_{11} - g_8\bar{\varepsilon}_{11} + 2g_7\overline{m}_{11}] \\
& + 2t_L^2[-\bar{e}_{15}(-g_8 + g_7) + \bar{q}_{15}(g_8 - g_7)] + t_L^4[\bar{\mu}_{11} - \overline{2m}_{11} + \bar{\varepsilon}_{11}]\} \\
& + g_1\{-g_6\bar{\mu}_{11} - g_8\bar{\varepsilon}_{11} - 2g_7\overline{m}_{11} + g_5[\bar{\varepsilon}_{11}\bar{\mu}_{11} - \overline{m}_{11}^2] - \bar{e}_{15}[g_8\bar{e}_{15} + 2t_L^2(\bar{\mu}_{11} - \overline{m}_{11})] \\
& + \bar{q}_{15}[-g_6\bar{q}_{15} + 2t_L^2 ik_3(\overline{m}_{11} - \bar{\varepsilon}_{11})] + 2g_7\bar{e}_{15}\bar{q}_{15}\} \\
& + g_2 t_L^2 ik_1\{g_2[-g_6\bar{\mu}_{11} - g_8\bar{\varepsilon}_{11} + 2g_7\overline{m}_{11}] - \bar{e}_{15}[g_3 g_8 - g_4 g_7] + \bar{q}_{15}[g_3 g_7 - g_4 g_6] \\
& + t_L^2 ik_3[-g_3\bar{\mu}_{11} - g_4\bar{\varepsilon}_{11} + g_4\overline{m}_{11} + g_3\overline{m}_{11}]\} \\
& - g_3 t_L^2 ik_1\{g_2[g_8\bar{e}_{15} - g_7\bar{q}_{15} + t_L^2(\bar{\mu}_{11} - \overline{m}_{11})] - g_3 g_8 - g_4 g_7 - g_5[-g_3\bar{\mu}_{11} + g_4\overline{m}_{11}] \\
& + g_{15} t_L^2[g_4 - g_3] - t_L^2[g_3\bar{q}_{15} - g_4\bar{e}_{15}]\} \\
& + g_4 t_L^2\overline{\beta}\{g_2[g_7\bar{e}_{15} - g_6\bar{q}_{15} + t_L^2[\overline{m}_{11} - \bar{\varepsilon}_{11}]] - g_3 g_7 - g_4 g_6 - g_5[g_4\bar{\varepsilon}_{11} - g_3\overline{m}_{11}] \\
& + \bar{e}_{15} t_L^2[g_4 - g_3] - t_L^2[g_3\bar{q}_{15} - g_4\bar{e}_{15}]\}
\end{aligned}
$$

$$
\begin{aligned}
D = {} & \bar{c}_{11}\left\{g_5\left[g_6 g_8 - g_7^2\right] + t_L^4[-g_8 + 2g_7 - g_6]\right\} \\
& + g_4 t_L^2\left\{g_2 t_L^2[-g_7 + g_6] - g_5 ik_1[g_3 g_7 - g_4 g_6] - t_L^4[-g_3 + g_4]\right\} \\
& + g_1\left\{g_6 g_8 - g_7^2 + g_5 ik_3[-g_6\bar{\mu}_{11} - g_8\bar{\varepsilon}_{11} + 2g_7\overline{m}_{11}]\right. \\
& \left. + 2t_L^2[-\bar{e}_{15}(-g_8 + g_7) + \bar{q}_{15}(-g_7 + g_8)] + t_L^4 ik_3[\bar{\mu}_{11} - \overline{2m}_{11} + \bar{\varepsilon}_{11}]\right\} \\
& + g_2 t_L^2\overline{\beta}\left\{g_2\left[g_6 g_8 - g_7^2\right] + t_L^2 ik_1[g_3 g_8 - g_4 g_7 - g_3 g_7 + g_4 g_6]\right\} \\
& - g_3 t_L^2\left\{g_2 t_L^2[-g_8 + g_7] - g_5[g_3 g_8 - g_4 g_7] - t_L^4[-g_3 + g_4]\right\}
\end{aligned}
$$

$$
E = g_1 ik_1\left\{g_5\left[g_6 g_8 - g_7^2\right] + t_L^4[-g_8 + 2g_7 - g_6]\right\}
$$

$$
F = \zeta^2 g_1 g_2 g_5 - \zeta^2\left(g_1 g_8 - g_1 g_7 + g_4^2 g_6 + \overline{\beta} g_8 g_8 - g_4 g_3 g_8\right).
$$

Solving equation (8.20) gives the solution for the symmetric mode

$$
\begin{aligned}
\phi_n^* &= \sum_{i=1}^{5}(A_{in}J_n(\alpha_i r) + B_{in}Y_n(\alpha_i r))\cos n\theta \\
W_n^* &= \sum_{i=1}^{5}a_i(A_{in}J_n(\alpha_i r) + B_{in}Y_n(\alpha_i r))\cos n\theta \\
E_n^* &= \sum_{i=1}^{5}b_i(A_{in}J_n(\alpha_i r) + B_{in}Y_n(\alpha_i r))\cos n\theta \\
H_n^* &= \sum_{i=1}^{5}d_i(A_{in}J_n(\alpha_i r) + B_{in}Y_n(\alpha_i r))\cos n\theta. \qquad (8.21)
\end{aligned}
$$

The solutions to the antisymmetric modes of vibrations $\overline{\phi}_n^*$, $\overline{W}_n^*$, $\overline{E}_n^*$, $\overline{H}_n^*$ are obtained by changing $\cos n\theta$ by $\sin n\theta$ in equation (8.21), and we get

$$\overline{\phi}_n^* = \sum_{i=1}^{5}(\overline{A}_{in}J_n(\alpha_i r) + \overline{B}_{in}Y_n(\alpha_i r))\sin n\theta$$

$$\overline{W}_n^* = \sum_{i=1}^{5}a_i(\overline{A}_{in}J_n(\alpha_i r) + \overline{B}_{in}Y_n(\alpha_i r))\sin n\theta$$

$$\overline{E}_n^* = \sum_{i=1}^{5}b_i(\overline{A}_{in}J_n(\alpha_i r) + \overline{B}_{in}Y_n(\alpha_i r))\sin n\theta$$

$$\overline{H}_n^* = \sum_{i=1}^{5}d_i(\overline{A}_{in}J_n(\alpha_i r) + \overline{B}_{in}Y_n(\alpha_i r))\sin n\theta \tag{8.22}$$

where J_n and Y_n are the Bessel function of first and second kind of order n.

The constants a_i, b_i, c_i and d_i defined in equations (8.21) and (8.22) are calculated using the following equations

$$-g_2 ik_1 t_L a_i - g_3 t_L b_i - g_4 t_L c_i = \overline{c}_{11} ik_1 \alpha_i^2 - g_1$$

$$-(\alpha_i^2 - g_1)a_i - (\overline{e}_{15}\alpha_i^2 - t_L^2)b_i - (\overline{q}_{15}\alpha_i^2 - t_L^2)ik_3 c_i = g_2 t_L \alpha_i^2$$

$$-(\overline{e}_{15}\alpha_i^2 - t_L^2)a_i + (\overline{\varepsilon}_{11}\alpha_i^2 + g_6)ik_1 b_i + (g_7 + \overline{m}_{11}\alpha_i^2)c_i = g_3 t_L \alpha_i^2$$

$$-(\overline{q}_{15}\alpha_i^2 + t_L^2)a_i + (\overline{m}_{11}\alpha_i^2 + g_7)b_i + (g_8 + \overline{\mu}_{11}\alpha_i^2)c_i = g_4 t_L \alpha_i^2. \tag{8.23}$$

Evaluvating equation (8.23), we get

$$a_i = \frac{-g_3(g_7 + \overline{m}_{11}\alpha_i^2) + g_4(\overline{\varepsilon}_{11}\alpha_i^2 + g_6)}{-g_2 ik_1(\overline{\varepsilon}_{11}\alpha_i^2 + g_6) - g_3(\overline{e}_{15}\alpha_i^2 - t_L^2)}$$

$$b_i = \frac{g_3 g_4 t_L^2 \alpha_i^2 + ik_1(\overline{c}_{11}\alpha_i^2 - g_1)(\overline{m}_{11}\alpha_i^2 + g_7)}{t_L\left[g_2(\overline{\varepsilon}_{11}\alpha_i^2 + g_6) + g_3(\overline{e}_{15}\alpha_i^2 + t_L^2)\right]}$$

$$c_i = \frac{ik_1(\overline{c}_{11}\alpha_i^2 - g_1)(\overline{e}_{15}\alpha_i^2 - t_L^2) - g_2 g_3 t_L^2 \alpha_i^2}{t_L\left[g_2(\overline{\varepsilon}_{11}\alpha_i^2 + g_6) + g_3(\overline{e}_{15}\alpha_i^2 + t_L^2)\right]}$$

$$d_i = \frac{((\alpha_i a)^2(g_4\overline{\beta} - \overline{c}_{11}) - g_1)}{(g_4 L + \overline{\beta})}$$

$$\mathrm{L} = (\overline{d} - i\overline{k}_1)(\alpha_i a)^2 - i\overline{k}_3\zeta^2.$$

Evaluating equation (8.13), the symmetric mode solution is defined as

$$\psi_n^* = (A_5 J_n(\alpha_5 r) + B_5 J_n(\alpha_5 r))\sin n\theta. \tag{8.24}$$

By changing cos $n\theta$ by sin $n\theta$ in equation (8.24), the antisymmetric mode solution is obtained as

$$\overline{\psi}_n^* = (\overline{A}_6 J_n(\alpha_6 r) + \overline{B}_6 J_n(\alpha_6 r)) \cos n\theta \tag{8.25}$$

where J_n and Y_n are the first and second kind of the Bessel function for order n, and

$$\alpha_6^2 = \frac{t_L^2 - \Omega^2}{c_{66}}.$$

If

$$(\alpha_i a)^2 < 0 (i = 1, 2, 3, 4, 5, 6),$$

then the Bessel functions J_n and Y_n are replaced by the modified Bessel function I_n and Y_n, respectively.

8.4 Equations of motion of the fluid

For an non-viscous fluid, the acoustic pressure and radial displacement equation of motion in cylindrical polar coordinates r, θ and z can be expressed as

$$p^f = -B^f (u_{r,\, r}^f + r^{-1}\left(u_r^f + u_{\theta,\, \theta}^f\right) + u_{z,\, z}^f \tag{8.26}$$

and

$$c_f^{-2} u_{r,\, tt}^f = \Delta_r. \tag{8.27}$$

Here B^f is the adiabatic bulk modulus, ρ^f is the density, $c^f = \sqrt{B^f/\rho^f}$ is the acoustic phase velocity in the fluid, and u_r^f, u_θ^f, u_z^f are the displacement vectors.

$$\Delta = (u_{r,\, r}^f + r^{-1}\left(u_r^f + u_{\theta,\, \theta}^f\right) + u_{z,\, z}^f). \tag{8.28}$$

Substituting

$$u_r^f = \phi_r^f,\ u_\theta^f = r^{-1}\phi_\theta^f \text{ and } u_z^f = \phi_z^f \tag{8.29}$$

and the solution of (8.26) is in the form

$$\phi^f(r, \theta, z, t) = \sum_{n=0}^{\infty} \phi^f(r) \cos n\theta e^{i(kz+\omega t)}. \tag{8.30}$$

The oscillating waves propagating in the inner fluid located in the annulus are given by

$$\phi^f = A_5 J_n(\alpha_5 a x) \tag{8.31}$$

where $\alpha_5 a = \Omega^2/\overline{\rho}_1^f \overline{B}_1^f - \varsigma^2$, in which $\overline{\rho}_1^f = \rho_1/\rho^f$, $\overline{B}_1^f = B_1^f/c_{44}$. If $(\alpha_5 a)^2 < 0$, the Bessel function J_n in (8.22) is to be replaced by modified Bessel function I_n. Similarly, for the outer fluid that represents the oscillatory waves propagating away

$$\phi^f = B_5 H_n^{(2)}(\alpha_6 ax) \tag{8.32}$$

where $(\alpha_6 a)^2 = \Omega^2/\overline{\rho}_2^f \overline{B}_2^f - \varsigma^2$, in which $\overline{\rho}_2^f = \rho_2/\rho^f$, $\overline{B}_2^f = B_2^f/c_{44}$, $H_n^{(2)}$ is the Hankel function of the second kind. If $(\alpha_6 a)^2 < 0$, then the Hankel function of second kind is to be replaced by K_n, where K_n is the modified Bessel function of the second kind. By substituting equation (8.30) in (8.26) along with (8.31) and (8.32), the acoustic pressure for the inner fluid can be expressed as

$$p_1^f = A_5 \Omega^2 \overline{\rho}_1 J_n(\alpha_5 ax) \cos n\theta \tag{8.33}$$

and for the outer fluid it is

$$p_2^f = B_5 \Omega^2 \overline{\rho}_2 H_n^2(\alpha_6 ax) \cos n\theta. \tag{8.34}$$

8.5 Solid–fluid interface conditions and frequency equations

In this problem, the free vibration of an arbitrary cross-sectional plate is considered. Since the boundary is irregular in shape, the FECM is used to satisfy the boundary conditions along the inner and outer the surface of the plate directly. For the plate, the normal stress σ'_{xx} and shear stresses σ'_{xy}, σ'_{xz}, the electric field D'_x, the magnetic field B'_x and the magnetic field T'_x are equal to zero for the stress-free inner boundary. Similarly, normal stress σ_{xx} and shear stresses σ_{xy}, σ_{xz}, the electric field D_x and the magnetic field B_x are equal to zero for the stress-free outer boundary. Thus the following traction-free boundary condition for the inner surface of the plate is obtained as

$$(\sigma'_{xx} + p_i^f)_i = (\sigma'_{xy})_i = (\sigma'_{xz})_i = (D'_x)_i = (B'_x)_i = (u'_x - u_x^f)_i = 0 \tag{8.35a}$$

and for the outer boundary, the boundary condition is obtained as

$$(\sigma_{xx} + p_2^f)_i = (\sigma_{xy})_i = \sigma_{xx}(\sigma_{xz})_i = (D_x)_i = (B_x)_i = (u - u^f)_i = 0 \tag{8.35b}$$

where $()_i$ is the value at the boundary Γ_i shown in figure 8.1 (geometry of segments). It is convenient to replace the coordinates r and θ instead of the coordinates x_i and y_i in boundary conditions.

The displacement relations of a straight line are defined for the ith segment as follows

$$u = u\cos(\theta - \gamma_i) - v\sin(\theta - \gamma_i)v = v\cos(\theta - \gamma_i) + u\sin(\theta - \gamma_i). \tag{8.36}$$

Since the angle γ_i between the reference axis and normal of the ith boundary has a constant value in a segment Γ_i, we obtain

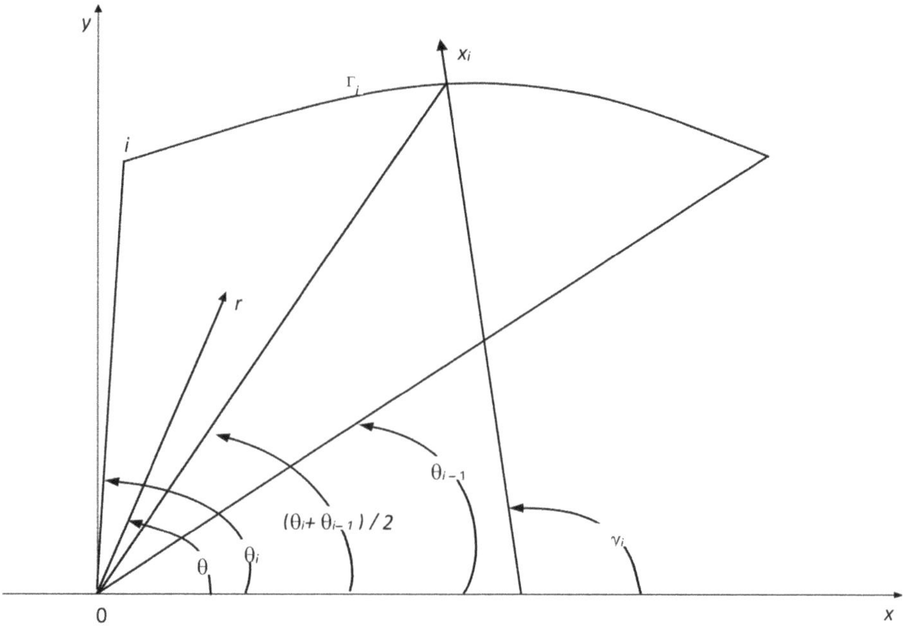

Figure 8.1. Geometry of a straight line segment.

$$\frac{\partial r}{\partial x_i} = \cos(\theta - \gamma_i), \; \frac{\partial \theta}{\partial x_i} = -\left(\frac{1}{r}\right)\sin(\theta - \gamma_i). \tag{8.37}$$

With the help of equations (8.36) and (8.37), the normal and shearing stresses can be transformed as

$$\begin{aligned}\sigma_{xx} = &(c_{11}\cos^2(\theta - \gamma_i) + c_{12}\sin^2(\theta - \gamma_i))u_r + r^{-1}(c_{11}\sin^2(\theta - \gamma_i)\\ &+ c_{12}\cos^2(\theta - \gamma_i))(u + v_\theta) + c_{66} - r^{-1}(v - u_\theta) - v_r)\\ &\sin^2(\theta - \gamma_i) + c_{13}W_z + e_{31}E_z + q_{31}H_z = 0\end{aligned}$$

$$\sigma_{xy} = c_{66}((u_r - r^{-1}(v_\theta + u))\sin^2(\theta - \gamma_i) + (r^{-1}(u_\theta - v) + v_r)\cos^2(\theta - \gamma_i)) = 0$$

$$\sigma_{xz} = c_{44}((u_z + W_r)\cos(\theta - \gamma_i) - (v_z + r^{-1}W_\theta)\sin(\theta - \gamma_i)) + e_{15}E_r + q_{15}H_r = 0$$

$$D_x = e_{15}(u_z + W_r) - \varepsilon_{11}E_r - m_{11}H_r = 0$$

$$B_x = q_{15}(u_z + W_r) - m_{11}E_r - \mu_{11}H_r = 0. \tag{8.38}$$

Substituting equations (8.36)–(8.38) in equations (8.35), we get

$$\left[\left(P^1_{xx}\right)_i + \left(\overline{P}^1_{xx}\right)_i\right]e^{i\Omega T_a} = 0, \left[\left(P^1_{xy}\right)_i + \left(\overline{P}^1_{xy}\right)_i\right]e^{i\Omega T_a} = 0, \left[\left(P^1_{xz}\right)_i + \left(\overline{P}^1_{xz}\right)_i\right]e^{i\Omega T_a} = 0,$$

$$\left[(Q_x^1)_i + \left(\overline{Q}_x^1\right)_i\right]e^{i\Omega T_a} = 0,\ \left[(R_x^1)_i + \left(\overline{R}_x^1\right)_i\right]e^{i\Omega T_a} = 0,\ \left[(S_t^1)_i + \left(\overline{S}_t^1\right)_i\right]e^{i\Omega T_a} = 0. \quad (8.39)$$

For the outer surface of the plate

$$[(P_{xx})_i + (\overline{P}_{xx})_i]e^{i\Omega T_a} = 0,\ [(P_{xy})_i + (\overline{P}_{xy})_i]e^{i\Omega T_a} = 0,\ [(P_{xz})_i + (\overline{P}_{xz})_i]e^{i\Omega T_a} = 0,$$

$$[(Q_x)_i + (\overline{Q}_x)_i]e^{i\Omega T_a} = 0,\ [(R_x)_i + (\overline{R}_x)_i]e^{i\Omega T_a} = 0,\ [(S_t)_i + (\overline{S}_t)_i]e^{i\Omega T_a} = 0 \quad (8.40)$$

where

$$\begin{aligned}P_{xx} = {} & 0.5(f_0^1 A_{10} + f_0^2 B_{10} + f_0^3 A_{20} + f_0^4 B_{20} + f_0^5 A_{30} \\ & + f_0^6 B_{30} + f_0^7 A_{40} + f_0^8 B_{40} + f_0^9 A_{50} + f_0^{10} B_{50}) \\ & + \sum_{n=1}^{\infty}\Big(f_n^1 A_{1n} + f_n^2 B_{1n} + f_n^3 A_{2n} + f_n^4 B_{2n} + f_n^5 A_{3n} + f_n^6 B_{3n} \\ & + f_n^7 A_{4n} + f_n^8 B_{4n} + f_n^9 A_{5n} + f_n^{10} B_{5n} + f_n^{11} A_{6n} + f_n^{12} B_{6n}\Big)\end{aligned}$$

$$\begin{aligned}P_{xy} = {} & 0.5(g_0^1 A_{10} + g_0^2 B_{10} + g_0^3 A_{20} + g_0^4 B_{20} + g_0^5 A_{30} \\ & + g_0^6 B_{30} + g_0^7 A_{40} + g_0^8 B_{40} + g_0^9 A_{50} + g_0^{10} B_{50}) \\ & + \sum_{n=1}^{\infty}\Big(g_n^1 A_{1n} + g_n^2 B_{1n} + g_n^3 A_{2n} + g_n^4 B_{2n} + g_n^5 A_{3n} + g_n^6 B_{3n} \\ & + g_n^7 A_{4n} + g_n^8 B_{4n} + g_n^9 A_{5n} + g_n^{10} B_{5n} + g_n^{11} A_{6n} + g_n^{12} B_{6n}\Big)\end{aligned}$$

$$\begin{aligned}P_{xz} = {} & 0.5(h_0^1 A_{10} + h_0^2 B_{10} + h_0^3 A_{20} + h_0^4 B_{20} + h_0^5 A_{30} \\ & + h_0^6 B_{30} + h_0^7 A_{40} + h_0^8 B_{40} + h_0^9 A_{50} + h_0^{10} B_{50}) \\ & + \sum_{n=1}^{\infty}\Big(h_n^1 A_{1n} + h_n^2 B_{1n} + h_n^3 A_{2n} + h_n^4 B_{2n} + h_n^5 A_{3n} + h_n^6 B_{3n} \\ & + h_n^7 A_{4n} + h_n^8 B_{4n} + h_n^9 A_{5n} + h_n^{10} B_{5n} + h_n^{11} A_{6n} + h_n^{12} B_{6n}\Big)\end{aligned}$$

$$\begin{aligned}Q_x = {} & 0.5(j_0^1 A_{10} + j_0^2 B_{10} + j_0^3 A_{20} + j_0^4 B_{20} + j_0^5 A_{30} \\ & + j_0^6 B_{30} + j_0^7 A_{40} + j_0^8 B_{40} + j_0^9 A_{50} + j_0^{10} B_{50}) \\ & + \sum_{n=1}^{\infty}\Big(j_n^1 A_{1n} + j_n^2 B_{1n} + j_n^3 A_{2n} + j_n^4 B_{2n} + j_n^5 A_{3n} + j_n^6 B_{3n} \\ & + j_n^7 A_{4n} + j_n^8 B_{4n} + j_n^9 A_{5n} + j_n^{10} B_{5n} + j_n^{11} A_{6n} + j_n^{12} B_{6n}\Big)\end{aligned}$$

$$\begin{aligned}R_x = 0.5(&k_0^1 A_{10} + k_0^2 B_{10} + k_0^3 A_{20} + k_0^4 B_{20} + k_0^5 A_{30}\\&+ k_0^6 B_{30} + k_0^7 A_{40} + k_0^8 B_{40} + k_0^9 A_{50} + k_0^{10} B_{50})\\&+ \sum_{n=1}^{\infty}\Big(k_n^1 A_{1n} + k_n^2 B_{1n} + k_n^3 A_{2n} + k_n^4 B_{2n} + k_n^5 A_{3n} + k_n^6 B_{3n}\\&+ k_n^7 A_{4n} + k_n^8 B_{4n} + k_n^9 A_{5n} + k_n^{10} B_{5n} + k_n^{11} A_{6n} + k_n^{12} B_{6n}\Big)\end{aligned}$$

$$\begin{aligned}S_t = 0.5(&l_0^1 A_{10} + l_0^2 B_{10} + l_0^3 A_{20} + l_0^4 B_{20} + l_0^5 A_{30} + l_0^6 B_{30}\\&+ l_0^7 A_{40} + l_0^8 B_{40} + l_0^9 A_{50} + l_0^{10} B_{50})\\&+ \sum_{n=1}^{\infty}\Big(l_n^1 A_{1n} + l_n^2 B_{1n} + l_n^3 A_{2n} + l_n^4 B_{2n} + l_n^5 A_{3n} + l_n^6 B_{3n} + l_n^7 A_{4n}\\&+ l_n^8 B_{4n} + l_n^9 A_{5n} + l_n^{10} B_{5n} + l_n^{11} A_{6n} + l_n^{12} B_{6n}\Big)\end{aligned}$$

$$\begin{aligned}P_{xx}^1 = 0.5(&f_0^1 A_{10} + f_0^2 B_{10} + f_0^3 A_{20} + f_0^4 B_{20} + f_0^5 A_{30}\\&+ f_0^6 B_{30} + f_0^7 A_{40} + f_0^8 B_{40} + f_0^9 A_{50} + f_0^{10} B_{50})\\&+ \sum_{n=1}^{\infty}\Big(f_n^1 A_{1n} + f_n^2 B_{1n} + f_n^3 A_{2n} + f_n^4 B_{2n} + f_n^5 A_{3n} + f_n^6 B_{3n} + f_n^7 A_{4n}\\&+ f_n^8 B_{4n} + f_n^9 A_{5n} + f_n^{10} B_{5n} + f_n^{11} A_{6n} + f_n^{12} B_{6n}\Big)\end{aligned}$$

$$\begin{aligned}P_{xy}^1 = 0.5(&g_0^1 A_{10} + g_0^2 B_{10} + g_0^3 A_{20} + g_0^4 B_{20} + g_0^5 A_{30} + g_0^6 B_{30}\\&+ g_0^7 A_{40} + g_0^8 B_{40} + g_0^9 A_{50} + g_0^{10} B_{50})\\&+ \sum_{n=1}^{\infty}\Big(g_n^1 A_{1n} + g_n^2 B_{1n} + g_n^3 A_{2n} + g_n^4 B_{2n} + g_n^5 A_{3n} + g_n^6 B_{3n} + g_n^7 A_{4n}\\&+ g_n^8 B_{4n} + g_n^9 A_{5n} + g_n^{10} B_{5n} + g_n^{11} A_{6n} + g_n^{12} B_{6n}\Big)\end{aligned}$$

$$\begin{aligned}P_{xz}^1 = 0.5(&h_0^1 A_{10} + h_0^2 B_{10} + h_0^3 A_{20} + h_0^4 B_{20} + h_0^5 A_{30}\\&+ h_0^6 B_{30} + h_0^7 A_{40} + h_0^8 B_{40} + h_0^9 A_{50} + h_0^{10} B_{50})\\&+ \sum_{n=1}^{\infty}\Big(h_n^1 A_{1n} + h_n^2 B_{1n} + h_n^3 A_{2n} + h_n^4 B_{2n} + h_n^5 A_{3n} + h_n^6 B_{3n}\\&+ h_n^7 A_{4n} + h_n^8 B_{4n} + h_n^9 A_{5n} + h_n^{10} B_{5n} + h_n^{11} A_{6n} + h_n^{12} B_{6n}\Big)\end{aligned}$$

$$\begin{aligned}Q_x^1 = 0.5(&j_0^1 A_{10} + j_0^2 B_{10} + j_0^3 A_{20} + j_0^4 B_{20} + j_0^5 A_{30}\\&+ j_0^6 B_{30} + j_0^7 A_{40} + j_0^8 B_{40} + j_0^9 A_{50} + j_0^{10} B_{50})\\&+ \sum_{n=1}^{\infty}\Big(j_n^1 A_{1n} + j_n^2 B_{1n} + j_n^3 A_{2n} + j_n^4 B_{2n} + j_n^5 A_{3n} + j_n^6 B_{3n}\\&+ j_n^7 A_{4n} + j_n^8 B_{4n} + j_n^9 A_{5n} + j_n^{10} B_{5n} + j_n^{11} A_{6n} + j_n^{12} B_{6n}\Big)\end{aligned}$$

$$\begin{aligned}
R_x^1 = {} & 0.5(k_0^1 A_{10} + k_0^2 B_{10} + k_0^3 A_{20} + k_0^4 B_{20} + k_0^5 A_{30} \\
& + k_0^6 B_{30} + k_0^7 A_{40} + k_0^8 B_{40} + k_0^9 A_{50} + k_0^{10} B_{50}) \\
& + \sum_{n=1}^{\infty}\Big(k_n^1 A_{1n} + k_n^2 B_{1n} + k_n^3 A_{2n} + k_n^4 B_{2n} + k_n^5 A_{3n} + k_n^6 B_{3n} \\
& + k_n^7 A_{4n} + k_n^8 B_{4n} + k_n^9 A_{5n} + k_n^{10} B_{5n} + k_n^{11} A_{6n} + k_n^{12} B_{6n}\Big)
\end{aligned}$$

$$\begin{aligned}
S_t^1 = {} & 0.5(l_0^1 A_{10} + l_0^2 B_{10} + l_0^3 A_{20} + l_0^4 B_{20} + l_0^5 A_{30} \\
& + l_0^6 B_{30} + l_0^7 A_{40} + l_0^8 B_{40} + l_0^9 A_{50} + l_0^{10} B_{50}) \\
& + \sum_{n=1}^{\infty}\Big(l_n^1 A_{1n} + l_n^2 B_{1n} + l_n^3 A_{2n} + l_n^4 B_{2n} + l_n^5 A_{3n} + l_n^6 B_{3n} \\
& + l_n^7 A_{4n} + l_n^8 B_{4n} + l_n^9 A_{5n} + l_n^{10} B_{5n} + l_n^{11} A_{6n} + l_n^{12} B_{6n}\Big).
\end{aligned} \tag{8.41}$$

For the antisymmetric mode

$$\begin{aligned}
\overline{P}_{xx} = {} & 0.5\Big(\overline{f}_0^{11}\,\overline{A}_{10} + \overline{f}_0^{12}\,\overline{B}_{10}\Big) \\
& + \sum_{n=1}^{\infty}\Big(\overline{f}_n^1\,\overline{A}_{1n} + \overline{f}_n^2\,\overline{B}_{1n} + \overline{f}_n^3\,\overline{A}_{2n} + \overline{f}_n^4\,\overline{B}_{2n} + \overline{f}_n^5\,\overline{A}_{3n} + \overline{f}_n^6\,\overline{B}_{3n} \\
& + \overline{f}_n^7\,\overline{A}_{4n} + \overline{f}_n^8\,\overline{B}_{4n} + \overline{f}_n^9\,\overline{A}_{5n} + \overline{f}_n^{10}\,\overline{B}_{5n} + \overline{f}_n^{11}\,\overline{A}_{6n} + \overline{f}_n^{12}\,\overline{B}_{6n}\Big)
\end{aligned}$$

$$\begin{aligned}
\overline{P}_{xy} = {} & 0.5\Big(\overline{g}_0^{11}\,\overline{A}_{10} + \overline{g}_0^{12}\,\overline{B}_{10}\Big) \\
& + \sum_{n=1}^{\infty}\Big(\overline{g}_n^1\,\overline{A}_{1n} + \overline{g}_n^2\,\overline{B}_{1n} + \overline{g}_n^3\,\overline{A}_{2n} + \overline{g}_n^4\,\overline{B}_{2n} + \overline{g}_n^5\,\overline{A}_{3n} + \overline{g}_n^6\,\overline{B}_{3n} \\
& + \overline{g}_n^7\,\overline{A}_{4n} + \overline{g}_n^8\,\overline{B}_{4n} + \overline{g}_n^9\,\overline{A}_{5n} + \overline{g}_n^{10}\,\overline{B}_{5n} + \overline{g}_n^{11}\,\overline{A}_{6n} + \overline{g}_n^{12}\,\overline{B}_{6n}\Big)
\end{aligned}$$

$$\begin{aligned}
\overline{P}_{xz} = {} & 0.5\Big(\overline{h}_0^{11}\,\overline{A}_{10} + \overline{h}_0^{12}\,\overline{B}_{10}\Big) \\
& + \sum_{n=1}^{\infty}\Big(\overline{h}_n^1\,\overline{A}_{1n} + \overline{h}_n^2\,\overline{B}_{1n} + \overline{h}_n^3\,\overline{A}_{2n} + \overline{h}_n^4\,\overline{B}_{2n} + \overline{h}_n^5\,\overline{A}_{3n} + \overline{h}_n^6\,\overline{B}_{3n} \\
& + \overline{h}_n^7\,\overline{A}_{4n} + \overline{h}_n^8\,\overline{B}_{4n} + \overline{h}_n^9\,\overline{A}_{5n} + \overline{h}_n^{10}\,\overline{B}_{5n} + \overline{h}_n^{11}\,\overline{A}_{6n} + \overline{h}_n^{12}\,\overline{B}_{6n}\Big)
\end{aligned}$$

$$\begin{aligned}
\overline{Q}_x = {} & 0.5\Big(\overline{j}_0^{11}\,\overline{A}_{10} + \overline{j}_0^{12}\,\overline{B}_{10}\Big) \\
& + \sum_{n=1}^{\infty}\Big(\overline{j}_n^1\,\overline{A}_{1n} + \overline{j}_n^2\,\overline{B}_{1n} + \overline{j}_n^3\,\overline{A}_{2n} + \overline{j}_n^4\,\overline{B}_{2n} + \overline{j}_n^5\,\overline{A}_{3n} + \overline{j}_n^6\,\overline{B}_{3n} \\
& + \overline{j}_n^7\,\overline{A}_{4n} + \overline{j}_n^8\,\overline{B}_{4n} + \overline{j}_n^9\,\overline{A}_{5n} + \overline{j}_n^{10}\,\overline{B}_{5n} + \overline{j}_n^{11}\,\overline{A}_{6n} + \overline{j}_n^{12}\,\overline{B}_{6n}\Big)
\end{aligned}$$

$$\overline{R}_x = 0.5\left(\overline{k}_0^{11}\overline{A}_{10} + \overline{k}_0^{12}\overline{B}_{10}\right) + \sum_{n=1}^{\infty}\left(\overline{k}_n^{1}\overline{A}_{1n} + \overline{k}_n^{2}\overline{B}_{1n} + \overline{k}_n^{3}\overline{A}_{2n} + \overline{k}_n^{4}\overline{B}_{2n} + \overline{k}_n^{5}\overline{A}_{3n} + \overline{k}_n^{6}\overline{B}_{3n} + \overline{k}_n^{7}\overline{A}_{4n} + \overline{k}_n^{8}\overline{B}_{4n} + \overline{k}_n^{9}\overline{A}_{5n} + \overline{k}_n^{10}\overline{B}_{5n} + \overline{k}_n^{11}\overline{A}_{6n} + \overline{k}_n^{12}\overline{B}_{6n}\right)$$

$$\overline{S}_t = 0.5\left(\overline{l}_0^{11}\overline{A}_{10} + \overline{l}_0^{12}\overline{B}_{10}\right) + \sum_{n=1}^{\infty}\left(\overline{l}_n^{1}\overline{A}_{1n} + \overline{l}_n^{2}\overline{B}_{1n} + \overline{l}_n^{3}\overline{A}_{2n} + \overline{l}_n^{4}\overline{B}_{2n} + \overline{l}_n^{5}\overline{A}_{3n} + \overline{l}_n^{6}\overline{B}_{3n} + \overline{l}_n^{7}\overline{A}_{4n} + \overline{l}_n^{8}\overline{B}_{4n} + \overline{l}_n^{9}\overline{A}_{5n} + \overline{l}_n^{10}\overline{B}_{5n} + \overline{l}_n^{11}\overline{A}_{6n} + \overline{l}_n^{12}\overline{B}_{6n}\right)$$

$$\overline{P}_{xx}^{1} = 0.5\left(\overline{f}_0^{11}\overline{A}_{10} + \overline{f}_0^{12}\overline{B}_{10}\right) + \sum_{n=1}^{\infty}\left(\overline{f}_n^{1}\overline{A}_{1n} + \overline{f}_n^{2}\overline{B}_{1n} + \overline{f}_n^{3}\overline{A}_{2n} + \overline{f}_n^{4}\overline{B}_{2n} + \overline{f}_n^{5}\overline{A}_{3n} + \overline{f}_n^{6}\overline{B}_{3n} + \overline{f}_n^{7}\overline{A}_{4n} + \overline{f}_n^{8}\overline{B}_{4n} + \overline{f}_n^{9}\overline{A}_{5n} + \overline{f}_n^{10}\overline{B}_{5n} + \overline{f}_n^{11}\overline{A}_{6n} + \overline{f}_n^{12}\overline{B}_{6n}\right)$$

$$\overline{P}_{xy}^{1} = 0.5\left(\overline{g}_0^{11}\overline{A}_{10} + \overline{g}_0^{12}\overline{B}_{10}\right) + \sum_{n=1}^{\infty}\left(\overline{g}_n^{1}\overline{A}_{1n} + \overline{g}_n^{2}\overline{B}_{1n} + \overline{g}_n^{3}\overline{A}_{2n} + \overline{g}_n^{4}\overline{B}_{2n} + \overline{g}_n^{5}\overline{A}_{3n} + \overline{g}_n^{6}\overline{B}_{3n} + \overline{g}_n^{7}\overline{A}_{4n} + \overline{g}_n^{8}\overline{B}_{4n} + \overline{g}_n^{9}\overline{A}_{5n} + \overline{g}_n^{10}\overline{B}_{5n} + \overline{g}_n^{11}\overline{A}_{6n} + \overline{g}_n^{12}\overline{B}_{6n}\right)$$

$$\overline{P}_{xz}^{1} = 0.5\left(\overline{h}_0^{11}\overline{A}_{10} + \overline{h}_0^{12}\overline{B}_{10}\right) + \sum_{n=1}^{\infty}\left(\overline{h}_n^{1}\overline{A}_{1n} + \overline{h}_n^{2}\overline{B}_{1n} + \overline{h}_n^{3}\overline{A}_{2n} + \overline{h}_n^{4}\overline{B}_{2n} + \overline{h}_n^{5}\overline{A}_{3n} + \overline{h}_n^{6}\overline{B}_{3n} + \overline{h}_n^{7}\overline{A}_{4n} + \overline{h}_n^{8}\overline{B}_{4n} + \overline{h}_n^{9}\overline{A}_{5n} + \overline{h}_n^{10}\overline{B}_{5n} + \overline{h}_n^{11}\overline{A}_{6n} + \overline{h}_n^{12}\overline{B}_{6n}\right)$$

$$\overline{Q}_{x}^{1} = 0.5\left(\overline{j}_0^{11}\overline{A}_{10} + \overline{j}_0^{12}\overline{B}_{10}\right) + \sum_{n=1}^{\infty}\left(\overline{j}_n^{1}\overline{A}_{1n} + \overline{j}_n^{2}\overline{B}_{1n} + \overline{j}_n^{3}\overline{A}_{2n} + \overline{j}_n^{4}\overline{B}_{2n} + \overline{j}_n^{5}\overline{A}_{3n} + \overline{j}_n^{6}\overline{B}_{3n} + \overline{j}_n^{7}\overline{A}_{4n} + \overline{j}_n^{8}\overline{B}_{4n} + \overline{j}_n^{9}\overline{A}_{5n} + \overline{j}_n^{10}\overline{B}_{5n} + \overline{j}_n^{11}\overline{A}_{6n} + \overline{j}_n^{12}\overline{B}_{6n}\right)$$

$$\overline{R}_x^1 = 0.5\left(\overline{k}_0^{11}\overline{A}_{10} + \overline{k}_0^{12}\overline{B}_{10}\right) + \sum_{n=1}^{\infty}\left(\overline{k}_n^1\overline{A}_{1n} + \overline{k}_n^2\overline{B}_{1n} + \overline{k}_n^3\overline{A}_{2n} + \overline{k}_n^4\overline{B}_{2n} + \overline{k}_n^5\overline{A}_{3n} + \overline{k}_n^6\overline{B}_{3n} + \overline{k}_n^7\overline{A}_{4n} + \overline{k}_n^8\overline{B}_{4n} + \overline{k}_n^9\overline{A}_{5n} + \overline{k}_n^{10}\overline{B}_{5n} + \overline{k}_n^{11}\overline{A}_{6n} + \overline{k}_n^{12}\overline{B}_{6n}\right)$$

$$\overline{S}_t^1 = 0.5\left(\overline{l}_0^{11}\overline{A}_{10} + \overline{l}_0^{12}\overline{B}_{10}\right) + \sum_{n=1}^{\infty}\left(\overline{l}_n^1\overline{A}_{1n} + \overline{l}_n^2\overline{B}_{1n} + \overline{l}_n^3\overline{A}_{2n} + \overline{l}_n^4\overline{B}_{2n} + \overline{l}_n^5\overline{A}_{3n} + \overline{l}_n^6\overline{B}_{3n} + \overline{l}_n^7\overline{A}_{4n} + \overline{l}_n^8\overline{B}_{4n} + \overline{l}_n^9\overline{A}_{5n} + \overline{l}_n^{10}\overline{B}_{5n} + \overline{l}_n^{11}\overline{A}_{6n} + \overline{l}_n^{12}\overline{B}_{6n}\right). \quad (8.42)$$

The coefficients $f_n^i - \overline{l}_n^i$ are given in appendix A.

Fourier expansion is used to satisfy the boundary conditions for the boundary line. For the convergence of solution, the straight line is taken as one segment and the curved line is taken as many segments. Hence, the symmetric boundary conditions along the inner boundary are expressed as follows

$$\sum_{m=0}^{\infty}\varepsilon_m[\hat{F}_{m0}^1 A_{10} + \hat{F}_{m0}^2 B_{10} + \hat{F}_{m0}^3 A_{20} + \hat{F}_{m0}^4 B_{20} + \hat{F}_{m0}^5 A_{30} + \hat{F}_{m0}^6 A_{30} + \hat{F}_{m0}^7 A_{40} + \hat{F}_{m0}^8 B_{40} + \hat{F}_{m0}^9 A_{50} + \hat{F}_{m0}^{10} B_{50} + \sum_{n=1}^{\infty}(\hat{F}_{mn}^1 A_{1n} + \hat{F}_{mn}^2 B_{1n} + \hat{F}_{mn}^3 A_{2n} + \hat{F}_{mn}^4 B_{2n} + \hat{F}_{mn}^5 A_{3n} + \hat{F}_{mn}^6 A_{3n} + \hat{F}_{mn}^7 A_{4n} + \hat{F}_{mn}^8 B_{4n} + \hat{F}_{mn}^9 A_{5n} + \hat{F}_{mn}^{10} B_{5n} + \hat{F}_{mn}^{11} A_{6n} + \hat{F}_{mn}^{12} B_{6n})]\cos m\theta = 0$$

$$\sum_{m=0}^{\infty}\varepsilon_m[\hat{G}_{m0}^1 A_{10} + \hat{G}_{m0}^2 B_{10} + \hat{G}_{m0}^3 A_{20} + \hat{G}_{m0}^4 B_{20} + \hat{G}_{m0}^5 A_{30} + \hat{G}_{m0}^6 A_{30} + \hat{G}_{m0}^7 A_{40} + \hat{G}_{m0}^8 B_{40} + \hat{G}_{m0}^9 A_{50} + \hat{G}_{m0}^{10} B_{50} + \sum_{n=1}^{\infty}(\hat{G}_{mn}^1 A_{1n} + \hat{G}_{mn}^2 B_{1n} + \hat{G}_{mn}^3 A_{2n} + \hat{G}_{mn}^4 B_{2n} + \hat{G}_{mn}^5 A_{3n} + \hat{G}_{mn}^6 A_{3n} + \hat{G}_{mn}^7 A_{4n} + \hat{G}_{mn}^8 B_{4n} + \hat{G}_{mn}^9 A_{5n} + \hat{G}_{mn}^{10} B_{5n} + \hat{G}_{mn}^{11} A_{6n} + \hat{G}_{mn}^{12} B_{6n})]\sin m\theta = 0$$

$$\begin{aligned}
&\sum_{m=0}^{\infty}\varepsilon_m[\hat{H}_{m0}^{1}A_{10}+\hat{H}_{m0}^{2}B_{10}+\hat{H}_{m0}^{3}A_{20}+\hat{H}_{m0}^{4}B_{20}+\hat{H}_{m0}^{5}A_{30}\\
&+\hat{H}_{m0}^{6}A_{30}+\hat{H}_{m0}^{7}A_{40}+\hat{H}_{m0}^{8}B_{40}+\hat{H}_{m0}^{9}A_{50}\\
&+\hat{H}_{m0}^{10}B_{50}+\sum_{n=1}^{\infty}(\hat{H}_{mn}^{1}A_{1n}+\hat{H}_{mn}^{2}B_{1n}+\hat{H}_{mn}^{3}A_{2n}+\hat{H}_{mn}^{4}B_{2n}\\
&+\hat{H}_{mn}^{5}A_{3n}+\hat{H}_{mn}^{6}A_{3n}+\hat{H}_{mn}^{7}A_{4n}+\hat{H}_{mn}^{8}B_{4n}\\
&+\hat{H}_{mn}^{9}A_{5n}+\hat{H}_{mn}^{10}B_{5n}+\hat{H}_{mn}^{11}A_{6n}+\hat{H}_{mn}^{12}B_{6n})]\cos m\theta=0
\end{aligned}$$

$$\begin{aligned}
&\sum_{m=0}^{\infty}\varepsilon_m[\hat{J}_{m0}^{1}A_{10}+\hat{J}_{m0}^{2}B_{10}+\hat{J}_{m0}^{3}A_{20}+\hat{J}_{m0}^{4}B_{20}+\hat{J}_{m0}^{5}A_{30}\\
&+\hat{J}_{m0}^{6}A_{30}+\hat{J}_{m0}^{7}A_{40}+\hat{J}_{m0}^{8}B_{40}+\hat{J}_{m0}^{9}A_{50}\\
&+\hat{J}_{m0}^{10}B_{50}+\sum_{n=1}^{\infty}(\hat{J}_{mn}^{1}A_{1n}+\hat{J}_{mn}^{2}B_{1n}+\hat{J}_{mn}^{3}A_{2n}+\hat{J}_{mn}^{4}B_{2n}\\
&+\hat{J}_{mn}^{5}A_{3n}+\hat{J}_{mn}^{6}A_{3n}+\hat{J}_{mn}^{7}A_{4n}+\hat{J}_{mn}^{8}B_{4n}\\
&+\hat{J}_{mn}^{9}A_{5n}+\hat{J}_{mn}^{10}B_{5n}+\hat{J}_{mn}^{11}A_{6n}+\hat{J}_{mn}^{12}B_{6n})]\cos m\theta=0
\end{aligned}$$

$$\begin{aligned}
&\sum_{m=0}^{\infty}\varepsilon_m[\hat{K}_{m0}^{1}A_{10}+\hat{K}_{m0}^{2}B_{10}+\hat{K}_{m0}^{3}A_{20}+\hat{K}_{m0}^{4}B_{20}\\
&+\hat{K}_{m0}^{5}A_{30}+\hat{K}_{m0}^{6}A_{30}+\hat{K}_{m0}^{7}A_{40}+\hat{K}_{m0}^{8}B_{40}+\hat{K}_{m0}^{9}A_{50}\\
&+\hat{K}_{m0}^{10}B_{50}+\sum_{n=1}^{\infty}(\hat{K}_{mn}^{1}A_{1n}+\hat{K}_{mn}^{2}B_{1n}+\hat{K}_{mn}^{3}A_{2n}+\hat{K}_{mn}^{4}B_{2n}\\
&+\hat{K}_{mn}^{5}A_{3n}+\hat{K}_{mn}^{6}A_{3n}+\hat{K}_{mn}^{7}A_{4n}+\hat{K}_{mn}^{8}B_{4n}\\
&+\hat{K}_{mn}^{9}A_{5n}+\hat{K}_{mn}^{10}B_{5n}+\hat{K}_{mn}^{11}A_{6n}+\hat{K}_{mn}^{12}B_{6n})]\cos m\theta=0
\end{aligned} \tag{8.43}$$

where

$$\hat{F}_{mn}^{j}=\left(\frac{2\varepsilon_n}{\pi}\right)\sum_{i=1}^{I}\int_{\theta_{i-1}}^{\theta_i}f_n^{j}(\hat{R}_i,\,\theta)\cos m\theta d\theta,\ \hat{G}_{mn}^{j}=\left(\frac{2\varepsilon_n}{\pi}\right)\sum_{i=1}^{I}\int_{\theta_{i-1}}^{\theta_i}g_n^{j}(\hat{R}_i,\,\theta)\sin m\theta d\theta,$$

$$\hat{H}_{mn}^{j}=\left(\frac{2\varepsilon_n}{\pi}\right)\sum_{i=1}^{I}\int_{\theta_{i-1}}^{\theta_i}h_n^{j}(\hat{R}_i,\,\theta)\cos m\theta d\theta,\ \hat{J}_{mn}^{j}=\left(\frac{2\varepsilon_n}{\pi}\right)\sum_{i=1}^{I}\int_{\theta_{i-1}}^{\theta_i}J_n^{j}(\hat{R}_i,\,\theta)\cos m\theta d\theta,$$

$$\hat{K}_{mn}^{j}=\left(\frac{2\varepsilon_n}{\pi}\right)\sum_{i=1}^{I}\int_{\theta_{i-1}}^{\theta_i}k_n^{j}(\hat{R}_i,\,\theta)\cos m\theta d\theta \tag{8.44}$$

and for the outer surface

$$\sum_{m=0}^{\infty}\varepsilon_m[F^1_{m0}A_{10} + F^2_{m0}B_{10} + F^3_{m0}A_{20} + F^4_{m0}B_{20}$$
$$+ F^5_{m0}A_{30} + F^6_{m0}A_{30} + F^7_{m0}A_{40} + F^8_{m0}B_{40} + F^9_{m0}A_{50}$$
$$+F^{10}_{m0}B_{50} + \sum_{n=1}^{\infty}(F^1_{mn}A_{1n} + F^2_{mn}B_{1n} + F^3_{mn}A_{2n} + F^4_{mn}B_{2n}$$
$$+ F^5_{mn}A_{3n} + F^6_{mn}A_{3n} + F^7_{mn}A_{4n} + F^8_{mn}B_{4n}$$
$$+ F^9_{mn}A_{5n} + F^{10}_{mn}B_{5n} + F^{11}_{mn}A_{6n} + F^{12}_{mn}B_{6n})]\cos m\theta = 0$$

$$\sum_{m=0}^{\infty}\varepsilon_m[G^1_{m0}A_{10} + G^2_{m0}B_{10} + G^3_{m0}A_{20} + G^4_{m0}B_{20}$$
$$+ G^5_{m0}A_{30} + G^6_{m0}A_{30} + G^7_{m0}A_{40} + G^8_{m0}B_{40} + G^9_{m0}A_{50}$$
$$+G^{10}_{m0}B_{50} + \sum_{n=1}^{\infty}(G^1_{mn}A_{1n} + G^2_{mn}B_{1n} + G^3_{mn}A_{2n} + G^4_{mn}B_{2n}$$
$$+ G^5_{mn}A_{3n} + G^6_{mn}A_{3n} + G^7_{mn}A_{4n} + G^8_{mn}B_{4n}$$
$$+ G^9_{mn}A_{5n} + G^{10}_{mn}B_{5n} + G^{11}_{mn}A_{6n} + G^{12}_{mn}B_{6n})]\sin m\theta = 0$$

$$\sum_{m=0}^{\infty}\varepsilon_m[H^1_{m0}A_{10} + H^2_{m0}B_{10} + H^3_{m0}A_{20} + H^4_{m0}B_{20}$$
$$+ H^5_{m0}A_{30} + H^6_{m0}A_{30} + H^7_{m0}A_{40} + H^8_{m0}B_{40} + H^9_{m0}A_{50}$$
$$+ H^{10}_{m0}B_{50} + \sum_{n=1}^{\infty}(H^1_{mn}A_{1n} + H^2_{mn}B_{1n} + H^3_{mn}A_{2n} + H^4_{mn}B_{2n}$$
$$+ H^5_{mn}A_{3n} + H^6_{mn}A_{3n} + H^7_{mn}A_{4n} + H^8_{mn}B_{4n}$$
$$+ H^9_{mn}A_{5n} + H^{10}_{mn}B_{5n} + H^{11}_{mn}A_{6n} + H^{12}_{mn}B_{6n})]\cos m\theta = 0$$

$$\sum_{m=0}^{\infty}\varepsilon_m[J^1_{m0}A_{10} + J^2_{m0}B_{10} + J^3_{m0}A_{20} + J^4_{m0}B_{20}$$
$$+ J^5_{m0}A_{30} + J^6_{m0}A_{30} + J^7_{m0}A_{40} + J^8_{m0}B_{40} + J^9_{m0}A_{50}$$
$$+ J^{10}_{m0}B_{50} + \sum_{n=1}^{\infty}(J^1_{mn}A_{1n} + J^2_{mn}B_{1n} + J^3_{mn}A_{2n} + J^4_{mn}B_{2n}$$
$$+ J^5_{mn}A_{3n} + J^6_{mn}A_{3n} + J^7_{mn}A_{4n} + J^8_{mn}B_{4n}$$
$$+ J^9_{mn}A_{5n} + J^{10}_{mn}B_{5n} + J^{11}_{mn}A_{6n} + J^{12}_{mn}B_{6n})]\cos m\theta = 0$$

$$\begin{aligned}
&\sum_{m=0}^{\infty}\varepsilon_m[K_{m0}^{1}A_{10} + K_{m0}^{2}B_{10} + K_{m0}^{3}A_{20} + K_{m0}^{4}B_{20} \\
&+ K_{m0}^{5}A_{30} + K_{m0}^{6}A_{30} + K_{m0}^{7}A_{40} + K_{m0}^{8}B_{40} + K_{m0}^{9}A_{50} \\
&+ K_{m0}^{10}B_{50} + \sum_{n=1}^{\infty}(K_{mn}^{1}A_{1n} + K_{mn}^{2}B_{1n} + K_{mn}^{3}A_{2n} + K_{mn}^{4}B_{2n} \\
&+ K_{mn}^{5}A_{3n} + K_{mn}^{6}A_{3n} + K_{mn}^{7}A_{4n} + K_{mn}^{8}B_{4n} \\
&+ K_{mn}^{9}A_{5n} + K_{mn}^{10}B_{5n} + K_{mn}^{11}A_{6n} + K_{mn}^{12}B_{6n})]\cos m\theta = 0
\end{aligned} \tag{8.45}$$

where

$$F_{mn}^{j} = \left(\frac{2\varepsilon_n}{\pi}\right)\sum_{i=1}^{I}\int_{\theta_{i-1}}^{\theta_i} f_n^{j}(R_i,\,\theta)\cos m\theta d\theta,\; G_{mn}^{j} = \left(\frac{2\varepsilon_n}{\pi}\right)\sum_{i=1}^{I}\int_{\theta_{i-1}}^{\theta_i} g_n^{j}(R_i,\,\theta)\sin m\theta d\theta,$$

$$H_{mn}^{j} = \left(\frac{2\varepsilon_n}{\pi}\right)\sum_{i=1}^{I}\int_{\theta_{i-1}}^{\theta_i} h_n^{j}(R_i,\,\theta)\cos m\theta d\theta,\; J_{mn}^{j} = \left(\frac{2\varepsilon_n}{\pi}\right)\sum_{i=1}^{I}\int_{\theta_{i-1}}^{\theta_i} J_n^{j}(R_i,\,\theta)\cos m\theta d\theta,$$

$$K_{mn}^{j} = \left(\frac{2\varepsilon_n}{\pi}\right)\sum_{i=1}^{I}\int_{\theta_{i-1}}^{\theta_i} k_n^{j}(R_i,\,\theta)\cos m\theta d\theta. \tag{8.46}$$

Similarly, the boundary conditions for the antisymmetric mode are given by

$$\begin{aligned}
&\sum_{m=0}^{\infty}[\overline{\hat{F}}_{m0}^{11}\overline{A}_{50} + \overline{\hat{F}}_{m0}^{12}\overline{B}_{50} \\
&+ \sum_{n=1}^{\infty}\overline{\hat{F}}_{mn}^{1}\overline{A}_{1n} + \overline{\hat{F}}_{mn}^{2}\overline{B}_{1n} + \overline{\hat{F}}_{mn}^{3}\overline{A}_{2n} + \overline{\hat{F}}_{mn}^{4}\overline{B}_{2n} + \overline{\hat{F}}_{mn}^{5}\overline{A}_{3n} \\
&+ \overline{\hat{F}}_{mn}^{6}\overline{B}_{3n} + \overline{\hat{F}}_{mn}^{7}\overline{A}_{4n} + \overline{\hat{F}}_{mn}^{8}\overline{B}_{4n} \\
&+ \overline{\hat{F}}_{mn}^{9}\overline{A}_{5n} + \overline{\hat{F}}_{mn}^{10}\overline{B}_{5n} + \overline{\hat{F}}_{mn}^{11}\overline{A}_{6n} + \overline{\hat{F}}_{mn}^{12}\overline{B}_{6n})]\sin m\theta = 0
\end{aligned}$$

$$\begin{aligned}
&\sum_{m=0}^{\infty}[\overline{\hat{G}}_{m0}^{11}\overline{A}_{50} + \overline{\hat{G}}_{m0}^{12}\overline{B}_{50} \\
&+ \sum_{n=1}^{\infty}(\overline{\hat{G}}_{mn}^{1}\overline{A}_{1n} + \overline{\hat{G}}_{mn}^{2}\overline{B}_{1n} + \overline{\hat{G}}_{mn}^{3}\overline{A}_{2n} + \overline{\hat{G}}_{mn}^{4}\overline{B}_{2n} \\
&+ \overline{\hat{G}}_{mn}^{5}\overline{A}_{3n} + \overline{\hat{G}}_{mn}^{6}\overline{B}_{3n} + \overline{\hat{G}}_{mn}^{7}\overline{A}_{4n} + \overline{\hat{G}}_{mn}^{8}\overline{B}_{4n} \\
&+ \overline{\hat{G}}_{mn}^{9}\overline{A}_{5n} + \overline{\hat{G}}_{mn}^{10}\overline{B}_{5n} + \overline{\hat{G}}_{mn}^{11}\overline{A}_{6n} + \overline{\hat{G}}_{mn}^{12}\overline{B}_{6n})]\cos m\theta = 0
\end{aligned}$$

$$
\begin{aligned}
&\sum_{m=0}^{\infty}[\overline{\hat{H}}{}^{11}_{m0}\overline{A}_{50} + \overline{\hat{H}}{}^{12}_{m0}\overline{B}_{50}\\
&+\sum_{n=1}^{\infty}(\overline{\hat{H}}{}^{1}_{mn}\overline{A}_{1n} + \overline{\hat{H}}{}^{2}_{mn}\overline{B}_{1n} + \overline{\hat{H}}{}^{3}_{mn}\overline{A}_{2n} + \overline{\hat{H}}{}^{4}_{mn}\overline{B}_{2n}\\
&+\overline{\hat{H}}{}^{5}_{mn}\overline{A}_{3n} + \overline{\hat{H}}{}^{6}_{mn}\overline{B}_{3n} + \overline{\hat{H}}{}^{7}_{mn}\overline{A}_{4n} + \overline{\hat{H}}{}^{8}_{mn}\overline{B}_{4n}\\
&+\overline{\hat{H}}{}^{9}_{mn}\overline{A}_{5n} + \overline{\hat{H}}{}^{10}_{mn}\overline{B}_{5n} + \overline{\hat{H}}{}^{11}_{mn}\overline{A}_{6n} + \overline{\hat{H}}{}^{12}_{mn}\overline{B}_{6n})]\sin m\theta = 0
\end{aligned}
$$

$$
\begin{aligned}
&\sum_{m=0}^{\infty}[\overline{\hat{J}}{}^{11}_{m0}\overline{A}_{50} + \overline{\hat{J}}{}^{12}_{m0}\overline{B}_{50}\\
&+\sum_{n=1}^{\infty}(\overline{\hat{J}}{}^{1}_{mn}\overline{A}_{1n} + \overline{\hat{J}}{}^{2}_{mn}\overline{B}_{1n} + \overline{\hat{J}}{}^{3}_{mn}\overline{A}_{2n} + \overline{\hat{J}}{}^{4}_{mn}\overline{B}_{2n}\\
&+\overline{\hat{J}}{}^{5}_{mn}\overline{A}_{3n} + \overline{\hat{J}}{}^{6}_{mn}\overline{B}_{3n} + \overline{\hat{J}}{}^{7}_{mn}\overline{A}_{4n} + \overline{\hat{J}}{}^{8}_{mn}\overline{B}_{4n}\\
&+\overline{\hat{J}}{}^{9}_{mn}\overline{A}_{5n} + \overline{\hat{J}}{}^{10}_{mn}\overline{B}_{5n} + \overline{\hat{J}}{}^{11}_{mn}\overline{A}_{6n} + \overline{\hat{J}}{}^{12}_{mn}\overline{B}_{6n})]\sin m\theta = 0
\end{aligned}
$$

$$
\begin{aligned}
&\sum_{m=0}^{\infty}[\overline{\hat{K}}{}^{11}_{m0}\overline{A}_{50} + \overline{\hat{K}}{}^{12}_{m0}\overline{B}_{50}\\
&+\sum_{n=1}^{\infty}(\overline{\hat{K}}{}^{1}_{mn}\overline{A}_{1n} + \overline{\hat{K}}{}^{2}_{mn}\overline{B}_{1n} + \overline{\hat{K}}{}^{3}_{mn}\overline{A}_{2n} + \overline{\hat{K}}{}^{4}_{mn}\overline{B}_{2n}\\
&+\overline{\hat{K}}{}^{5}_{mn}\overline{A}_{3n} + \overline{\hat{K}}{}^{6}_{mn}\overline{B}_{3n} + \overline{\hat{K}}{}^{7}_{mn}\overline{A}_{4n} + \overline{\hat{K}}{}^{8}_{mn}\overline{B}_{4n}\\
&+\overline{\hat{K}}{}^{9}_{mn}\overline{A}_{5n} + \overline{\hat{K}}{}^{10}_{mn}\overline{B}_{5n} + \overline{\hat{K}}{}^{11}_{mn}\overline{A}_{6n} + \overline{\hat{K}}{}^{12}_{mn}\overline{B}_{6n})]\sin m\theta = 0
\end{aligned}
\tag{8.47}
$$

where

$$
\overline{F}{}^{j}_{mn} = \left(\frac{2\varepsilon_n}{\pi}\right)\sum_{i=1}^{I}\int_{\theta_{i-1}}^{\theta_i} f_n^j(\overline{R}_i, \theta)\sin m\theta d\theta,\ \overline{\hat{G}}{}^{j}_{mn} = \left(\frac{2\varepsilon_n}{\pi}\right)\sum_{i=1}^{I}\int_{\theta_{i-1}}^{\theta_i} g_n^j(\overline{R}_i, \theta)\cos m\theta d\theta,
$$

$$
\overline{\hat{H}}{}^{j}_{mn} = \left(\frac{2\varepsilon_n}{\pi}\right)\sum_{i=1}^{I}\int_{\theta_{i-1}}^{\theta_i} h_n^j(\overline{\hat{R}}_i, \theta)\sin m\theta d\theta,\ \overline{\hat{J}}{}^{j}_{mn} = \left(\frac{2\varepsilon_n}{\pi}\right)\sum_{i=1}^{I}\int_{\theta_{i-1}}^{\theta_i} J_n^j(\overline{\hat{R}}_i, \theta)\sin m\theta d\theta,
$$

$$
\overline{\hat{K}}{}^{j}_{mn} = \left(\frac{2\varepsilon_n}{\pi}\right)\sum_{i=1}^{I}\int_{\theta_{i-1}}^{\theta_i} k_n^j(\overline{\hat{R}}_i, \theta)\sin m\theta d\theta \tag{8.48}
$$

and for the outer surface

$$
\begin{aligned}
&\sum_{m=1}^{\infty}[\overline{F}^{5}_{m0}\overline{A}_{30} + \overline{F}^{6}_{m0}\overline{B}_{30} \\
&+ \sum_{n=1}^{\infty}(\overline{F}^{1}_{mn}\overline{A}_{1n} + \overline{F}^{2}_{mn}\overline{B}_{1n} + \overline{F}^{3}_{mn}\overline{A}_{2n} + \overline{F}^{4}_{mn}\overline{B}_{2n} \\
&+ \overline{F}^{5}_{mn}\overline{A}_{3n} + \overline{F}^{6}_{mn}\overline{B}_{3n} + \overline{F}^{7}_{mn}\overline{A}_{4n} + \overline{F}^{8}_{mn}\overline{B}_{4n} \\
&+ \overline{F}^{9}_{mn}\overline{A}_{5n} + \overline{F}^{10}_{mn}\overline{B}_{5n} + \overline{F}^{11}_{mn}\overline{A}_{6n} + \overline{F}^{12}_{mn}\overline{B}_{6n})]\sin m\theta = 0
\end{aligned}
$$

$$
\begin{aligned}
&\sum_{m=1}^{\infty}[\overline{G}^{5}_{m0}\overline{A}_{30} + \overline{G}^{6}_{m0}\overline{B}_{30} \\
&+ \sum_{n=1}^{\infty}(\overline{G}^{1}_{mn}\overline{A}_{1n} + \overline{G}^{2}_{mn}\overline{B}_{1n} + \overline{G}^{3}_{mn}\overline{A}_{2n} + \overline{G}^{4}_{mn}\overline{B}_{2n} \\
&+ \overline{G}^{5}_{mn}\overline{A}_{3n} + \overline{G}^{6}_{mn}\overline{B}_{3n} + \overline{G}^{7}_{mn}\overline{A}_{4n} + \overline{G}^{8}_{mn}\overline{B}_{4n} \\
&+ \overline{G}^{9}_{mn}\overline{A}_{5n} + \overline{G}^{10}_{mn}\overline{B}_{5n} + \overline{G}^{11}_{mn}\overline{A}_{6n} + G^{12}_{mn}\overline{B}_{6n})]\cos m\theta = 0
\end{aligned}
$$

$$
\begin{aligned}
&\sum_{m=1}^{\infty}[\overline{H}^{5}_{m0}\overline{A}_{30} + \overline{H}^{6}_{m0}\overline{B}_{30} \\
&+ \sum_{n=1}^{\infty}(\overline{H}^{1}_{mn}\overline{A}_{1n} + \overline{H}^{2}_{mn}\overline{B}_{1n} + \overline{H}^{3}_{mn}\overline{A}_{2n} + \overline{H}^{4}_{mn}\overline{B}_{2n} \\
&+ \overline{H}^{5}_{mn}\overline{A}_{3n} + \overline{H}^{6}_{mn}\overline{B}_{3n} + \overline{H}^{7}_{mn}\overline{A}_{4n} + \overline{H}^{8}_{mn}\overline{B}_{4n} \\
&+ \overline{H}^{9}_{mn}\overline{A}_{5n} + \overline{H}^{10}_{mn}\overline{B}_{5n} + \overline{H}^{11}_{mn}\overline{A}_{6n} + \overline{H}^{12}_{mn}\overline{B}_{6n})]\sin m\theta = 0
\end{aligned}
$$

$$
\begin{aligned}
&\sum_{m=1}^{\infty}[\overline{J}^{5}_{m0}\overline{A}_{30} + \overline{J}^{6}_{m0}\overline{B}_{30} \\
&+ \sum_{n=1}^{\infty}(\overline{J}^{1}_{mn}\overline{A}_{1n} + \overline{J}^{2}_{mn}\overline{B}_{1n} + \overline{J}^{3}_{mn}\overline{A}_{2n} + \overline{J}^{4}_{mn}\overline{B}_{2n} \\
&+ \overline{J}^{5}_{mn}\overline{A}_{3n} + \overline{J}^{6}_{mn}\overline{B}_{3n} + \overline{J}^{7}_{mn}\overline{A}_{4n} + \overline{J}^{8}_{mn}\overline{B}_{4n} \\
&+ \overline{J}^{9}_{mn}\overline{A}_{5n} + \overline{J}^{10}_{mn}\overline{B}_{5n} + \overline{J}^{11}_{mn}\overline{A}_{6n} + \overline{J}^{12}_{mn}\overline{B}_{6n})]\sin m\theta = 0
\end{aligned}
$$

$$
\begin{aligned}
&\sum_{m=1}^{\infty}[\overline{K}^{5}_{m0}\overline{A}_{30} + \overline{K}^{6}_{m0}\overline{B}_{30} \\
&+ \sum_{n=1}^{\infty}(\overline{K}^{1}_{mn}\overline{A}_{1n} + \overline{K}^{2}_{mn}\overline{B}_{1n} + \overline{K}^{3}_{mn}\overline{A}_{2n} + \overline{K}^{4}_{mn}\overline{B}_{2n} \\
&+ \overline{K}^{5}_{mn}\overline{A}_{3n} + \overline{K}^{6}_{mn}\overline{B}_{3n} + \overline{K}^{7}_{mn}\overline{A}_{4n} + \overline{K}^{8}_{mn}\overline{B}_{4n} \\
&+ \overline{K}^{9}_{mn}\overline{A}_{5n} + \overline{K}^{10}_{mn}\overline{B}_{5n} + \overline{K}^{11}_{mn}\overline{A}_{6n} + \overline{K}^{12}_{mn}\overline{B}_{6n})]\sin m\theta = 0
\end{aligned}
\tag{8.49}
$$

where

$$\overline{F}_{mn}^{j} = \left(\frac{2\varepsilon_n}{\pi}\right)\sum_{i=1}^{I}\int_{\theta_{i-1}}^{\theta_i} f_n^j(\overline{R}_i,\ \theta)\sin m\theta d\theta,\ \overline{G}_{mn}^{j} = \left(\frac{2\varepsilon_n}{\pi}\right)\sum_{i=1}^{I}\int_{\theta_{i-1}}^{\theta_i} g_n^j(\overline{R}_i,\ \theta)\cos m\theta d\theta,$$

$$\overline{H}_{mn}^{j} = \left(\frac{2\varepsilon_n}{\pi}\right)\sum_{i=1}^{I}\int_{\theta_{i-1}}^{\theta_i} h_n^j(\overline{R}_i,\ \theta)\sin m\theta d\theta,\ \overline{J}_{mn}^{j} = \left(\frac{2\varepsilon_n}{\pi}\right)\sum_{i=1}^{I}\int_{\theta_{i-1}}^{\theta_i} J_n^j(\overline{R}_i,\ \theta)\sin m\theta d\theta,$$

$$\overline{K}_{mn}^{j} = \left(\frac{2\varepsilon_n}{\pi}\right)\sum_{i=1}^{I}\int_{\theta_{i-1}}^{\theta_i} k_n^j(\overline{R}_i,\ \theta)\sin m\theta d\theta \tag{8.50}$$

where $j = 1,2,3,4,5,6,7,8,9$ and 10, $\varepsilon_m = 1/2$ for $m = 0$ and $\varepsilon_m = 1$ for $\geqslant 0$, I is the number of segments, $\hat{\overline{R}}_i$ is the coordinate r at the inner boundary, and R_i is the coordinate r at the outer boundary.

For the symmetric and antisymmetric mode, the frequency equations are obtained for inner and outer boundary conditions of equations (8.43) and (8.45). By truncating the series to $N + 1$ term, the frequency equations are obtained from equations (8.47) and (8.49) and equating the determinant of the coefficients of the amplitudes A_{in}, B_{in}, $\overline{A}_{in}$ and $\overline{B}_{in}$ ($i = 1,2,3,4,5,6$) to zero. Thus the frequency equation for the symmetric mode is obtained in the form of a matrix as follows:

$$\begin{bmatrix}
\bar{\hat{F}}^1_{00} & \bar{\hat{F}}^2_{00} & \cdots & \cdots & \bar{\hat{F}}^8_{00} & \bar{\hat{F}}^1_{01} & \cdots & \bar{\hat{F}}^1_{0N} & \bar{\hat{F}}^2_{01} & \cdots & \bar{\hat{F}}^2_{0N} & \cdots & \cdots & \bar{\hat{F}}^{12}_{01} & \cdots & \bar{\hat{F}}^{12}_{0N} \\
\vdots & \vdots & \vdots & \vdots & \vdots & \vdots & \vdots & \vdots & \vdots & \vdots & \vdots & \cdots & \cdots & \vdots & \vdots & \vdots \\
\bar{\hat{F}}^1_{N0} & \bar{\hat{F}}^2_{N0} & \cdots & \cdots & \bar{\hat{F}}^8_{N0} & \bar{\hat{F}}^1_{N1} & \cdots & \bar{\hat{F}}^1_{NN} & \bar{\hat{F}}^2_{N1} & \cdots & \bar{\hat{F}}^2_{NN} & \cdots & \cdots & \bar{\hat{F}}^{12}_{N1} & \cdots & \bar{\hat{F}}^{12}_{NN} \\
\vdots & \vdots & \vdots & \vdots & \vdots & \vdots & \cdots & \cdots & \cdots & \cdots & \cdots & \cdots & \cdots & \vdots & \cdots & \vdots \\
\vdots & \vdots & \vdots & \vdots & \vdots & \vdots & \cdots & \cdots & \cdots & \cdots & \cdots & \cdots & \cdots & \vdots & \cdots & \vdots \\
\bar{\hat{K}}^1_{00} & \bar{\hat{K}}^2_{00} & \cdots & \cdots & \bar{\hat{K}}^8_{00} & \bar{\hat{K}}^1_{01} & \cdots & \bar{\hat{K}}^1_{0N} & \bar{\hat{K}}^2_{01} & \cdots & \bar{\hat{K}}^2_{0N} & \cdots & \cdots & \bar{\hat{K}}^{12}_{01} & \cdots & \bar{\hat{K}}^{12}_{0N} \\
\vdots & \vdots & \vdots & \vdots & \vdots & \vdots & \vdots & \vdots & \vdots & \vdots & \vdots & \vdots & \vdots & \vdots & \vdots & \vdots \\
\bar{\hat{K}}^1_{N0} & \bar{\hat{K}}^2_{N0} & \cdots & \cdots & \bar{\hat{K}}^8_{N0} & \bar{\hat{K}}^1_{N1} & \cdots & \bar{\hat{K}}^1_{NN} & \bar{\hat{K}}^2_{N1} & \cdots & \bar{\hat{K}}^2_{NN} & \cdots & \cdots & \bar{\hat{K}}^{12}_{N1} & \cdots & \bar{\hat{K}}^{12}_{NN} \\
F^1_{00} & F^2_{00} & \cdots & \cdots & F^8_{00} & F^1_{01} & \cdots & F^1_{0N} & F^2_{01} & \cdots & F^2_{0N} & \cdots & \cdots & F^2_{01} & \cdots & F^2_{0N} \\
\vdots & \vdots & \vdots & \vdots & \vdots & \vdots & \vdots & \vdots & \vdots & \vdots & \vdots & \vdots & \vdots & \vdots & \vdots & \vdots \\
F^1_{N0} & F^2_{N0} & \cdots & \cdots & F^8_{N0} & F^1_{N1} & \cdots & F^1_{NN} & F^2_{N1} & \cdots & F^2_{NN} & \cdots & \cdots & F^2_{N1} & \cdots & F^2_{NN} \\
\vdots & \vdots & \vdots & \vdots & \vdots & \vdots & \vdots & \vdots & \vdots & \vdots & \vdots & \vdots & \vdots & \vdots & \vdots & \vdots \\
\vdots & \vdots & \vdots & \vdots & \vdots & \vdots & \vdots & \vdots & \vdots & \vdots & \vdots & \vdots & \vdots & \vdots & \vdots & \vdots \\
K^1_{00} & K^2_{00} & \cdots & \cdots & K^8_{00} & K^1_{00} & \cdots & K^1_{0N} & K^2_{00} & \cdots & K^1_{N0} & \cdots & \cdots & K^{12}_{00} & \cdots & K^{12}_{0N} \\
\vdots & \vdots & \vdots & \vdots & \vdots & \vdots & \vdots & \vdots & \vdots & \vdots & \vdots & \vdots & \vdots & \vdots & \vdots & \vdots \\
K^1_{N0} & K^2_{N0} & \cdots & \cdots & K^8_{N0} & K^1_{N1} & \cdots & K^1_{NN} & K^2_{N1} & \cdots & K^2_{NN} & \cdots & \cdots & K^{12}_{N1} & \cdots & K^{12}_{NN}
\end{bmatrix}
\begin{bmatrix}
A_{10} \\ B_{10} \\ \vdots \\ A_{80} \\ B_{80} \\ A_{11} \\ \vdots \\ A_{1n} \\ B_{11} \\ \vdots \\ B_{1N} \\ \vdots \\ \vdots \\ B_{121} \\ \vdots \\ \vdots \\ B_{12N}
\end{bmatrix} = 0 \tag{8.51}$$

The frequency equation for the antisymmetric mode is obtained in the form of a matrix as follows:

$$
\begin{bmatrix}
\bar{\hat{F}}^{9}_{00} & \bar{\hat{F}}^{10}_{00} & \cdots & \cdots & \cdots & \bar{\hat{F}}^{1}_{01} & \cdots & \bar{\hat{F}}^{1}_{0N} & \bar{\hat{F}}^{2}_{01} & \cdots & \bar{\hat{F}}^{2}_{0N} & \cdots & \cdots & \bar{\hat{F}}^{12}_{01} & \cdots & \bar{\hat{F}}^{12}_{0N} \\
\vdots & \vdots & \vdots & \vdots & \vdots & \vdots & \vdots & \vdots & \vdots & \vdots & \vdots & \cdots & \cdots & \vdots & \vdots & \vdots \\
\bar{\hat{F}}^{9}_{N0} & \bar{\hat{F}}^{10}_{N0} & \cdots & \cdots & \cdots & \bar{\hat{F}}^{1}_{N1} & \cdots & \bar{\hat{F}}^{1}_{NN} & \bar{\hat{F}}^{2}_{N1} & \cdots & \bar{\hat{F}}^{2}_{NN} & \cdots & \cdots & \bar{\hat{F}}^{12}_{N1} & \cdots & \bar{\hat{F}}^{12}_{NN} \\
\vdots & \vdots & \vdots & \vdots & \vdots & \vdots & \cdots & \cdots & \cdots & \cdots & \cdots & \cdots & \cdots & \vdots & \cdots & \vdots \\
\vdots & \vdots & \vdots & \vdots & \vdots & \vdots & \cdots & \cdots & \cdots & \cdots & \cdots & \cdots & \cdots & \vdots & \cdots & \vdots \\
\bar{\hat{K}}^{1}_{00} & \bar{\hat{K}}^{2}_{00} & \cdots & \cdots & \cdots & \bar{\hat{K}}^{1}_{01} & \cdots & \bar{\hat{K}}^{1}_{0N} & \bar{\hat{K}}^{2}_{01} & \cdots & \bar{\hat{K}}^{2}_{0N} & \cdots & \cdots & \bar{\hat{K}}^{12}_{01} & \cdots & \bar{\hat{K}}^{12}_{0N} \\
\vdots & \vdots & \vdots & \vdots & \vdots & \vdots & \vdots & \vdots & \vdots & \vdots & \vdots & \vdots & \vdots & \vdots & \vdots & \vdots \\
\bar{\hat{K}}^{1}_{N0} & \bar{\hat{K}}^{2}_{N0} & \cdots & \cdots & \cdots & \bar{\hat{K}}^{1}_{N1} & \cdots & \bar{\hat{K}}^{1}_{NN} & \bar{\hat{K}}^{2}_{N1} & \cdots & \bar{\hat{K}}^{2}_{NN} & \cdots & \cdots & \bar{\hat{K}}^{12}_{N1} & \cdots & \bar{\hat{K}}^{12}_{NN} \\
\bar{F}^{1}_{00} & \bar{F}^{2}_{00} & \cdots & \cdots & \cdots & \bar{F}^{1}_{01} & \cdots & \bar{F}^{1}_{0N} & \bar{F}^{2}_{01} & \cdots & \bar{F}^{2}_{0N} & \cdots & \cdots & \bar{F}^{2}_{01} & \cdots & \bar{F}^{2}_{0N} \\
\vdots & \vdots & \vdots & \vdots & \vdots & \vdots & \vdots & \vdots & \vdots & \vdots & \vdots & \vdots & \vdots & \vdots & \vdots & \vdots \\
\bar{F}^{1}_{N0} & \bar{F}^{2}_{N0} & \cdots & \cdots & \cdots & \bar{F}^{1}_{N1} & \cdots & \bar{F}^{1}_{NN} & \bar{F}^{2}_{N1} & \cdots & \bar{F}^{2}_{NN} & \cdots & \cdots & \bar{F}^{2}_{N1} & \cdots & \bar{F}^{2}_{NN} \\
\vdots & \vdots & \vdots & \vdots & \vdots & \vdots & \vdots & \vdots & \vdots & \vdots & \vdots & \vdots & \vdots & \vdots & \vdots & \vdots \\
\vdots & \vdots & \vdots & \vdots & \vdots & \vdots & \vdots & \vdots & \vdots & \vdots & \vdots & \vdots & \vdots & \vdots & \vdots & \vdots \\
\bar{K}^{1}_{00} & \bar{K}^{2}_{00} & \cdots & \cdots & \cdots & \bar{K}^{1}_{00} & \cdots & \bar{K}^{1}_{0N} & \bar{K}^{2}_{00} & \cdots & \bar{K}^{1}_{N0} & \cdots & \cdots & \bar{K}^{12}_{00} & \cdots & \bar{K}^{12}_{0N} \\
\vdots & \vdots & \vdots & \vdots & \vdots & \vdots & \vdots & \vdots & \vdots & \vdots & \vdots & \vdots & \vdots & \vdots & \vdots & \vdots \\
\bar{K}^{1}_{N0} & \bar{K}^{2}_{N0} & \cdots & \cdots & \cdots & \bar{K}^{1}_{N1} & \cdots & \bar{K}^{1}_{NN} & \bar{K}^{2}_{N1} & \cdots & \bar{K}^{2}_{NN} & \cdots & \cdots & \bar{K}^{12}_{N1} & \cdots & \bar{K}^{12}_{NN}
\end{bmatrix}
\begin{bmatrix}
\bar{A}_{50} \\ \bar{B}_{50} \\ \vdots \\ \vdots \\ \vdots \\ \bar{A}_{11} \\ \vdots \\ \bar{A}_{1N} \\ \bar{B}_{11} \\ \vdots \\ \bar{B}_{1N} \\ \vdots \\ \vdots \\ \vdots \\ \bar{B}_{61} \\ \vdots \\ \vdots \\ \bar{B}_{6N}
\end{bmatrix}
= 0. \tag{8.52}
$$

8.6 Results and discussions

The dimensions of each plate used in the numerical calculation are shown in figures 8.2 and 8.3, respectively. The axis of symmetry is denoted by the lines in figures 8.2 and 8.3. The computation of Fourier coefficients given in (8.44), (8.46), (8.48), and (8.50) is carried out using the five-point Gaussian quadrature. The secant method is used to obtain the roots of the frequency equation, for the complex roots. The material properties are considered from Peng-Fei *et al* (2008)

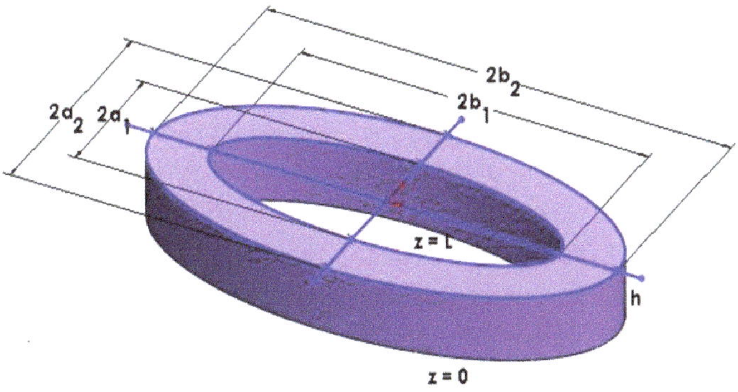

Figure 8.2. Geometry of a ring-shaped elliptical plate.

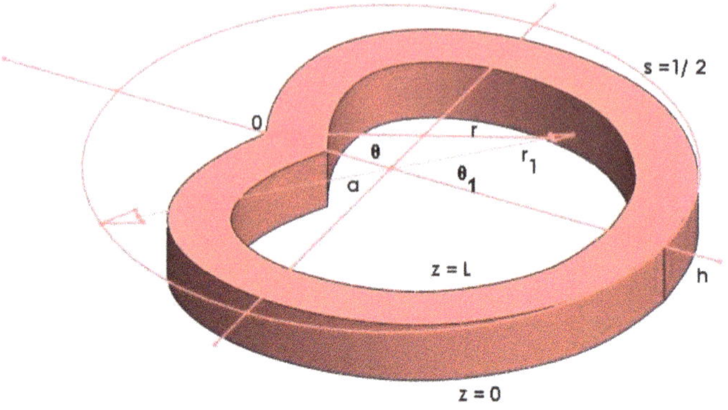

Figure 8.3. Geometry of a ring-shaped cardioid plate.

Table 8.1. Comparison of non-dimensional wave number $|\varsigma|$ of elliptical cross-sectional plate.

	FECM			FEM		
Mode	$\frac{b_1}{b_2} = 0.5$	$\frac{b_1}{b_2} = 1.0$	$\frac{b_1}{b_2} = 1.5$	$\frac{b_1}{b_2} = 0.5$	$\frac{b_1}{b_2} = 1.0$	$\frac{b_1}{b_2} = 1.5$
S1	0.1628	0.1650	0.1677	0.1673	0.1669	0.1665
S2	0.2509	0.2624	0.2527	0.2525	0.2524	0.2523
S3	0.4209	0.4337	0.4193	0.4192	0.4192	0.4193
S4	0.5897	0.6005	0.5860	0.5861	0.5863	0.5865
S5	0.8519	0.8588	0.8354	0.8366	0.8374	0.8382

$c_{11} = 7.41 \times 10^{10}\,\mathrm{N\,m^{-2}}$, $c_{12} = 4.52 \times 10^{10}\,\mathrm{N\,m^{-2}}$,
$c_{13} = 3.93 \times 10^{10}\,\mathrm{N\,m^{-2}}$, $c_{33} = 8.36 \times 10^{10}\,\mathrm{N\,m^{-2}}$,
$c_{44} = 1.322 \times 10^{10}\,\mathrm{N\,m^{-2}}$, $c_{66} = 49 \times 10^{9}\,\mathrm{N\,m^{-2}}$, $e_{31} = -0.16\,\mathrm{Cm^{-2}}$, $e_{33} = 0.347\,\mathrm{Cm^{-2}}$,
$e_{15} = -0.138\,\mathrm{Cm^{-2}}$, $q_{31} = 580.3\,\mathrm{N\,A^{-1}m^{-1}}$, $q_{33} = 699.7\,\mathrm{N\,A^{-1}m^{-1}}$, $q_{15} = 550\,\mathrm{N\,A^{-1}m^{-1}}$,
$\varepsilon_{11} = 8.26 \times 10^{-11}\mathrm{C^{-2}\,N^{-1}m^{-2}}$, $\varepsilon_{33} = 9.03 \times 10^{-11}\mathrm{C^{-2}\,N^{-1}m^{-2}}$, $\mu_{11} = -5 \times 10^{-6}\,\mathrm{N\,s^2\,C^{-2}}$,
$\mu_{33} = 10 \times 10^{-6}\,\mathrm{N\,s^2\,C^{-2}}$, $m_{11} = -3612 \times 10^{-11}\mathrm{Ns^{-1}(VC)^{-1}}$ and
$m_{33} = -2.4735 \times 10^{-11}\mathrm{Ns^{-1}(VC)^{-1}}$.

In the numerical calculation, the angle θ is taken as an independent variable and the coordinate R_i at the ith segment of the boundary is expressed in terms of θ. Substituting R_i and the angle γ_i, between the reference axis and the normal to the ith boundary line, the integrations of the Fourier coefficients f_n^i, g_n^i, h_n^i, j_n^i, k_n^i, $\bar{f}_n^i$, $\bar{g}_n^i$, $\bar{h}_n^i$, $\bar{j}_n^i$ and $\bar{k}_n^i$ can be expressed in terms of the angle θ. Using these coefficients in equations (8.43) and (8.49), we obtained the frequency equations.

The notation S_1, S_2, S_3, S_4, and S_5 used in the table represents the vibration of symmetric modes and the subscripts 1, 2, 3,4 and 5 represent the vibration modes. In table 8.1, a comparison is made between the non-dimensional wave number $|\varsigma|$ of

Table 8.2. Comparison of non-dimensional wave number $|\varsigma|$ of cardioid cross-sectional plate

	FECM			FEM		
Mode	$s = 0.0$	$s = 0.25$	$s = 0.5$	$s = 0.0$	$s = 0.25$	$s = 0.5$
S1	0.1683	0.1683	0.1692	0.1680	0.1678	0.1682
S2	0.2531	0.2531	0.2890	0.2528	0.2527	0.2789
S3	0.4189	0.4189	0.4190	0.4188	0.4188	0.4192
S4	0.5837	0.5837	0.5849	0.5842	0.5846	0.5849
S5	0.8226	0.8226	0.8248	0.8282	0.8307	0.8327

FECM—Fourier expansion collocation method.
FEM—finite element method.

elliptical cross-sectional plates using FECM and FEM with different ratios of semi-minor axis $b_1/b_2 = 0.5,1.0,1.5$. From this, the vibration modes increase the dimensionless wave numbers in the ratios $b_1/b_2 = 0.5,1.0,1.5$. Another spectrum of the non-dimensional wave number $|\varsigma|$ of a cardioid cross-sectional plate with different eccentric values $s = 0.0,0.25,0.5$ is discussed in table 8.2 using FECM and FEM. From table 8.2, it is observed that the modes increase the non-dimensional wave number. Compared with circular cross-section, the dispersion modes get high value in the cardioid cross-section ($s = 0.5$) ($s = 0$). The amplitude of all the modes of a non-dimensional wave number possess high energy and also converge faster in FECM than FEM. The notations used in the figures, namely LMWF, LMWOF, FASMWF and FASMWOF denote the longitudinal mode with fluid, longitudinal mode without fluid, flexural antisymmetric mode with fluid, flexural antisymmetric mode without fluid, respectively.

8.6.1 Elliptic cross-sectional plate

The geometry of an elliptic cross-sectional ring-shaped plate is shown in figure 8.2. The geometrical relations of an elliptic ring-shaped plate given by Nagaya (1981a) are used for numerical calculation and are given below:

$$R_i/b_1 = (a_2/b_1)/[\cos^2\theta + (a_2/b_2)^2\sin^2\theta]^{1/2}$$

$$\gamma_i = \pi/2 - \tan^{-1}[(b_2/a_2)^2/\tan\theta_i^*], \quad \text{for } \theta_i^* < \pi/2$$

$$\gamma_i = \pi/2, \ \theta_i^* = \pi/2$$

$$\gamma_i = \pi/2 + \tan^{-1}[(b_2/a_2)^2/|\tan\theta_i^*|], \quad \text{for } \theta_i^* > \pi/2 \tag{8.53a}$$

for the outer surface and

$$R_i/b_1 = (a_1/b_1)/[\cos^2\theta + (a_1/b_2)^2\sin^2\theta]^{1/2}$$

$$\gamma_i = \pi/2 - \tan^{-1}[(b_1/a_1)^2/\tan\theta_i^*], \quad \text{for } \theta_i^* < \pi/2$$

$$\gamma_i = \pi/2,\ \theta_i^* = \pi/2$$

$$\gamma_i = \pi/2 + \tan^{-1}[(b_1/a_1)^2/|\ \tan\theta_i^*\ |],\quad \text{for}\ \theta_i^* > \pi/2 \tag{8.53b}$$

for the inner surface, where a_1 and a_2 are the length of inner and outer semi-major axis, and b_1 and b_2 are the length of semi-minor axis of an elliptic cross-section. Also $\theta_l^* = (\theta_l + \theta_{l-1})/2$ and R_i is the coordinate r at the $i-$ thboundary, γ_i is the angle between the reference axis and the normal to the segment.

The frequency equation is obtained by choosing both terms of n and m as 0,2,4,6, … in equation (8.51) for the numerical calculations. Since the boundary of the cross-sections namely, elliptic and cardioid are irregular in shape, it is difficult to satisfy the boundary conditions along the curved surface, and hence FECM is applied. In this method, the curved surface, in the range $\theta = 0$ and $\theta = \pi$ is divided into 20 segments, such that the distance between any two segments is negligible and the integration is performed for each segment numerically by using the Gauss five-point formula.

The influence of hoop stresses $\sigma_{\theta\theta}$ along with aspect ratio (a/b) is shown in figures 8.4 and 8.5 for longitudinal, and flexural antisymmetric modes with and without fluid environment of elliptic and cardioid cross-sectional plates. The propagation behaviour is, however, decreasing exponentially in figures 8.4 and 8.5, but in figure 8.5 it becomes dispersed due to the existence of various interacting fields such as magnetic and fluid. Moreover, it is observed that the magnitude of dispersion is higher in lower modes of aspect ratio (*a*/*b*) and also higher in flexural antisymmetric modes, which might happen because the radiation of the energy is high in the transverse direction of the elliptical system. Also, increasing the aspect ratio increases the coupling effect.

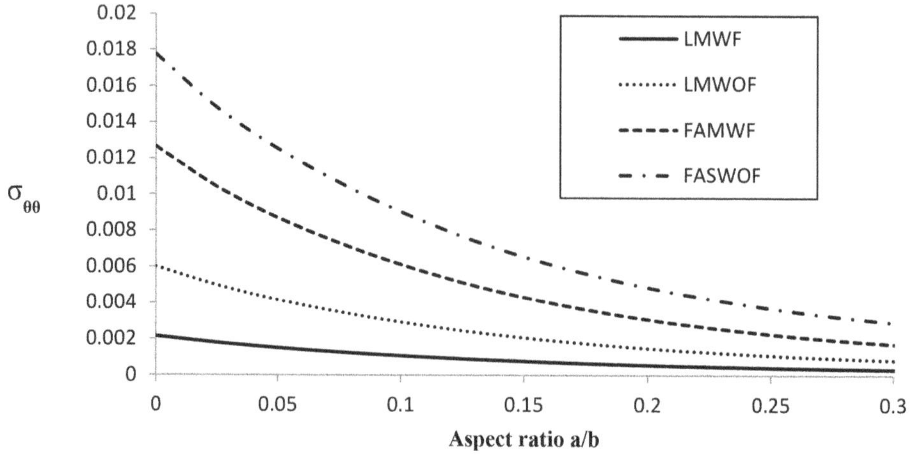

Figure 8.4. Variation of hoop stress $\sigma_{\theta\theta}$ versus aspect ratio (*a*/*b*) of an elliptic plate.

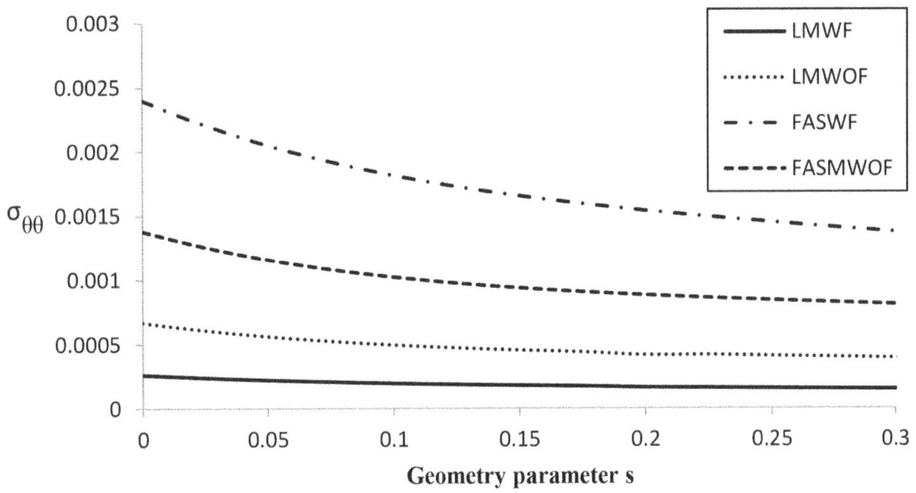

Figure 8.5. Variation of hoop stress $\sigma_{\theta\theta}$ versus geometry parameter s of a cardioid plate.

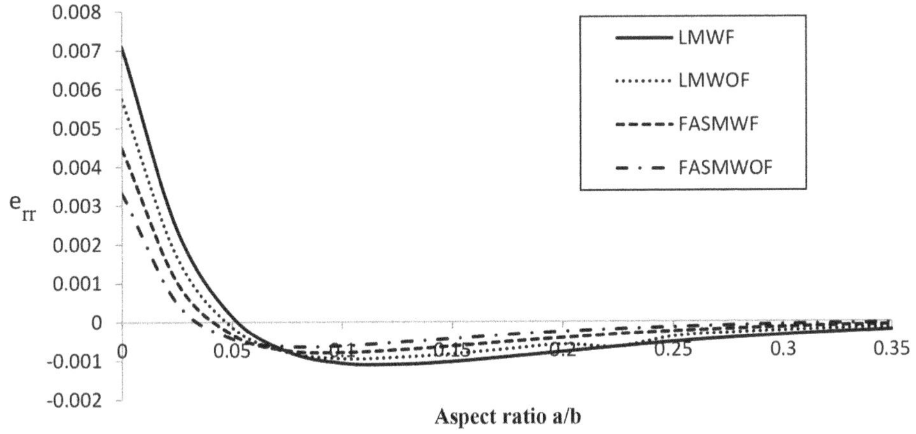

Figure 8.6. Variation of radial strain e_{rr} versus aspect ratio (a/b) of an elliptic plate.

The effect of radial strain e_{rr} along with aspect ratio a/b of magneto-electroelastic elliptical cross-section is considered in figure 8.6 for longitudinal, flexural symmetric and flexural antisymmetric modes with and without fluid. From this, it is observed that the radial strain energy is decreasing by increasing the aspect ratio. In figure 8.7, a graph is drawn between radial strain e_{rr} and geometry of magneto-electroelastic cardioid cross-section for longitudinal, flexural symmetric and flexural antisymmetric modes with and without fluid. It is observed that as the aspect ratio and geometry s increases, the radial strain e_{rr} decreases linearly, but for flexural modes, these curves become dispersive in the range $0 \leqslant s \leqslant 0.05$.

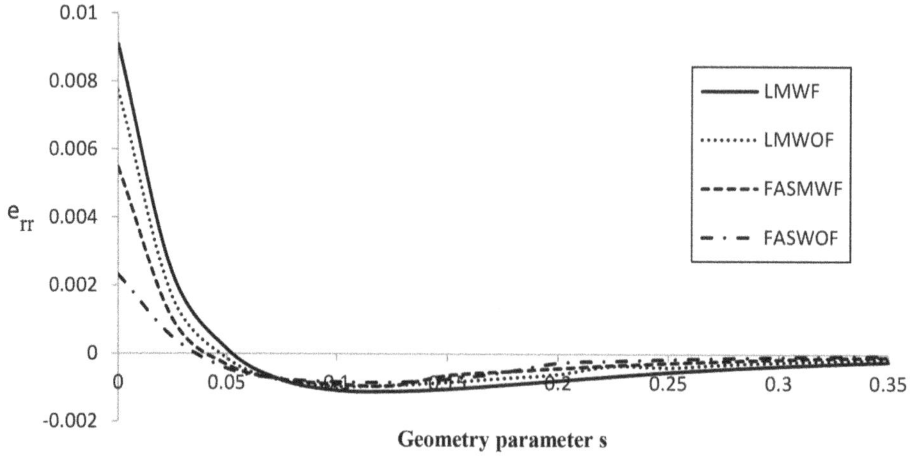

Figure 8.7. Variation of radial strain e_{rr} versus geometry parameter s of a cardioid plate.

8.6.2 Cardioid cross-sectional plate

The cardioid cross-section of the plate is shown in figure 8.3, and its relation used for the numerical calculations is given below taken from equations (8.24) and (8.26) of Nagaya (1983b) as:

$$R_i/a = 1 + s^2 + 2s\cos\theta_1/1 + s \tag{8.54a}$$

$$\theta = \cos^{-1}\cos\theta_1 + s\cos 2\theta_1/(1 + s^2 + 2s\cos\theta_1)^{1/2}$$

where a is the radius of the circumscribing circle and:

$$G(\theta_1) = \cos\theta_1 + 2s\cos 2\theta_1/-\sin\theta_1 - 2s\sin 2\theta_1$$

$$\gamma_i = \pi/2, \quad \text{for} \quad G(\theta_i^*) = 0.$$

$$\gamma_i = \pi/2 - \tan^{-1}(-G(\theta_i^*)), \quad \text{for} \quad G(\theta_i^*) < 0$$

$$\gamma_i = \pi/2 + \tan^{-1}(-G(\theta_i^*)), \quad \text{for} \quad G(\theta_i^*) > 0 \tag{8.54b}$$

where $\theta_i^* = (\theta_i - \theta_{i-1})/2$ is the mean angle of the segment i and R_i is the coordinate r at the boundary, γ_i is the angle between the normal to the segment and the reference axis at the ith boundary. Parameter s in the above relations represents a circle when $s = 0$ and represents a cardioid when $s = 0.5$. In the present problem, there are three kinds of basic independent modes of wave propagation considered, namely the longitudinal and two flexural (symmetric and antisymmetric) modes of vibrations.

The same problem is solved for cardioidal cross-sectional plate using the geometrical relations given in equations (8.24) and (8.26) of Nagaya (1987), which are functions depending on a parameter s. Since a cardioid is symmetric about only one axis, the longitudinal and flexural symmetric modes are carried out by choosing

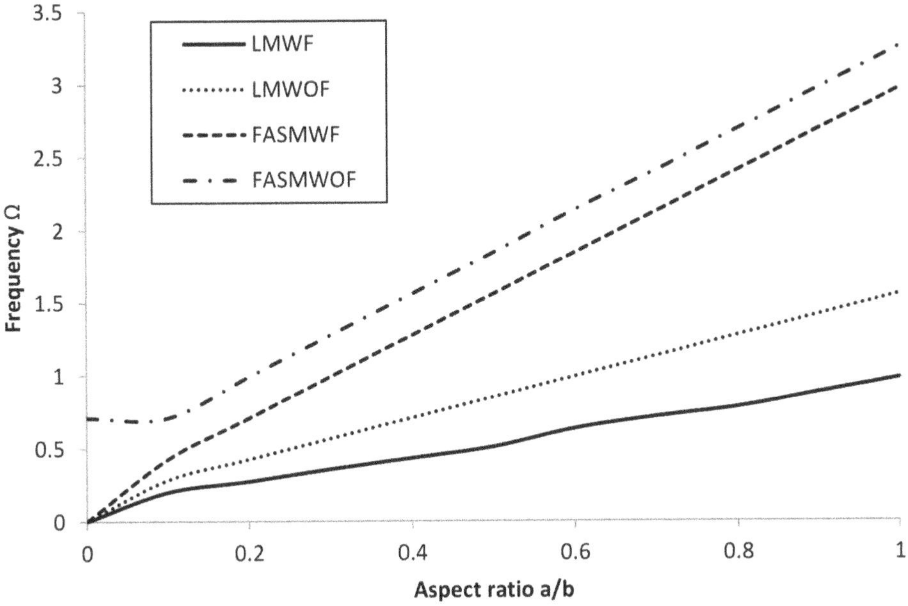

Figure 8.8. Variation of frequency Ω versus aspect ratio (*a*/*b*) of an elliptic plate.

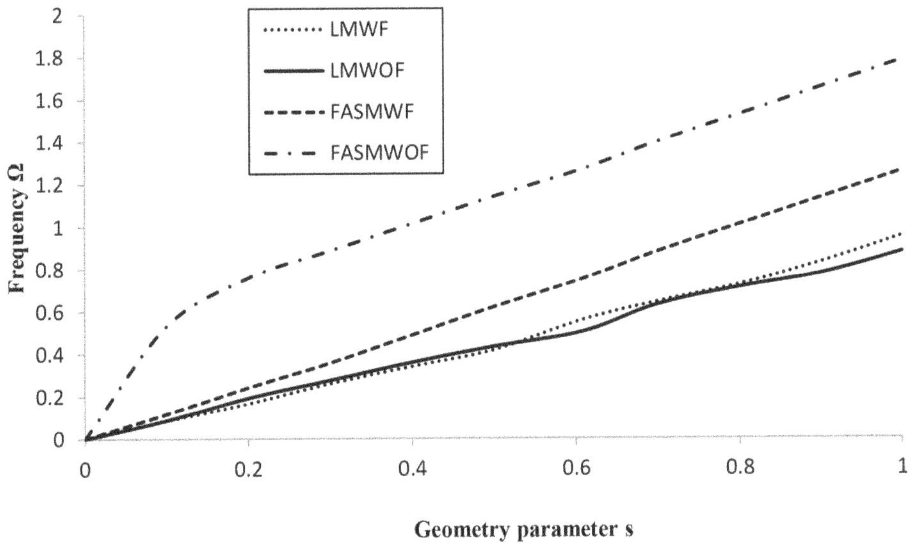

Figure 8.9. Variation of frequency Ω versus geometry parameter *s* of a cardioid plate.

n, $m = 0,1,2,3\ldots$ in equation (8.51) and flexural (antisymmetric) modes are obtained by choosing n, $m = 1,2,3\ldots$ in equation (8.52).

The variation of dimensionless frequency Ω along with aspect ratio (a/b) are shown in figure 8.8 for longitudinal, and flexural antisymmetric modes of elliptic cross-sectional plates and between the non-dimensional frequency Ω and geometry *s* of cardioid cross-sectional plates in figure 8.9. From this it is observed that the

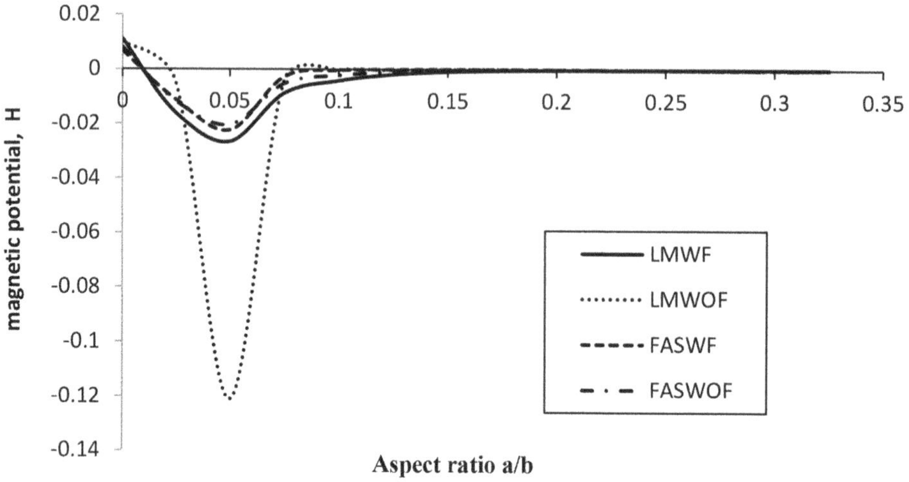

Figure 8.10. Variation of magnetic potential H versus aspect ratio (a/b) of an elliptic plate.

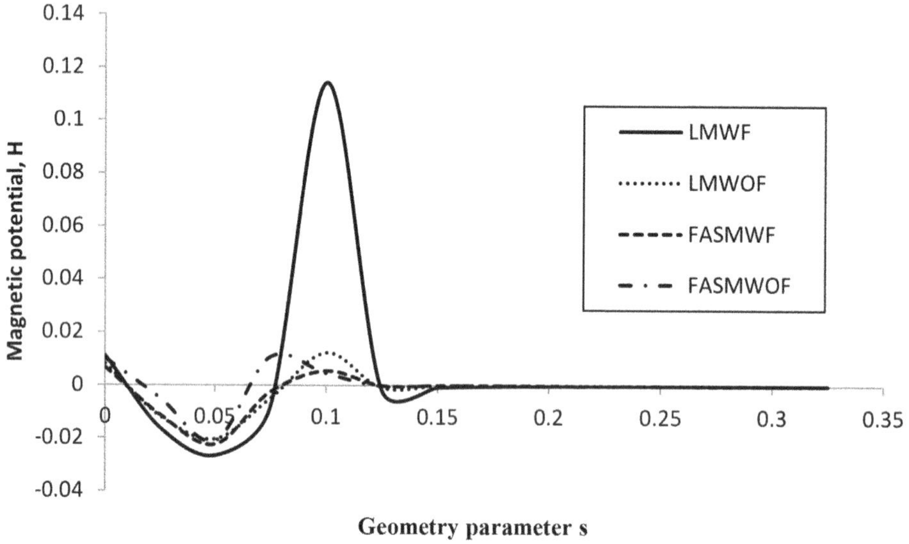

Figure 8.11. Variation of magnetic potential H versus geometry parameter s of a cardioid plate.

dimensionless frequency is increased by increasing the aspect ratio (a/b) and the geometry s.

The effect of magnetic potential H and aspect ratio (a/b) is shown in figures 8.10 and 8.11 for elliptic cross-section and geometry s for cardioids cross-sectional plates for longitudinal, flexural symmetric and flexural antisymmetric modes of vibrations with and without fluid loading. It is clear that the magnetic potential is decreased by the act of aspect ratio (a/b) and the geometry s of elliptic and cardioid cross-sections. In figure 8.10 all the vibration modes are oscillating in the lower range of $0 \leqslant a/b \leqslant 0.1$ aspect ratio, while in figure 8.11 it is $0 \leqslant s \leqslant 0.15$. The transfer of

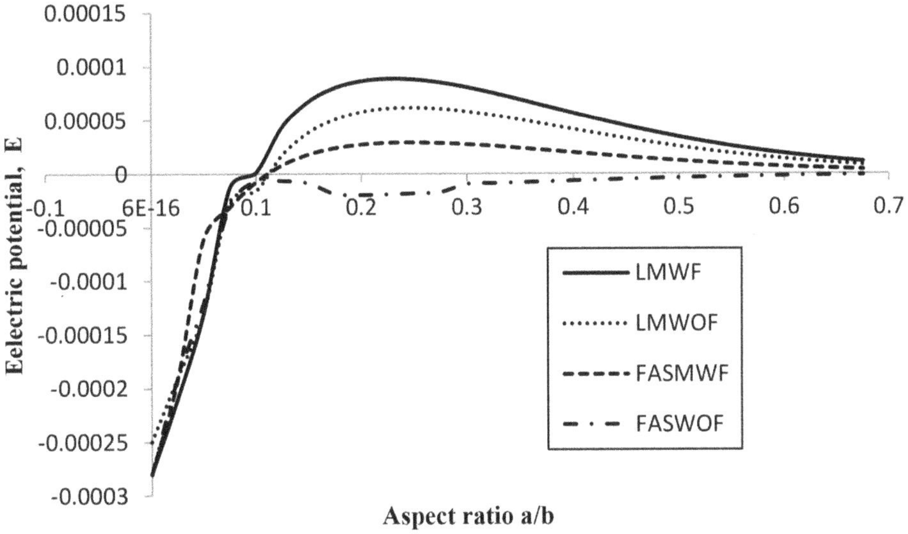

Figure 8.12. Variation of electric potential E versus aspect ratio (a/b) of an elliptic plate.

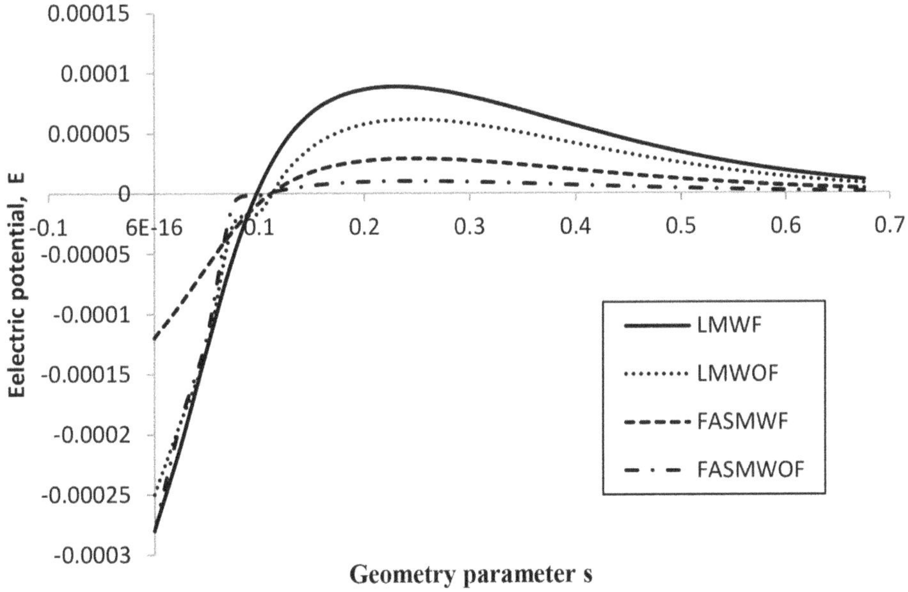

Figure 8.13. Variation of electric potential E versus geometry parameter s of a cardioid plate.

energy is highly oscillated in the cardiod cross-section as compared to the elliptic cross-section. This cross-over point represents the transfer of energy between modes of vibration of elliptic and cardioid plates.

The graphs are plotted between electric potential E and the aspect ratio (a/b) in figures 8.12 and 8.13. The electric potential is less at lower aspect ratio (a/b) and geometry s and attains maximum in the range $0.1 \leqslant a/b \leqslant 0.3$ for an elliptical plate and $0 \leqslant s \leqslant 0.4$ for cardioid cross-sections. From this the electric potential profile gets higher value in longitudinal modes. It is found that the electric potential E of the waves is modified due to the surrounding fluid medium and anisotropic effects.

Appendix A

$$\begin{aligned} f_n^i = & [-2\bar{c}_{66}\cos^2(\theta-\gamma_i)\{n(n-1)J_n(\alpha_i ax)+(\alpha_i ax)J_{n+1}(\alpha_i ax)\}+x^2(\alpha_i a)^2 \\ & \times(\bar{c}_{11}\cos^2(\theta-\gamma_i)+\bar{c}_{12}\sin^2(\theta-\gamma_i))+t_L(\bar{c}_{13}a_i+\bar{e}_{31}b_i+\bar{q}_{31}c_i)\times J_n(\alpha_i ax)] \\ & \cos(m\pi\zeta)\cos n\theta-2n\bar{c}_{66}\{n(n-1)J_n(\alpha_i ax)-(\alpha_i ax)J_{n+1}(\alpha_i ax)\} \\ & \times\sin 2(\theta-\gamma_i)\cos(m\pi\zeta)\sin n\theta,\ i=1,2,3,4 \end{aligned} \tag{A1}$$

$$\begin{aligned} f_n^5 = & [2\bar{c}_{66}\cos^2(\theta-\gamma_i)\{n(n-1)J_n(\alpha_5 ax)-(\alpha_5 ax)J_{n+1}(\alpha_5 ax)\}]\cos(m\pi\zeta) \\ & \cos n\theta+\bar{c}_{66}[2\{n(n-1)J_n(\alpha_5 ax)+(\alpha_5 ax)J_{n+1}(\alpha_5 ax)\}-(\alpha_5 ax)^2J_n(\alpha_5 ax)] \\ & \times\sin 2(\theta-\gamma_i)\cos(m\pi\zeta)\sin n\theta \end{aligned} \tag{A2}$$

$$f_n^6 = \Omega^2\bar{\rho}H_n^{(2)}(\delta ax)\cos n\theta \tag{A3}$$

$$\begin{aligned} f_n^i = & [-2\bar{c}_{66}\cos^2(\theta-\gamma_i)\{n(n-1)Y_n(\alpha_i ax)+(\alpha_i ax)Y_{n+1}(\alpha_i ax)\}+x^2(\alpha_i a)^2\times \\ & (\bar{c}_{11}\cos^2(\theta-\gamma_i)+\bar{c}_{12}\sin^2(\theta-\gamma_i))+t_L(\bar{c}_{13}a_i+\bar{e}_{31}b_i+\bar{q}_{31}c_i)\times Y_n(\alpha_i ax)] \\ & \cos(m\pi\zeta)\cos n\theta-2n\bar{c}_{66}\{n(n-1)Y_n(\alpha_i ax)-(\alpha_i ax)Y_{n+1}(\alpha_i ax)\} \\ & \times\sin^2(\theta-\gamma_i)\cos(m\pi\zeta)\sin n\theta,\ i=6,7,8,9,10,11 \end{aligned} \tag{A4}$$

$$\begin{aligned} f_n^{12} = & [2\bar{c}_{66}\cos^2(\theta-\gamma_i)\{n(n-1)Y_n(\alpha_{12}ax)-(\alpha_{12}ax)Y_{n+1}(\alpha_{12}ax)\}] \\ & \cos(m\pi\zeta)\cos n\theta+\bar{c}_{66}[2\{n(n-1)Y_n(\alpha_{12}ax)+(\alpha_{12}ax)Y_{n+1}(\alpha_{12}ax)\}-(\alpha_{12}ax)^2Y_n(\alpha_{12}ax)] \\ & \times\sin 2(\theta-\gamma_i)\cos(m\pi\zeta)\sin n\theta \end{aligned} \tag{A5}$$

$$\begin{aligned} g_n^i = & [-2\{n(n-1)J_n(\alpha_i ax)+(\alpha_i ax)J_{n+1}(\alpha_i ax)\}+(\alpha_i ax)^2J_n(\alpha_i ax)] \\ & \cos(m\pi\zeta)\cos n\theta\sin 2(\theta-\gamma_i)+2n\{(n-1)J_n(\alpha_i ax)-(\alpha_i ax)J_{n+1}(\alpha_i ax)\} \\ & \times\cos^2(\theta-\gamma_i)\cos(m\pi\zeta)\sin n\theta,\ i=1,2,3,4 \end{aligned} \tag{A6}$$

$$\begin{aligned} g_n^5 = & [2n\{(n-1)J_n(\alpha_5 ax)+(\alpha_5 ax)J_{n+1}(\alpha_5 ax)\}]\cos(m\pi\zeta)\cos n\theta\sin 2(\theta-\gamma_i) \\ & +[-2\{n(n-1)J_n(\alpha_5 ax)+(\alpha_5 ax)J_{n+1}(\alpha_5 ax)\}+(\alpha_5 ax)^2J_n(\alpha_5 ax)] \\ & \cos^2(\theta-\gamma_i)\cos(m\pi\zeta)\sin n\theta \end{aligned} \tag{A7}$$

$$\begin{aligned} g_n^6 = & [2n\{(n-1)J_n(\alpha_5 ax)-(\alpha_5 ax)J_{n+1}(\alpha_5 ax)\}]\cos(m\pi\zeta) \\ & \cos n\theta\sin 2(\theta-\gamma_i)+[-2\{n(n-1)J_n(\alpha_5 ax)+(\alpha_5 ax)J_{n+1}(\alpha_5 ax)\} \\ & +(\alpha_5 ax)^2J_n(\alpha_5 ax)]\cos 2(\theta-\gamma_i)\cos(m\pi\zeta)\sin n\theta \end{aligned} \tag{A8}$$

$$\begin{aligned} g_n^i = & [-2\{n(n-1)Y_n(\alpha_i ax)+(\alpha_i ax)Y_{n+1}(\alpha_i ax)\}+(\alpha_i ax)^2Y_n(\alpha_i ax)] \\ & \cos(m\pi\zeta)\cos n\theta\sin 2(\theta-\gamma_i)+2n\{(n-1)Y_n(\alpha_i ax)-(\alpha_i ax)Y_{n+1}(\alpha_i ax)\} \\ & \cos 2(\theta-\gamma_i)\cos(m\pi\zeta)\sin n\theta \end{aligned} \tag{A9}$$

$$g_n^{12} = [2n\{(n-1)Y_n(\alpha_{12}ax) - (\alpha_{12}ax)Y_{n+1}(\alpha_{12}ax)\}]$$
$$\cos(m\pi\zeta)\cos n\theta \sin 2(\theta - \gamma_i) + [-2\{n(n-1)Y_n(\alpha_{12}ax) + (\alpha_{12}ax)Y_{n+1}(\alpha_{12}ax)\}] \quad \text{(A10)}$$
$$\cos 2(\theta - \gamma_i)\cos(m\pi\zeta)\sin n\theta$$

$$h_n^i = [(t_L + a_i)\cos(\theta - \gamma_i) + \bar{e}_{15}b_i + \bar{q}_{31}c_i][\{nJ_n(\alpha_i ax) + (\alpha_i ax)J_{n+1}(\alpha_i ax)\}]$$
$$\sin(m\pi\zeta)\cos n\theta + (t_L + a_i)n\{J_n(\alpha_i ax)\} \times \sin(\theta - \gamma_i)\sin(m\pi\zeta)\sin n\theta \; i = 1, 2, 3, 4 \quad \text{(A11)}$$

$$h_n^5 = [-nt_L J_n(\alpha_{10}ax)\sin(m\pi\zeta)$$
$$\cos n\theta \cos(\theta - \gamma_i) - \{nJ_n(\alpha_{10}ax) - (\alpha_{10}ax)J_{n+1}(\alpha_{10}ax)\}t_L \sin(\theta - \gamma_i)\sin(m\pi\zeta)\sin n\theta \quad \text{(A12)}$$

$$h_n^i = [(t_L + a_i)\cos(\theta - \gamma_i) + \bar{e}_{15}b_i + \bar{q}_{31}c_i][\{nY_n(\alpha_i ax) - (\alpha_i ax)Y_{n+1}(\alpha_i ax)\}]$$
$$\sin(m\pi\zeta)\cos n\theta + (t_L + a_i)n\{Y_n(\alpha_i ax)\} \times \sin(\theta - \gamma_i)\sin(m\pi\zeta)\sin n\theta \; i = 6, 7, 8, 9, 10, 11 \quad \text{(A13)}$$

$$h_n^{12} = [-nt_L Y_n(\alpha_{12}ax)\sin(m\pi\zeta)$$
$$\cos n\theta \cos(\theta - \gamma_i) - \{nY_n(\alpha_{12}ax) - (\alpha_{12}ax)Y_{n+1}(\alpha_{12}ax)\}t_L \sin(\theta - \gamma_i)\sin(m\pi\zeta)\sin n\theta \quad \text{(A14)}$$

$$j_n^i = [\bar{e}_{15}(t_L + a_i) - \bar{\varepsilon}_{11}b_i - \bar{m}_{11}c_i][\{nJ_n(\alpha_i ax) - (\alpha_i ax)J_{n+1}(\alpha_i ax)\}]$$
$$\sin(m\pi\zeta)\cos n\theta, \; i = 1, 2, 3, 4 \quad \text{(A15)}$$

$$j_n^5 = -\bar{e}_{15}nt_L J_n(\alpha_5 ax) \quad \text{(A16)}$$

$$j_n^i = [\bar{e}_{15}(t_L + a_i) - \bar{\varepsilon}_{11}b_i - \bar{m}_{11}c_i][\{nY_n(\alpha_i ax) - (\alpha_i ax)Y_{n+1}(\alpha_i ax)\}]$$
$$\sin(m\pi\zeta)\cos n\theta, \; i = 6, 7, 8, 9, 10, 11 \quad \text{(A17)}$$

$$j_n^{12} = -\bar{e}_{15}nt_L Y_n(\alpha_{12}ax) \quad \text{(A18)}$$

$$k_n^i = [\bar{q}_{15}(t_L + a_i) - \bar{m}_{11}b_i - \bar{\mu}_{11}c_i]$$
$$[\{nJ_n(\alpha_i ax) - (\alpha_i ax)J_{n+1}(\alpha_i ax)\}], \; i = 1, 2, 3, 4 \quad \text{(A19)}$$

$$k_n^5 = -\bar{q}_{15}nt_L J_n(\alpha_5 ax) \quad \text{(A20)}$$

$$k_n^6 = \Omega^2 \rho'\{nH_n^{(2)}(\delta ax) - (\delta ax)H_{n+1}^{(2)}(\delta ax)\}\sin n\theta \quad \text{(A21)}$$

$$k_n^i = [\bar{q}_{15}(t_L + a_i) - \bar{m}_{11}b_i - \bar{\mu}_{11}c_i]$$
$$[\{nY_n(\alpha_i ax) - (\alpha_i ax)Y_{n+1}(\alpha_i ax)\}], \; i = 6, 7, 8, 9, 10, 11 \quad \text{(A22)}$$

$$k_n^{12} = -\bar{q}_{15}nt_L Y_n(\alpha_{12}ax). \quad \text{(A23)}$$

This chapter was used with permission of Emerald Publishing Ltd, from Selvamani (2018); permission conveyed through Copyright Clearance Center, Inc.

References

Annigeri A R, Ganesan N and Swarnamani S 2006 Free vibration of clamped–clamped magneto-electro-elastic cylindrical shells *J. Sound Vib.* **292** 300–14

Annigeri A R, Ganesan N and Swarnamani S 2007a Free vibration behavior of multiphase and layered magneto-electro-elastic beam *J. Sound Vib.* **299** 44–63

Annigeri A R, Ganesan N and Swarnamani S 2007b Free vibrations of simply supported layered and multiphase magneto-electro-elastic cylindrical shells *Smart Mater. Struct.* **15** 459–67

Easwaran V and Munjal M L 1995 A note on the effect of wall compliance on lowest-order mode propagation in a fluid-filled/submerged impedance tubes *J. Acoust. Soc. Am.* **97** 3494–501

Ech-Cherif El-Kettani M, Luppe F and Guillet A 2004 Guided waves in a plate with linearly varying thickness: experimental and numerical results *Ultrasonics* **42** 807–12

Feng W J and Pan E 2008 Dynamic fracture behavior of an internal interfacial crack between two dissimilar magneto-electro-elastic plates *J. Eng. Fract. Mech.* **75** 1468–87

Guo J H, Chen J Y and Pan E 2016 Analytical three-dimensional solutions of anisotropic multilayered composite plates with modified couple-stress effect *Compos. Struct.* **153** 321–31

Guo S H 2011 The thermo-electromagnetic waves in piezoelectric solids *Acta Mech.* **219** 231–40

Hon P F, Leung A Y and Ding H J 2008 A point heat source on the surface of a semi-infinite transversely isotropic electro-magneto-thermo-elastic material *Int. J. Eng. Sci.* **46** 273–85

Jeong H and Park M C 2005 Finite-element analysis of laser-generated ultrasounds for wave propagation and interaction with surface breaking cracks *Res. Nondestr. Eval.* **16** 1–14

Kuang Z B 2009 Variational principles for generalized dynamical theory of thermo piezo-electricity *Acta Mech.* **203** 1–11

Moser F, Jacobs L J and Qu J 1999 Modeling elastic wave propagation in waveguides with the finite element method *NDT & E Int.* **32** 225–34

Nagaya K 1981a Simplified method for solving problems of plates of doubly connected arbitrary shape. Part I: derivation of the frequency equation *J. Sound Vib.* **74** 543–51

Nagaya K 1981b Simplified method for solving problems of plates of doubly connected arbitrary shape, Part II: applications and experiments *J. Sound Vib.* **74** 553–64

Nagaya K 1983a Dispersion of elastic waves in bar with polygonal cross-section *J. Acoust. Soc. Am.* **70** 763–70

Nagaya Y 1983b Dynamic response of a transversely isotropic solid bar of cardioid cross section immersed in a fluid *Appl. Math. Inf. Sci.* **8** 2909–19

Nagaya K 1983c Vibration of a thick walled pipe or ring of arbitrary shape in its plane *J. Appl. Mech.* **50** 757–64

Nagaya Y and Nakamura K 1987 Artificial radionuclides in the western Northwest Pacific (II): 137Cs and 239, 240 Pu inventories in water and sediment columns observed from 1980 to 1986 *J. Oceanogr. Soc. Japan* **43** 345–55

Nagaya K 1983d Vibration of a thick polygonal ring in its plane *J. Acoust. Soc. Am.* **74** 1441–7

Pan E and Heyliger P R 2003 Exact solutions for magneto-electro-elastic laminates in cylindrical bending *Int. J. Solids Struct.* **40** 6859–76

Peng-Fei H, Leung A Y T and Ding H-J 2008 Point heat source on the surface of a semi-infinite transversely isotropic electro-magneto-thermo-elastic material *Int. J. Eng. Sci.* **46** 273–85

Ponnusamy P 2007 Wave propagation in a generalized thermo elastic solid cylinder of arbitrary cross-section *Int. J. Solids Struct.* **44** 5336–48

Ponnusamy P and Selvamani R 2012 Dispersion analysis of generalized magneto-thermoelastic waves in a transversely isotropic cylindrical panel *J. Therm. Stresses* **35** 1119–42

Selvamani R and Ponnusamy P 2015 Wave propagation in a generalized piezothermoelastic rotating bar of circular cross section Multidiscip. Model MA**11** 216–37

Selvamani R 2017 Stress waves in a generalized thermo elastic polygonal plate of inner and outer cross sections *J. Solid Mech.* **9** 263–75

Selvamani R 2017 Modeling of elastic waves in a fluid loaded and immersed piezoelectric hollow fiber *Int. J. Appl. Comput. Math.* **3** 3263–77

Selvamani R 2018 Dispersion analysis of magneto-electro elastic plate of arbitrary cross-sections immersed in fluid *World J. Eng.* **15** 130–47

Chapter 9

Dispersion of thermomechanical waves in a non-homogeneous piezoelectric doubly connected polygonal resonator plate using a dual-phase lagging model

The wave propagation of non-homogeneous waves in a visco-thermoelastic piezoelectric resonator plate is investigated. The stress–strain equations are formulated using the non-homogeneous form of three-dimensional linear elasticity theory. The solution of the problem is derived via the first and second types of Bessel function. The irregular boundary conditions are evaluated using the Fourier expansion collocation method. The numerical results are carried out for various shapes of resonator plates such as triangle, square, pentagon and hexagon. The dispersion curves are presented for the physical variables.

9.1 Introduction

This section was reproduced from Infant Sujitha and Selvamani (2020). Copyright IOP Publishing Ltd. CC BY 3.0.

Resonators vibrate at high frequency, and resonator plates are used in vehicle engines to control the high frequency of sound. The only disadvantage of this is that it is not adjustable. They are also used in biological sensing, motion sensing, signal filtering and a microelectromechanical system oscillator. The vibration analysis of a non-homogeneous polygonal plate was analysed by Ponnusamy and Amuthalakshmi (2014). The behaviour of stress waves in a thermoelastic polygonal plate was considered by Selvamani (2017). The behaviour of a magneto-electroelastic plate was reported by Wu *et al* (2008). The fundamental solution of thermo-visco-elasticity was derived using the finite element method by Othman and Abbas (2012). In ancient

doi:10.1088/978-0-7503-5638-1ch9

days, the effect of thermoelasticity and the irreversibility on thermodynamics were introduced by Biot (1956). The decay and growth of an anistropic non-homogeneous thermo-viscoelastic medium was derived by Tarhan (1980). The transient waves were discussed in Shahin and Asad (2013). Rajneesh *et al* (2012) referred to Othman and Abbas (2012) and derived the effect of viscocity on the medium considered by Tarhan (1980) using the three-phase-log model. The mathematical model for thermoelctroelastic waves in a circular fibre was developed by Ponnusamy and Amuthalakshmi (2016). The wave propagation of thermo-viscoelasticity in a circular micro-plate was studied by Guo *et al* (2014). In this chapter, the visco-thermoelasticity of the non-homogeneous piezoelectric resonator plate of polygonal shape is discussed using the linear elasticity theory and the numerical computations are carried out for the polygonal shape plates.

9.2 Formulation of the problem

The stress–strain model of linear elasticity theory is given in cylindrical variables (r, θ, z) as

$$\prod_{rr,\,r} + r^{-1}\prod_{r\theta,\,\theta} + r^{-1}\left(\prod_{rr} - \prod_{\theta\theta}\right) = \rho u_{r,\,tt} \tag{9.1}$$

$$\prod_{r\theta,\,r} + r^{-1}\prod_{\theta\theta,\,\theta} + 2r^{-1}\prod_{r\theta} = \rho u_{\theta,\,tt} \tag{9.2}$$

$$D_{r,\,r} + r^{-1}D_r + r^{-1}D_{\theta,\,\theta} = 0. \tag{9.3}$$

The generalized heat conduction equation in visco-thermoelasticity by Guo *et al* (2014) is,

$$\begin{aligned} &K\left[\begin{array}{l} (T_{rr} + 1/rT_{,r} + 1/r^2T_{,\theta\theta}) + \\ \tau_T\dfrac{\partial}{\partial t}(T_{,rr} + 1/rT_{,r} + 1/r^2T_{,\theta\theta}) \end{array}\right] - \rho C_v(T_{,r} + \tau_q T_{,tt}) \\ &+ \beta T_0\,\nabla^2\left[\begin{array}{l} \dfrac{\partial}{\partial t}\left(u_{r,\,r} + \dfrac{1}{r(u_{\theta,\,\theta} + u_r)}\right) \\ +\tau_q\dfrac{\partial^2}{\partial t^2}\left(u_{r,\,r} + 1/r(u_{\theta,\,\theta} + u_r)\right) \end{array}\right] \end{aligned} \tag{9.4}$$

where

$$\begin{aligned} &\prod_{rr} = (\lambda + 2\mu)e_{rr} + \lambda e_{\theta\theta} - \beta T; \\ &\prod_{r\theta} = 2\mu e_{r\theta}; \prod_{\theta\theta} = \lambda e_{rr} + (\lambda + 2\mu)e_{\theta\theta} - \beta T. \end{aligned} \tag{9.5}$$

Here $\prod_{rr}$, $\prod_{r\theta}$, $\prod_{\theta\theta}$ and e_{rr}, $e_{r\theta}$, $e_{\theta\theta}$ denotes stress and strain variables, T denotes temperature, ρ stands for mass density, C_v denotes specific heat capacity, β is thermal capacity, K stands for thermal conductivity of the material, λ, μ denotes Lamé's constants. Equation (9.3) represents the electric conduction of the considered plate with the components D_r, D_θ. In (9.4) τ_T, τ_q are the temperature phase lagging and the heat flux parameter.

The strain components are given by,

$$e_{rr} = u_{r,\,r};\;\; e_{r\theta} = u_{\theta,\,r} - \left(\frac{1}{r}\right)(u_\theta - u_{r,\,\theta});\, e_{\theta\theta} = (1/r)(u_r + u_{\theta,\,\theta}). \tag{9.6}$$

The electric conduction equation (9.3) over the electric field $\Im_r$, $\Im_{,\theta}$ are related to the electric potential $\Im$ as

$$\Im = -\Im_{,r};\; \Im_\theta = -r^{-1}\Im_{,\theta}. \tag{9.7}$$

All the above constants and the mass density ρ can be expressed for the non-homogeneity of the material with the rational number m as the functions of radial coordinates and are as follows

$$\lambda = \lambda r^{2m},\; \mu = \mu r^{2m},\;\; \varepsilon_{11} = \varepsilon_{11} r^{2m},\; \rho = \rho r^{2m},\; \beta = \beta r^{2m},\; K = Kr^{2m}. \tag{9.8}$$

Using (9.5)–(9.8) and applying the non-homogeneity via (9.1)–(9.4), we get

$$\begin{aligned}(\lambda + 2\mu)&\begin{bmatrix} u_{r,\,r} \\ +1/r(u_{r,\,r} + u_{\theta,\,r\theta}) \\ -1/r^2 u_r \end{bmatrix} + 1/r^2\begin{bmatrix} 2\mu u_{\theta,\,r\theta} \\ -(\lambda + 4\mu)u_{\theta,\,\theta} \end{bmatrix} \\ &+ 2m\begin{bmatrix} (\lambda + 2\mu)u_{r,\,r} \\ +\dfrac{\lambda}{r(2u_r + u_{\theta,\,\theta})} \end{bmatrix} - \beta T_{,r} = \rho u_{r,\,tt}\end{aligned} \tag{9.9}$$

$$\begin{aligned}2\mu&\begin{bmatrix} u_{\theta,\,r} \\ +1/r u_{\theta,\,r} \\ -1/r^2(u_\theta - u_{\theta,\,\theta\theta} \end{bmatrix} + 1/r^2\begin{bmatrix} \lambda u_{\theta,\,\theta\theta} + (\lambda + 4\mu)u_{r,\,\theta} \\ +r(\lambda + 2\mu)u_{r,\,r\theta} \end{bmatrix} \\ &+ 4m/r\mu\begin{bmatrix} u_{e,\,r} - \\ \dfrac{1}{r(u_\theta - u_{r,\,\theta})} \end{bmatrix} - \beta T_{,\theta} = \rho u_{\theta,\,tt}\end{aligned} \tag{9.10}$$

$$\varepsilon_{11}(\Im_{rr} + 1/r\Im_{,r} + 1/r^2\Im_{,\theta\theta}) + 2m/r\varepsilon_{11}\Im_{,r} = 0 \tag{9.11}$$

$$\begin{aligned}K&\begin{bmatrix} (T_{rr} + 1/rT_{,r} + 1/r^2 T_{,\theta\theta}) + \\ \tau_T\dfrac{\partial}{\partial t}(T_{,rr} + 1/rT_{,r} + 1/r^2 T_{,\theta\theta}) \end{bmatrix} - \rho C_v(T_{,r} + \tau_q T_{,tt}) \\ &+ \beta T_0\,\nabla^2\begin{bmatrix} \dfrac{\partial}{\partial t}\left(u_{r,\,r} + \dfrac{1}{r(u_{\theta,\,\theta} + u_r)}\right) \\ +\tau_q\dfrac{\partial^2}{\partial t^2}\left(u_{r,\,r} + \dfrac{1}{r(u_{\theta,\,\theta} + u_r)}\right) \end{bmatrix} = 0.\end{aligned} \tag{9.12}$$

9.3 Solution of the problem

The above coupled solutions can be uncoupled by considering the following form:

$$u_{r,\theta}(r,\theta) = \sum_{n=0}^{\infty} \varepsilon_n[((1/r)\psi_{n,\theta} - \phi_{n,r}) + (1/r)\overline{\psi}_{n,\theta} - \overline{\phi}_{n,r})\ ]e^{i\omega t}$$

$$u_\theta(r,\theta) = \sum_{n=0}^{\infty} \varepsilon_n[((1/r)\phi_{n,\theta} - \psi_{n,r}) + (1/r)\overline{\phi}_{n,\theta} - \overline{\psi}_{n,r})]e^{i\omega t}$$

$$\mathfrak{I}(r,\theta) = \sum_{n=0}^{\infty} \varepsilon_n(\mathfrak{I}_n + \overline{\mathfrak{I}}_n)e^{i\omega t}$$

$$T(r,\theta) = (\lambda + 2\mu/a^2\beta)\sum_{n=0}^{\infty} \varepsilon_n(T_n + \overline{T}_n)e^{i\omega t} \tag{9.13}$$

where $\varepsilon_n = 1/2$ for $n = 0$ and $\varepsilon_n = 1$ for $n \geqslant 1$ and ω is the angular frequency, $\phi_n(r,\theta)$, $\psi_n(r,\theta)E_n(r,\theta)$, $T_n(r,\theta)$ are the displacement potentials. The bar symbol of the displacement potentials denote the antisymmetric modes of vibrations.

To get the solution substituting (9.13) in (9.9)–(9.12), we get

$$\lambda + 2\mu)(\nabla^2\phi_n - a^{-2}T_n) + 2m\left(\left(\frac{1}{r}\right)(\lambda + 2\mu)\phi_{n,r} - \left(\frac{1}{r^{-2}}\right)\lambda\phi_n\right) + \rho\phi_{n,tt} = 0 \tag{9.14}$$

$$\mu(\nabla^2\psi_n + a^{-2}T_n) + 2\mu m\left(\left(\frac{1}{r}\right)(\psi_{n,r} - \psi_n)\right) = 0 \tag{9.15}$$

$$\varepsilon_{11}\ \nabla^2\ \mathfrak{I}_n + 2mr^{-1}\varepsilon_{11}\mathfrak{I}_{n,r} = 0 \tag{9.16}$$

$$K[(1 + \tau_T)\ \nabla^2\ T_n] - i\omega\rho C_v(1 + i\tau_q)T_n + \beta T_0 i\omega\ (1 + i\tau_q)\ \nabla^4\ \phi_n = 0. \tag{9.17}$$

We consider the following solution which is suitable for the non-homogeneous resonator plate of polygonal shape.

$$\phi_n(r,\theta,t) = r^{-m}\phi_n(r)\cos n\theta e^{i\omega t}$$

$$\psi_n(r,\theta,t) = r^{-m}\psi_n(r)\cos n\theta e^{i\omega t}$$

$$E_n(r,\theta,t) = r^{-m}E_n(r)\cos n\theta e^{i\omega t}$$

$$T_n(r,\theta,t) = r^{-m}T_n(r)\cos n\theta e^{i\omega t}. \tag{9.18}$$

By substituting (9.18) in (9.14)–(9.17) and eliminating T_n and we get,

$$\begin{aligned}&\phi_n''(r) + 1/r\phi_n'(r) + (\omega^2\rho(\lambda + 2\mu)^{-1}\\ &- 1/r^2((m^2 + n^2) + 2m\lambda(\lambda + 2\mu)^{-1}))\phi_n(r) = 0\end{aligned} \tag{9.19}$$

$$\phi_n''(r) + 1/r\phi_n'(r) + (k^2r^2 - \zeta^2)\phi_n(r) = 0. \tag{9.20}$$

Equation (9.20) is a Bessel equation of order ζ

$$\phi_n(r) = [P_{1n}J_\zeta(kr) + P_{1n}'Y_\zeta(kr)]\cos n\theta \tag{9.21}$$

$$\psi_n''(r) + 1/r\psi_n'(r) + (\mu^{-1}\rho\omega^2 - 1/r^2(4m^2 + 4m + n^2))\psi_n(r) = 0 \tag{9.22}$$

$$\psi_n''(r) + \frac{1}{r\psi_n(r)} + (\gamma^2r^2 - \xi^2)\psi_n(r) = 0. \tag{9.23}$$

Equation (9.23) is a Bessel equation of order ξ

$$\psi_n(r) = [P_{2n}J_\xi(\gamma r) + P_{2n}'Y_\xi(\gamma r)]\cos n\theta \tag{9.24}$$

$$\Im_n''(r) + 1/r\Im_n'(r) - (1/r^2(m^2 + n^2))\Im_n(r) = 0. \tag{9.25}$$

The solution of (9.25) is,

$$\Im_n(r) = (P_{3n}r^p + P_{3n}'r^{-p})\cos n\theta \tag{9.26}$$

where $p^2 = m^2 + n^2$

$$C_1\,\nabla^4\,\phi_n(r) + C_2\,\nabla^2\,T_n(r) - C_3T_n(r) = 0 \tag{9.27}$$

where

$$C_1 = \frac{\beta T_0 i\beta\omega(1 + i\tau_q)}{K},\ C_2 = \frac{1 + \tau_T}{K},\ C_3 = \frac{i\omega\rho C_v(1 + i\tau_q)}{K}. \tag{9.28}$$

The solution of the thermal effect is

$$T_n(r) = [P_{4n}J_n(\alpha r) + P_{4n}'Y_n(\alpha r)]\cos n\theta, \tag{9.29}$$

where $\alpha = \frac{\beta T_0(1 + \tau T)}{\rho C_v}$.

The solution of the non-homogeneous solid plate of polygonal cross-sections can be considered as

$$\phi_n(r,\,\theta,\,t) = P_{1n}J_\zeta(kr)\cos n\theta \tag{9.30}$$

$$\psi_n(r,\,\theta,\,t) = P_{2n}J_\xi(\gamma r)\sin n\theta \tag{9.31}$$

$$\Im_n(r,\,\theta,\,t) = P_{3n}r^p\cos n\theta \tag{9.32}$$

$$T_n(r,\,\theta,\,t) = P_{4n}J_n(\alpha r)\cos n\theta. \tag{9.33}$$

9.4 Boundary conditions and frequency equations

The considered material has an irregular boundary condition (figure 9.1). Hence we are using the method of Nagaya *et al* (2001) to obtain the boundary conditions as,

$$\left(\Pi_{xx}\right)_j = \left(\Pi_{xy}\right)_j = (D_r)_j = (T_r)_j = 0. \tag{9.34}$$

Transforming the vibration displacements into the Cartesian coordinates x_i and y_i the relation between the displacements for the ith segment of straight line boundaries are (figure 9.2)

$$u_r = u_r \cos(\theta - \gamma_i) - u_\theta \sin(\theta - \gamma_i) \tag{9.35}$$

$$u_\theta = u_\theta \cos(\theta - \gamma_i) - u_r \sin(\theta - \gamma_i) \tag{9.36}$$

$$\frac{\partial r}{\partial x_i} = \cos(\theta - \gamma_i), \; \frac{\partial \theta}{\partial x_i} = -r^{-1} \sin(\theta - \gamma_i) \tag{9.37}$$

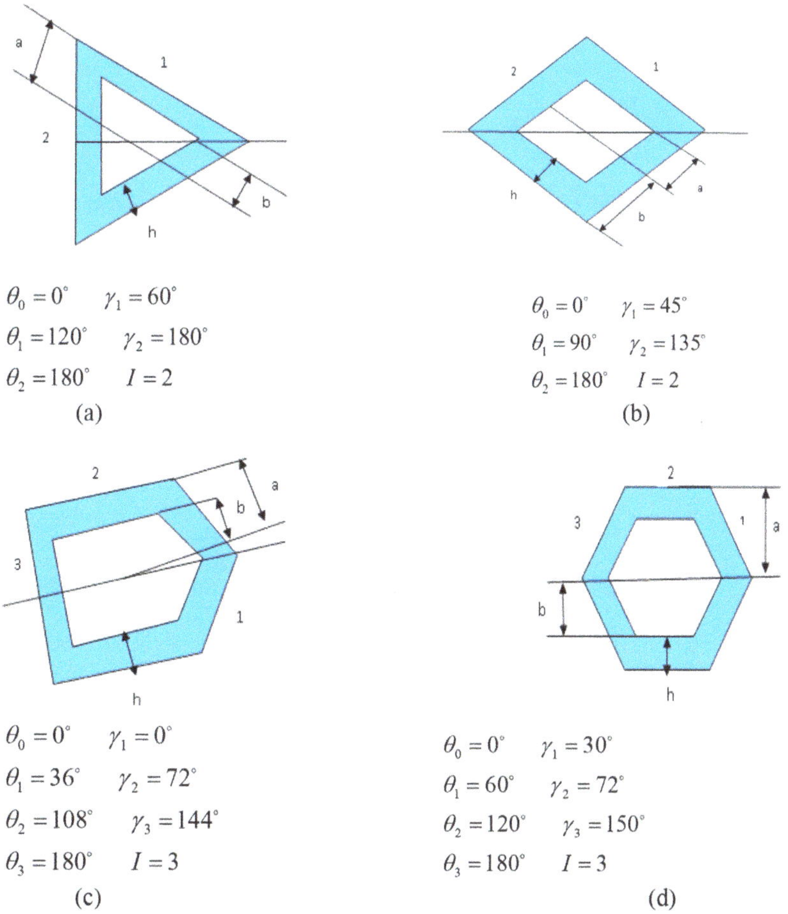

Figure 9.1. Polygonal resonator plates (a) triangular (b) square (c) pentagon (d) hexagon.

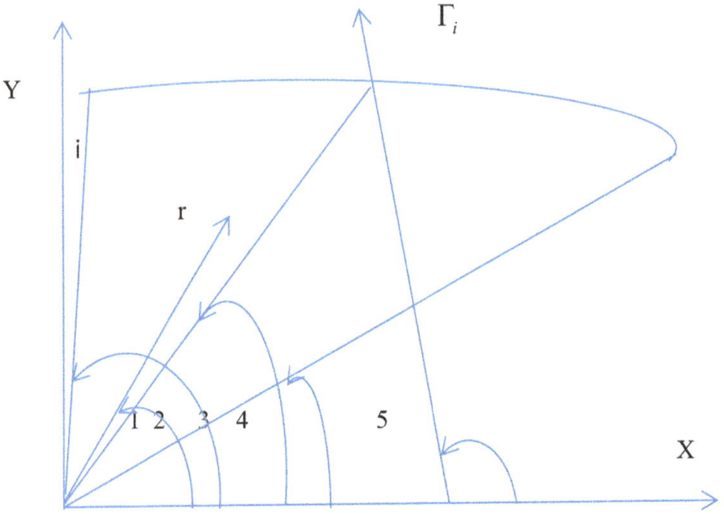

Figure 9.2. Line segment.

$$\frac{\partial r}{\partial y_i} = \sin(\theta - \gamma_i), \; \frac{\partial \theta}{\partial y_i} = r^{-1}\cos(\theta - \gamma_i). \tag{9.38}$$

Biot (1956) and (9.35)–(9.38) help to derive the stress equations for the non-homogeneity as

$$\begin{aligned}\prod_{xx} &= ((\lambda + 2\mu)\cos^2(\theta - \gamma_i) + \lambda\cos^2((\theta - \gamma_i))u_{r,\, r} \\ &+ \frac{1}{r}\begin{pmatrix}\lambda + 2\mu)\sin^2(\theta - \gamma_i) \\ \lambda\cos^2((\theta - \gamma_i))\end{pmatrix}(u_r + u_{\theta,\, \theta}) \\ &+ \frac{u_\theta}{2}\left(\frac{1}{r}(u_\theta - u_{r,\, \theta}) - u_{\theta,\, r}\right)\sin^2(\theta - \gamma_i) - T\beta = 0\end{aligned} \tag{9.39}$$

$$\prod_{xy} = \mu\begin{pmatrix}(u_{r,\, r} - r^{-1}u_{\theta,\, \theta} - r^{-1}u_r)\sin^2(\theta - \gamma_i) + \\ (r^{-1}u_{r,\, \theta} + u_{\theta,\, r} - r^{-1}u_\theta)\sin^2(\theta - \gamma_i)\end{pmatrix} = 0 \tag{9.40}$$

$$\Im_x = -\mathcal{E}_{11}\Im_r = 0. \tag{9.41}$$

The transformed solution equations via boundary conditions (9.34) are

$$[(S_{xx})_j + (\overline{S}_{xx})_j]e^{i\omega t} = 0$$

$$[(S_{xy})_j + (\overline{S}_{xy})_j]e^{i\omega t} = 0$$

$$[(\Im_x)_j + (\overline{\Im}_x)_j]e^{i\omega t} = 0$$

$$[(T_x)_j + (\overline{T}_x)_j]e^{i\omega t} = 0$$

where

$$S_{xx} = 0.5(P_{10}e_o^1 + P_{20}e_o^2 + P_{30}e_o^3) + \sum_{n=1}^{\infty}(P_{1n}e_n^1 + P_{2n}e_n^2 + P_{3n}e_n^3 + P_{4n}e_n^4)$$

$$S_{xy} = 0.5(P_{10}f_o^1 + P_{20}f_o^2 + P_{30}f_o^3) + \sum_{n=1}^{\infty}(P_{1n}f_n^1 + P_{2n}f_n^2 + P_{3n}f_n^3 + P_{4n}f_n^4)$$

$$\Im_x = 0.5(P_{10}l_o^1 + P_{20}l_o^2 + P_{30}l_o^3) + \sum_{n=1}^{\infty}(P_{1n}l_n^1 + P_{2n}l_n^2 + P_{3n}l_n^3 + P_{4n}l_n^4)$$

$$T_x = 0.5(P_{10}h_o^1 + P_{20}h_o^2 + P_{30}h_o^3) + \sum_{n=1}^{\infty}(P_{1n}h_n^1 + P_{2n}h_n^2 + P_{3n}h_n^3 + P_{4n}h_n^4).$$

For antisymmetric mode

$$\bar{S}_{xx} = 0.5\left(\bar{P}_{40}\bar{e}_0^4\right) + \sum_{n=1}^{\infty}(\bar{P}_{1n}\bar{e}_n^1 + \bar{P}_{2n}\bar{e}_n^2 + \bar{P}_{3n}\bar{e}_n^3 + \bar{P}_{4n}\bar{e}_n^4)$$

$$\bar{S}_{xy} = 0.5\left(\bar{P}_{40}\bar{f}_0^4\right) + \sum_{n=1}^{\infty}(\bar{P}_{1n}\bar{f}_n^1 + \bar{P}_{2n}\bar{f}_n^2 + \bar{P}_{3n}\bar{f}_n^3 + \bar{P}_{4n}\bar{f}_n^4)$$

$$\bar{\Im}_x = 0.5\left(\bar{P}_{40}\bar{l}_0^4\right) + \sum_{n=1}^{\infty}(\bar{P}_{1n}\bar{l}_n^1 + \bar{P}_{2n}\bar{l}_n^2 + \bar{P}_{3n}\bar{l}_n^3 + \bar{P}_{4n}\bar{l}_n^4)$$

$$\bar{T}_x = 0.5\left(\bar{P}_{40}\bar{h}_0^4\right) + \sum_{n=1}^{\infty}(\bar{P}_{1n}\bar{h}_n^1 + \bar{P}_{2n}\bar{h}_n^2 + \bar{P}_{3n}\bar{h}_n^3 + \bar{P}_{4n}\bar{h}_n^4).$$

Appendix A indicates the coefficients $e_i^n - \bar{h}_i^n$.

9.5 Numerical computation

The above analytical model is validated by the following numerical illustration. The physical constants for the numerical computation are taken from copper at 42 K, Poisson ratio $\upsilon = 0.3$, density $\rho = 8.96 \times 10^3\,\text{kg m}^{-3}$ Young's modulus $E = 2.139 \times 10^{11}\,\text{N m}^{-2}$, $\lambda = 8.20 \times 10^{11}\,\text{kg m}^{-2}$, $\mu = 4.20 \times 10^{10}\,\text{kg}$, $C_\upsilon = 9.1 \times 10^{-2}\,\text{m}^2(\text{ks})^{-2}$ and $K = 112 \times 10^{-2}\,\text{kg m (ks)}^{-2}$. The material properties of the magneto-electroelastic material based on Hou and Leung (2004) are given by $\varepsilon_{11} = 8.26 \times 10^{-11}\,\text{C}^2\,\text{N}^{-1}\,\text{m}^{-2}$, $\mu_{11} = -5 \times 10^{-6}\,\text{N s}^2\,\text{C}^{-2}$, $m_{11} = -3612 \times 10^{-11}\,\text{Ns (VC)}^{-1}$. Nagaya *et al* (2000) has given the geometric relations for the polygonal cross-sections as $R_i a^{-1} = [\cos(\theta - \gamma_i)]^{-1}, R_i b^{-1} = [\cos(\theta - \gamma_i)]^{-1}, \gamma_i = \gamma_i$. The thermoelastic damping factor is defined as $Q^{-1} = 2\left|\frac{Im(\omega)}{\text{Re}(\omega)}\right|$.

The dispersion curves are plotted in figures 9.3 and 9.4 for thermoelastic damping with the aspect ratio for different non-circular cross-sectional plates such as triangular, square, pentagonal and hexagonal resonator plates. Figures 9.3 and

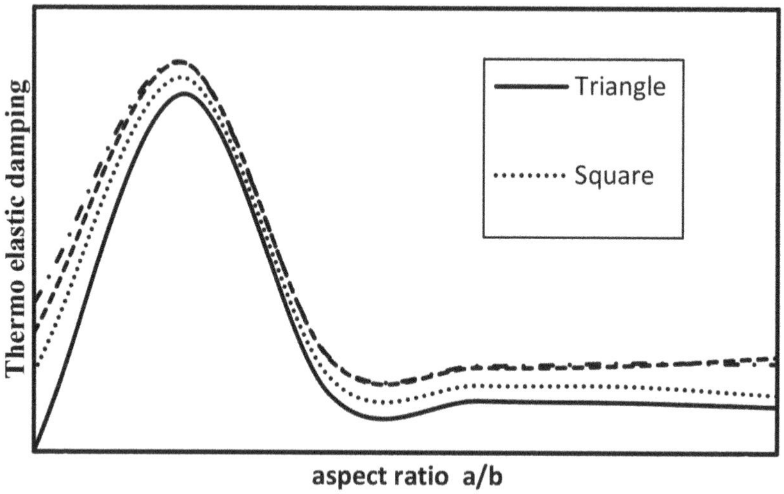

Figure 9.3. Thermoelastic damping with thickness of triangular resonator plate.

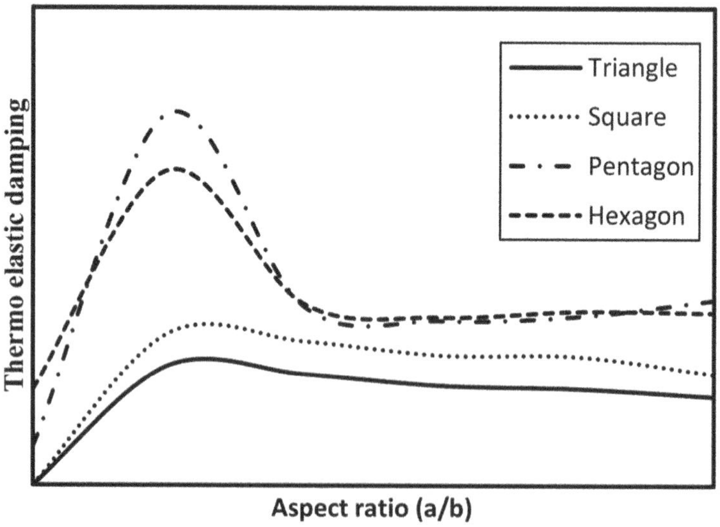

Figure 9.4. Thermoelastic damping with thickness of square resonator plate.

9.4 show that the increase in aspect ratio will lead to the reduction in thermoelastic damping for the different cross-sections of the resonator plate. Also in figures 9.3 and 9.4, the hexagon cross-section experiences high values of damping in the wave propagation compared with other cross-sections. It can be noted that the effect of temperature and non-circular cross-section shows increasing effect on the damping magnitude.

A comparative illustration is made among the thermoelastic damping, radial and circumferential distance of the piezoelectric pentagon and hexagon resonator plates in figures 9.5 and 9.6, respectively. From figures 9.5 and 9.6, it is clear that, for the

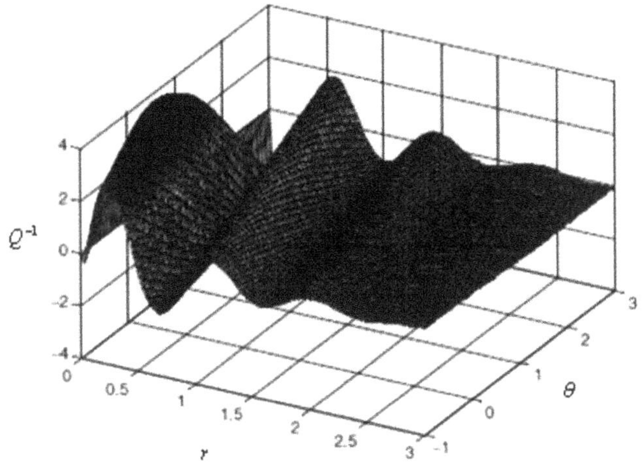

Figure 9.5. 3D curve of thermoelastic damping of pentagonal resonator plate.

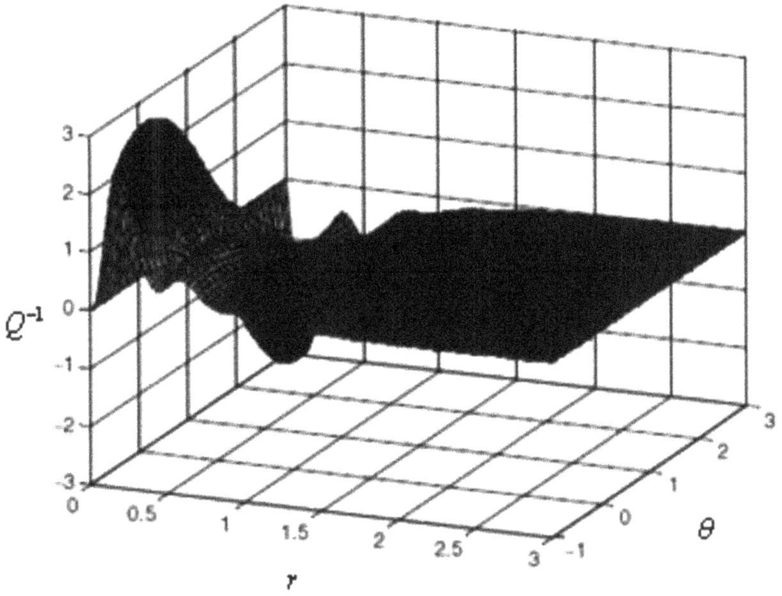

Figure 9.6. 3D curve of thermoelastic damping of hexagonal resonator plate.

initial range of radial and circumferential distances, the damping attains a maximum wave propagation trend in both types of polygonal plates, after which, in the higher values of radial and circumference distance, the damping becomes linear and constant. Also, it is observed that the hexagonal resonator plate attains maximum damping values in a smaller distances compared with pentagonal plate. These curves explain the dependence of distances on the damping of thermoelasticity.

Appendix A

$$e_n^1 = 2\{\zeta(\zeta - 1)J_\zeta(kr) + (kr)J_{\zeta+1}(kr)\}\cos^2(\theta - \gamma_i)\cos n\theta - r^2\{k^2(\overline{\lambda} + 2\cos^2(\theta - \gamma_i))\}J_\zeta(kr)\cos n\theta$$

$$\overline{e_n^1} = 2\{\zeta(\zeta - 1)J_\zeta(kr) + (kr)J_{\zeta+1}(kr)\}\sin^2(\theta - \gamma_i)\sin n\theta - r^2\{k^2(\overline{\lambda} + 2\cos^2(\theta - \gamma_i))\}J_\zeta(kr)\sin n\theta$$

$$e_n^2 = \{n(\zeta - 1)J_\zeta(\gamma r) + (\gamma r)J_\zeta(\gamma r)\}\cos^2(\theta - \gamma_i)\cos n\theta - \left\{\begin{matrix}(\xi(\xi + 2)) + (n^2 - (\gamma r)^2)\frac{J_\xi(\gamma r)}{2} \\ -(\gamma r)J_{\xi+1}(\gamma r)\end{matrix}\right\}\sin n\theta\sin^2(\theta - \gamma_i)$$

$$\overline{e_n^2} = \{n(\zeta - 1)J_\zeta(\gamma r) - (\gamma r)J_\zeta(\gamma r)\}\sin^2(\theta - \gamma_i)\sin n\theta - \left\{\begin{matrix}(\xi(\xi + 2)) + (n^2 - (\gamma r)^2)\frac{J_\xi(\gamma r)}{2} \\ -(\gamma r)J_{\xi+1}(\gamma r)\end{matrix}\right\}\cos n\theta\cos^2(\theta - \gamma_i)^{e_n^3 = 0,\ e_n^4 = 0}$$

$$f_n^1 = \begin{bmatrix}2\{\zeta J_\zeta(kr) - (kr)J_{\zeta+1}(kr)\} + \\ ((kr)^2 - \zeta^2 - n^2)J_\zeta(kr)\end{bmatrix}\cos n\theta\sin^2(\theta - \gamma_i) + 2n\{(\zeta - 1)J_\zeta(kr) - (kr)J_{\zeta+1}(kr)\}\sin n\theta\cos^2(\theta - \gamma_i)$$

$$\overline{f_n^1} = \begin{bmatrix}2\{\zeta J_\zeta(kr) - (kr)J_{\zeta+1}(kr)\} + \\ ((kr)^2 - \zeta^2 - n^2)J_\zeta(kr)\end{bmatrix}\sin n\theta\cos^2(\theta - \gamma_i) + 2n\{(\zeta - 1)J_\zeta(kr) - (kr)J_{\zeta+1}(kr)\}\cos n\theta\sin^2(\theta - \gamma_i)$$

$$f_n^2 = 2n[\xi J_\zeta(\gamma r) - (\gamma r)J_{\zeta+1}(\gamma r)]\cos n\theta\sin^2(\theta - \gamma_i) + 2\left\{\begin{matrix}[\xi J_\zeta(\gamma r) - (\gamma r)J_{\zeta+1}(\gamma r)] + \\ [(\gamma r)^2 - \xi^2 - n^2]J_\zeta(\gamma r)\end{matrix}\right\}\sin n\theta\cos^2(\theta - \gamma_i)$$

$$\overline{f_n^2} = 2n[\xi J_\zeta(\gamma r) - (\gamma r)J_{\zeta+1}(\gamma r)]\sin n\theta\cos^2(\theta - \gamma_i) + 2\left\{\begin{matrix}[\xi J_\zeta(\gamma r) - (\gamma r)J_{\zeta+1}(\gamma r)] + \\ [(\gamma r)^2 - \xi^2 - n^2]J_\zeta(\gamma r)\end{matrix}\right\}\cos n\theta\sin^2(\theta - \gamma_i)$$

$$f_n^3 = 0, f_n^4 = 0, h_n^3 = 0 = h_n^4$$

$$h_n^1 = \{nJ_n(\alpha r) - (\alpha r)J_{n+1}(\alpha r)\}\cos n\theta$$

$$\overline{h_n^1} = \{nJ_n(\alpha r) - (\alpha r)J_{n+1}(\alpha r)\}\sin n\theta$$

$$h_n^1 = nJ_n(\alpha r)\cos n\theta$$

$$\overline{h_n^2} = nJ_n(\alpha r)\sin n\theta.$$

References

Biot M 1956 Thermo elasticity and irreversible thermo dynamics *J. Appl. Phys.* **27** 240–53

Guo F L, Song J, Wang G O and Zhou Y F 2014 Analysis of thermoelastic dissipation in circular micro-plate resonators using the generalized thermoelasticity theory of dual-phase-lagging model *J. Sound Vib.* **333** 2465–74

Hou P-F and Leung Y T 2004 The transient responses of magneto-electro-elastic hollow cylinders *Smart Mater. Struct.* **13** 755–62

Infant Sujitha G and Selvamani R 2020 Visco thermo elastic waves in a nonhomogeneous piezoelectric resonator plate of polygonal shape *J. Phys.: Conf. Ser.* **1597** 012014

Nagaya, N, Toshio N, Masaaki U, Toru S, Shingo K, Fumio S and Mikio K 2000 Plasma brain natriuretic peptide as a prognostic indicator in patients with primary pulmonary hypertension Circulation 102 865–70

Nagaya N, Masaaki U, Masayasu K, Yoshihiko I, Fumiki Y, Wataru S and Hiroshi H 2001 Chronic administration of ghrelin improves left ventricular dysfunction and attenuates development of cardiac cachexia in rats with heart failure Circulation **104** 1430–5

Othman M I A and Abbas I A 2012 Fundamental solution of generalized thermo-visco elasticity using the finite element method *Comput. Math. Model.* **23** 158–67

Ponnusamy P and Amuthalakshmi A 2014 Dispersion anlysis of non-homogeneous transversely isotropic elctro-magneto-elastic palte of polygonal cross section *J. Appl. Math.* **72** 25473–86

Ponnusamy P and Amuthalakshmi A 2016 Modelling of thermo electro elastic waves in a transversely isotropic circular fiber *Mech. Res. Commun.* **73** 47–57

Rajneesh K, Vijay C and Abbas I A 2012 Effect of viscosity on the wave propagation in anistropic thermoelastic medium with three-phase-lag model *J. Theor. Appl. Mech.* **39** 313–41

Selvamani R 2017 Stress waves in a generalized thermo elastic polygonal plate of inner and outer cross sections *J. Solid Mech.* **9** 263–75

Shahin N A and Asad E 2013 Transient wave propagation in non-homogeneous viscoelastic media *Int. Rev. Mech. Eng.* **7** 847–56

Tarhan D 1980 Wave propagation in anistropic non-homogeneous thermo viscoelastic media *J. Sound Vib.* **69** 569–82

Wu B, Jiangong C and He C 2008 Wave propagation in magneto-electro-elastic plates *J. Sound Vib.* **317** 250–64

IOP Publishing

Farzad Ebrahimi and Rajendran Selvamani

Chapter 10

Assessment of hydrostatic stress and thermopiezoelectricity in a laminated multilayered rotating hollow cylinder

10.1 Introduction

Composite materials are generally utilized in engineering structures because of their predominance over the basic materials in applications requiring high quality and solidity in lightweight parts. Thus, the study of their mechanical conduct is taking an imperative part in basic planning. Procedures that encourage transversely isotropic flexible properties in them make most cylindrical parts, for example, poles, wires, cylinders, funnels and strands. Displaying the proliferation of waves in these parts is significant in different applications, including ultrasonic nondestructive assessment systems, progression of room sensors and numerous others.

Honarvar *et al* (2007) studied the characteristics analysis transversely isotropic cylinders. Akbarov and Guz (2010) formed a model for axisymmetric longitudinal wave propagation in pre-stressed compound circular cylinders. Mirsky (1965) studied wave propagation in transversely isotropic circular cylinders. White and Tongtaow (1981) thoroughly discussed cylindrical waves in transversely isotropic media. Paria (1967) studied magneto-elasticity and magneto-thermoelasticity. De and Sengupta (1972) discussed magneto-elastic waves and disturbances in initially stressed conducting media. Acharya and Sengupta (1978) constructed magneto-thermoelastic surface waves in initially stressed conducting media. Placidi and Hutter (2004) developed the thermodynamics of polycrystalline materials treated by the theory of mixtures with continuous diversity. Altenbach and Eremeyev (2014) studied vibration analysis of non-linear six-parameter pre-stressed shells. Sengupta and Nath (2011) carried out a study of surface waves in fibre-reinforced anisotropic

doi:10.1088/978-0-7503-5638-1ch10

elastic media. In transversely isotropic media, the generalized thermoelastic waves are investigated by Singh and Sharma (1985). Chadwick and Windle (1984) thoroughly discussed propagation of Rayleigh waves along isothermal and insulated boundaries. Iesan (2008) showed a theory of pre-stressed thermoelastic Cosserat continua. Othman and Song (2007) studied reflection of plane waves from an elastic solid half-space under hydrostatic initial stress without energy dissipation. Abbas and Othman (2012) generalized thermoelastic interaction in a fibre-reinforced anisotropic half-space under hydrostatic initial stress. Othman and Lotfy (2015) describe a model for the influence of gravity on a two-dimensional problem of two-temperature generalized thermoelastic medium with thermal relaxation. Abo-Dahab *et al* (2015) investigated rotation and magnetic field effect on surface waves propagation in an elastic layer lying over a generalized thermoelastic diffusive half-space with imperfect boundary. Hobinyand and Abbas (2015) discovered an analytical solution of magneto-thermoelastic interaction in a fibre-reinforced anisotropic material. Abouelregal and Abo-Dahab (2012) studied a dual-phase lag model of magneto-thermoelasticity infinite nonhomogeneous solid having a spherical cavity. Othman *et al* (2014) analysed gravitational effect and initial stress on generalized magneto-thermomicrostretch elastic solid for different theories. Abo-Dahab and Lotfy (2017) studied the two-temperature plane strain problem in a semiconducting medium under photothermal theory. Abo-Dahab (2011) investigated the reflection of P and SV waves from a stress-free surface elastic half-space under influence of magnetic field and hydrostatic initial stress. El-Sirafy *et al* (2012) reported on the wave propagation in a thermoelastic half-space under the influence of voids and rotation using Green–Naghdi theory.

The mathematical model is constructed for longitudinal wave propagation in a hollow multilayered composite cylinder in a thermal environment under the influence of initial hydrostatic stress. Lord and Shulman (1967) introduced a theory, generalized thermoelasticity with one relaxation time for an isotropic body. The cylinder is made of tetragonal system material, such as PZT-5A. After applying suitable boundary conditions, the coefficient of Bessel functions are presented in the frequency equation. The numerical computations obtained the roots of frequency equations. The dispersion curve carried for various parameters (figure 10.1).

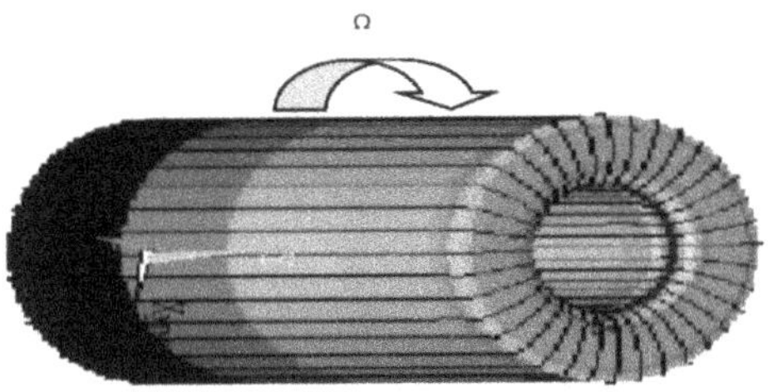

Figure 10.1. Geometry of the problem.

10.2 Formulation of the problem and basic equations

Longitudinal wave propagation in a homogeneous, transversely isotropic cylinder of tetragonal elastic material of inner and outer radius x and subjected to an axial thermal and electric field is considered. The cylinder is treated as a perfect conductor and the regions inside and outside the elastic material are assumed to be vacuumed. The medium is assumed to be rotating with uniform angular velocity $\bar{\Omega}$.

In cylindrical coordinates (r, θ, z), the displacement field can be expressed as

$$\sigma^l{}_{rr,r} + \sigma^l{}_{rz,z} + r^{-1}(\sigma^l{}_{rr}) + \rho(\bar{\Omega} \times (\bar{\Omega} \times \bar{u}) + 2(\bar{\Omega} \times \bar{u}_{,t})) = \rho u_{,tt} \tag{10.1a}$$

$$\sigma^l{}_{rz,r} + \sigma^l{}_{zz,z} + r^{-1}\sigma^l{}_{rz} + \rho(\bar{\Omega} \times (\bar{\Omega} \times \bar{u}) + 2(\bar{\Omega} \times \bar{u}_{,t})) = \rho w_{,tt}. \tag{10.1b}$$

The electric displacement equation is given by

$$\frac{1}{r}\frac{\partial}{\partial r}(rD^l{}_r) + \frac{\partial D^l{}_z}{\partial z} = 0. \tag{10.1c}$$

The heat conduction equation is

$$K_1(T^l{}_{,rr} + r^{-1}T^l{}_{,r} + r^{-2}T^l{}_{,\theta\theta}) + K_3 T^l{}_{,zz} - \rho c_v T^l{}_{,t} = T_0\frac{\partial}{\partial t}[\beta_1(e^l{}_{rr} + e^l{}_{\theta\theta}) + \beta_3 e^l{}_{,zz} - p_3\,\phi_{,Z}]. \tag{10.1e}$$

The stress–strain relations are given as follows

$$\sigma^l{}_{rr} = c_{11}e^l{}_{rr} + c_{12}e^l{}_{\theta\theta} + c_{13}e^l{}_{zz} - \beta_1 T^l - e_{31}\varphi^l_z$$

$$\sigma^l{}_{zz} = c_{13}e^l{}_{rr} + c_{13}e^l{}_{\theta\theta} + c_{33}e^l{}_{zz} - \beta_3 T^l - e_{33}\varphi^l_z$$

$$\sigma^l{}_{rz} = c_{44}e^l{}_{rz}$$

$$D_r = e_{15}e^l{}_{rz} + \varepsilon_{11}\varphi^l{}_r$$

$$D_z = e_{31}(e^l{}_{rr} + e^l{}_{\theta\theta}) + e_{33}e^l{}_{zz} + \varepsilon_{33}\varphi^l{}_z + p_3 T \tag{10.2}$$

where $\sigma^l{}_{rr}$, $\sigma^l{}_{r\theta}$, $\sigma^l{}_{rz}$, $\sigma^l{}_{\theta\theta}$, $\sigma^l{}_{zz}$, $\sigma^l{}_{\theta z}$ are the stress and $e^l{}_{rr}$, $e^l{}_{zz}$, $e^l{}_{\theta\theta}$, $e^l{}_{r\theta}$, $e^l{}_{z\theta}$, $e^l{}_{rz}$ stands for strain components, T denotes the temperature change, c_{11}, c_{12}, c_{13}, c_{33}, c_{44}, c_{66} are the five elastic constants, β_1, β_3 and K_1, K_3 denote the coefficients of thermal expansion and thermal conductivities along and perpendicular to the symmetry, ρ stands for mass density, c_v denotes specific heat capacity, and p_3 denotes pyroelectric effect.

The strains $e^l{}_{ij}$ related to the displacements are expressed as

$$e^l{}_{rr} = u^l{}_{,r} e^l{}_{\theta\theta} = r^{-1}(u^l + v^l{}_{,\theta}),$$
$$e^l{}_{zz} = w^l{}_{,z} e^l{}_{r\theta} = v^l{}_{,r} - r^{-1}(v^l - u^l{}_{,\theta}),$$

$$e^l{}_{z\theta} = v^l{}_{,z} + r^{-1}w^l{}_{,\theta}, \quad e^l{}_{rz} = w^l{}_{,r} + u^l{}_{,z}. \tag{10.3}$$

Substitution of equations (10.3) and (10.2) into equation (10.1) results in the following three-dimensional equation of motion, heat and electric conduction. We note that the first two equations under the influence of hydrostatical stress become:

$$\begin{aligned} &c_{11}(u^l{}_{,rr} + r^{-1}u^l{}_{,r} + r^{-2}u^l) + c_{13}w^l{}_{,rz} + c_{44}(u^l{}_{,zz} + w^l{}_{,rz}) \\ &+ (e_{15} + e_{31})\varphi^l_{,rz} - p_0(u^l{}_{,rr} + r^{-1}u^l{}_{,r} + r^{-2}u^l + u^l{}_{,zz}) \\ &- \beta_1 T^l{}_{,r} + \rho(\Omega^2 u + 2\Omega w_{,t}) = \rho u_{,tt} \end{aligned} \tag{10.4a}$$

$$\begin{aligned} &(c_{44} + c_{13})(u^l{}_{,rz} + r^{-1}u^l{}_{,z}) + c_{33}(w^l{}_{,zz}) \\ &+ c_{44}(w^l{}_{,rr} + r^{-1}w^l{}_{,r}) + e_{15}\left(\varphi^l_{,rr} + r^{-1}\varphi_{,r}\right) \\ &- p_0(w^l{}_{,rr} + r^{-1}w^l{}_{,r} + w^l{}_{,zz}) - \beta_3 T^l{}_{,r} + \rho(\Omega^2 w + 2\Omega u_{,t}) = \rho w_{,tt} \end{aligned} \tag{10.4b}$$

$$\begin{aligned} &r^{-1}e_{15}\left(w^l_{,rr} + u^l_{,rz}\right) + \varepsilon_{11}\varphi^l_{,r} + e_{13}u^l_{,rz} + r^{-1}\left(u^l_{,r} + v^l_{,\theta z}\right) \\ &+ e_{33}w^l_{,zz} + \varepsilon_{33}\varphi^l_{,z} + p_3 T = 0 \end{aligned} \tag{10.4c}$$

$$k_{11}(T^l{}_{,rr} + r^{-1}T^l{}_{,r}) + k_{33}T^l{}_{,zz} - T_0 dT^l{}_{,t} = T_0[\beta_1(u^l{}_{,rt} + r^{-1}u^l{}_{,t}) + \beta_3 w^l{}_{,z}] \tag{10.4d}$$

where k_{ij} is heat conduction coefficient, $d = \frac{\rho C_v}{T_0}$, u and w are the displacements along the r- *and* z-direction. The time is denoted as 't'. The solutions of equation (10.4) are considered in the form

$$\left.\begin{aligned} u^l &= U^l{}_{,r}\exp\{i(kz + pt)\} \\ w^l &= \left(\frac{i}{h}\right)W^l \exp\{i(kz + pt)\} \\ \varphi^l &= \left(\frac{ic_{44}}{ae_{33}}\right)E^l \exp\{i(kz + pt)\} \\ T^l &= \left(\frac{c_{44}}{\beta_3}\right)\left(\frac{T^l}{h^2}\right)\exp\{i(kz + pt)\} \end{aligned}\right\} \tag{10.5}$$

where, u^l, w^l, φ^l, T^l are displacement potentials, k is the wave number, p is the angular frequency and $i = \sqrt{-1}$. The non-dimensional quantities are given as $x = \frac{r}{a}$, $\varepsilon = ka$, $c = \rho p$. 'a' is the geometrical parameter of the composite hollow cylinder. $\overline{c}_{11} = c_{11}/c_{44}$, $\overline{c}_{13} = c_{13}/c_{44}$, $\overline{c}_{33} = c_{33}/c_{44}$, $\overline{c}_{66} = c_{66}/c_{44}$, $\overline{\beta} = \beta_1/\beta_3$, $\overline{k}_i = \frac{(\rho c_{44})^{\frac{1}{2}}}{\beta_3^2 T_0 a\Omega}$. Substituting equation (10.5) in equation (10.4), we obtain:

$$\begin{aligned} &[(\overline{c}_{11} - p_0)\nabla^2 - (1 - p_0)\varepsilon^2 + \zeta^2 + \chi^2 + \daleth]U^l \\ &- [\varepsilon(1 + \overline{c}_{13})]W^l - \varepsilon(\overline{e}_{31} + \overline{e}_{15})E^l - \overline{\beta}T^l = 0 \end{aligned} \tag{10.6a}$$

$$[\varepsilon(1+\bar{c}_{13})\nabla^2]U^l + [(1-p_0)\nabla^2 - (\bar{c}_{33}-p_0)\varepsilon^2 + \zeta^2 + \chi^2 + ℶ]W^l + (\bar{e}_{15}\nabla^2 - \varepsilon^2)E^l - \varepsilon T^l = 0 \tag{10.6b}$$

$$\varepsilon((\bar{e}_{31}+\bar{e}_{15})\nabla^2 U^l + (\bar{e}_{15}\nabla^2 + \varepsilon^2)W^l - (K_{11}^2\nabla^2 + K_{33}\varepsilon^2)E^l - p\varepsilon\, T^l = 0 \tag{10.6c}$$

$$\bar{\beta}\,\nabla^2 U^l + \varepsilon W^l - p\varepsilon\, E^l + (i\bar{K}_1\nabla^2 + i\bar{K}_3\varepsilon^2 - \bar{d})T^l = 0. \tag{10.6d}$$

The above relations (10.6a)–(10.6d) are rearranged in the following form as

$$\begin{vmatrix} (\bar{c}_{11}-p_0)\nabla^2 - s_1 + A_1 & -A_2 & A_3 & -A_4 \\ A_2 & (1-p_0)\nabla^2 - s_2 + A_1 & e_{15}^2\nabla^2 + A_5 & A_6 \\ A_3\nabla^2 & e_{15}^2\nabla^2 + A_5 & K_{11}^2\nabla^2 + A_7 & -A_8 \\ A_4\nabla^2 & A_6 & -A_8 & i\bar{K}_1\nabla^2 + A_9 \end{vmatrix} \times (U^l, W^l, E^l, T^l) = 0 \tag{10.7}$$

where $A_1 = \zeta^2 + \chi^2 + ℶ$, $A_2 = \varepsilon(1+\bar{c}_{13})$, $A_3 = \varepsilon(\bar{e}_{31}+\bar{e}_{15})$, $A_4 = -\beta$, $A_5 = \varepsilon^2$, $A_6 = \varepsilon$, $A_7 = K_{33}^2\varepsilon^2 A_8 = -p^l\varepsilon$, $A_9 = i\bar{k}_3\varepsilon^2 - \bar{d}$, $s_1 = (1-p_0)\varepsilon^2$, $s_2 = (\bar{c}_{33}-p_0)\varepsilon^2$.

By solving the above determinant, we get the following partial differential equation as:

$$(A\,\nabla^8 + B\,\nabla^6 + C\,\nabla^4 + D\,\nabla^2 + E)(U^lW^lE^lT^l)^T = 0. \tag{10.8}$$

Factorizing the relation given in equation (10.8) into the biquadratic equation for $(\alpha^l{}_j a)^2$, $i = 1, 2, 3, 4$ the symmetric mode solutions are given by

$$U^l = \sum_{j=1}^{4}[A_jJ_n(\alpha_jx) + B_jy_n(\alpha_jx)],$$

$$W^l = \sum_{j=1}^{4}a^l{}_j\;[A_jJ_n(\alpha_jx) + B_jy_n(\alpha_jx)],$$

$$E^l = \sum_{j=1}^{4}b^l{}_j[A_jJ_n(\alpha_jx) + B_jy_n(\alpha_jx)],$$

$$T^l = \sum_{j=1}^{4}c^l{}_j[A_jJ_n(\alpha_jx) + B_jy_n(\alpha_jx)],$$

here $(\alpha_i^l ax) > 0$, for $(i = 1,2,3,4)$ are the roots of the algebraic equation

$$(A(\alpha^l{}_ja)^8 + B(\alpha^l{}_ja)^6 + C(\alpha^l{}_ja)^4 + D(\alpha^l{}_ja)^2 + E)(U^l, W^l, E^l, T^l) = 0. \tag{10.10}$$

The solutions corresponding to the root $(\alpha_ia)^2 = 0$ are not considered here, since $J_n(0)$ is zero, except $n = 0$. The Bessel function J_n is used when the roots $(\alpha_ia)^2$, $(i = 1,2,3,4)$ are real or complex and the modified Bessel function I_n is used when the roots $(\alpha_ia)^2$, $(i = 1,2,3,4)$ are imaginary.

The constants $a^l{}_j$, $b^l{}_j$ and $c^l{}_j$ defined in equation (10.10) can be calculated from the following equations

$$[(\bar{c}_{11} - p_0)\,\nabla^2 - (1 - p_0)\varepsilon^2 + \zeta^2 + \chi^2 + \sqsupset] - \varepsilon(1 + \bar{c}_{13})a^l{}_j - \varepsilon(\bar{e}_{31} + \bar{e}_{15})b^l{}_j - \bar{\beta}c^l{}_j = 0$$

$$\varepsilon(1 + \bar{c}_{13})\,\nabla^2 + [(1 - p_0)\,\nabla^2 - (\bar{c}_{33} - p_0)\varepsilon^2 + \zeta^2 + \chi^2 + \sqsupset]a^l{}_j + (\bar{e}_{15}\,\nabla^2 - \varepsilon^2)b^l{}_j - \varepsilon c^l{}_j = 0$$

$$\varepsilon((\bar{e}_{31} + \bar{e}_{15})\,\nabla^2 + (\bar{e}_{15}\,\nabla^2 + \varepsilon^2)a^l{}_j - \left(K_{11}^2\,\nabla^2 + K_{33}\varepsilon^2\right)b^l{}_j - p\varepsilon\, c^l{}_j = 0$$

$$\bar{\beta}\,\nabla^2 + \varepsilon a^l{}_j - p\varepsilon\, b^l{}_j + (i\bar{K}_1\,\nabla^2 + i\bar{K}_3\varepsilon^2 - \bar{d})c^l{}_j = 0.$$

10.3 Equation of motion for linear elastic materials with voids (LEMV)

The displacement equations of motion and equation of equilibrated inertia for an isotropic LEMV are

$$(\lambda + 2\mu)(u_{,rr} + r^{-1}u_{,r} - r^{-2}u) + \mu u_{,zz} + (\lambda + \mu)w_{,zz} + \beta E_{,r} = \rho u_{tt}$$

$$(\lambda + \mu)(u_{,rz} + r^{-1}u_{,z}) + \mu(w_{,rr} + r^{-1}w_{,r}) + (\lambda + 2\mu)w_{,zz} + \beta E_{,z} = \rho w_{,tt}$$

$$-\beta(u_{,r} + r^{-1}u) - \beta w_{,z} + \alpha(E_{,rr} + r^{-1}E_{,r} + \phi_{zz}) - \delta k E_{,tt} - \omega E_{,t} - \xi E = 0. \quad (10.11)$$

The stress in the LEMV core materials are

$$\sigma_{,rr} = (\lambda + 2\mu)u_{,r} + \lambda r^{-1}u + \lambda w_{,z} + \beta\phi$$

$$\sigma_{rz} = \mu(u_{,t} + w_{,r}).$$

The solution for equation (10.11) is taken as

$$u = U_{,r}\exp i(kz + pt)$$

$$w = \left(\frac{i}{h}\right)W\exp i(kz + pt)$$

$$E = \left(\frac{1}{h^2}\right)E\exp i(kz + pt). \quad (10.12)$$

The above solution in (10.11) and dimensionless variables x and ε, equation can be simplified as

$$\begin{vmatrix} (\lambda + 2\mu)\,\nabla^2 + M_1 & -M_2 & M_3 \\ M_2\nabla^2 & \bar{\mu}\,\nabla^2 + M_4 & M_5 \\ -M_3\nabla^2 & M_5 & \alpha\,\nabla^2 + M_6 \end{vmatrix} \times (u, w, E) = 0 \quad (10.13)$$

where $\nabla^2 = \frac{\partial^2}{\partial x^2} + \frac{1}{x}\frac{\partial}{\partial x}$

$$M_1 = \frac{\rho}{\rho^1}(ch)^2 - \overline{\mu}\varepsilon^2, \quad M_2 = (\overline{\lambda} + \overline{\mu})\varepsilon,$$
$$M_3 = \overline{\beta}, \; M_4 = \frac{\rho}{\rho^1}(ch)^2 - (\overline{\lambda} + \overline{\mu})\varepsilon^2, \quad M_5 = \overline{\beta}\varepsilon$$

$$M_6 = \frac{\rho}{\rho^1}(ch)^2\overline{k} - \overline{\alpha}\varepsilon^2 - i\overline{\omega}(ch) - \overline{\xi}.$$

Equation (10.13) can specified as,

$$(\nabla^6 + P\,\nabla^4 + Q\,\nabla^2 + R)(U,\, W,\, E) = 0. \tag{10.14}$$

Thus the solution of equation (10.14) is as follows,

$$U = \sum_{j=1}^{3}[A_j J_0(\alpha_j x) + B_j y_0(\alpha_j x)],$$

$$W = \sum_{j=1}^{3} a_j[A_j J_0(\alpha_j x) + B_j y_0(\alpha_j x)],$$

$$E = \sum_{j=1}^{3} b_j[A_j J_0(\alpha_j x) + B_j y_0(\alpha_j x)],$$

$(\alpha_j x)^2$ are the roots of the equation when replacing $\nabla^2 = -(\alpha_j x)^2$. The arbitrary constants a_j and b_j are obtained from

$$M_2\,\nabla^2 + (\overline{\mu}\,\nabla^2 + M_4)a_j + M_5 b_j = 0,$$

$$-M_3\,\nabla^2 + M_5 a_j + (\alpha\,\nabla^2 + M_6)b_j = 0.$$

By taking the void volume fraction $E = 0$, and the Lamé's constants as $\lambda = c_{12}$, $\mu = \frac{c_{11} - c_{12}}{2}$ in equation (10.11) we get the governing equation for carbon fibre-reinforced polymer (CFRP) core material.

10.4 Boundary conditions and frequency equations

The frequency equations can obtain for the following boundary condition

- On the traction-free inner and outer surface $\sigma^l{}_{rr} = \sigma^l_{rz} = E^l = T^l = 0$ with $l = 1{,}3$
- At the interface $\sigma^l_{rr} = \sigma_{rr}; \sigma^l_{rz} = \sigma_{rz}; E^l = 0; T^l = 0; D^l = 0.$

Substituting the above boundary condition, we obtain as a 22 × 22 determinant equation

$$|(Y_{ij})| = 0, \; (i, j = 1, 2, 3, \ldots .22) \tag{10.15}$$

at $x = x_0$ where $j = 1, 2, 3, 4$

$$Y_{1j} = 2\overline{c}_{66}\left(\frac{\alpha^1_j}{x_0}\right)J_1(\alpha^1{}_j x_0) - [(\alpha^1{}_j a)^2\overline{c}_{11} + \zeta\overline{c}_{13}a^l{}_j + \overline{e}_{31}\zeta b^l{}_j + \overline{\beta}c^l{}_j]J_0(\alpha^1{}_j a x_0)$$

$$Y_{2j} = \left(\zeta + a^1_j + \overline{e}_{15}b^1_j\right)(\alpha^1_j)\ J_1(\alpha^1{}_j x_0)$$

$$Y_{3j} = b^1_j J_0(\alpha^1{}_j x_0)$$

$$Y_{4j} = \frac{h^1_j}{x_0}J_0(\alpha^1{}_j x_0) - (\alpha^1_j)\ J_1(\alpha^1{}_j x_0).$$

In addition, the other nonzero elements $Y_{1,j+4}$, $Y_{2,j+4}$, $Y_{3,j+4}$ and $Y_{4,j+4}$ are obtained by replacing J_0 by J_1 and Y_0 by Y_1.

At $x = x_1$

$$Y_{5j} = 2\overline{c}_{66}\left(\frac{\alpha^1_j}{x_1}\right)J_1(\alpha^1{}_j x_1) - [(\alpha^1{}_j a)^2\overline{c}_{11} + \zeta\overline{c}_{13}a^l{}_j + \overline{e}_{31}\zeta b^l{}_j + \overline{\beta}c^l{}_j]J_0(\alpha^1{}_j a x_1)$$

$$Y_{5,j+8} = -[2\overline{\mu}\left(\frac{\alpha_j}{x_1}\right)J_1(\alpha x_1) + \{-(\overline{\lambda} + \overline{\mu})(\alpha_j)^2 + \overline{\beta}b_j - \overline{\lambda}\zeta a_j\}J_0(\alpha_j x_1)$$

$$Y_{6j} = \left(\zeta + a^1_j + \overline{e}_{15}b^1_j\right)(\alpha^1_j)\ J_1(\alpha^1{}_j a x_1)$$

$$Y_{6,j+8} = -\overline{\mu}(\zeta + a_j)(\alpha_j)J_1(\alpha_j x_1)$$

$$Y_{7j} = \left(\alpha^l_j\right)J_1\left(\alpha^l_j x_1\right)$$

$$Y_{7,j+8} = -(\alpha_j)J_1(\alpha^l{}_j x_1)$$

$$Y_{8j} = a^l_j J_0\left(\alpha^l_j x_1\right)$$

$$Y_{8,j+8} = -a^l_j J_0\left(\alpha^l_j x_1\right)$$

$$Y_{9j} = b^l_j J_0\left(\alpha^l_j x_0\right)$$

$$Y_{10j} = e_j(\alpha_j)J_1(\alpha^1_j x_1)$$

$$Y_{11j} = \frac{c^l_j}{x_1}J_0\left(\alpha^l_j x_1\right) - (\alpha^l_j)J_1\left(\alpha^l_j x_1\right)$$

and the other nonzero elements at the interfaces $x = x_1$ can be obtained on replacing J_0 by J_1 and Y_0 by Y_1 in the above elements. They are $Y_{i,j+4}$, $Y_{i,j+8}$, $Y_{i,j+11}$, $Y_{i,j+14}$, $(i = 5,6,7,8)$ and $Y_{9,j+4}$, $Y_{10,j+4}$, $Y_{11,j+4}$,. At the interface $x = x_2$, nonzero elements along the following rows Y_{ij}, ($i = 12,13, \ldots, 18$ and $j = 8,9,\ldots,20$) are obtained on replacing x_1 by x_2 and superscript 1 by 2 in order. Similarly, at the outer surface $x = x_3$, the nonzero elements Y_{ij}, ($i = 19,20,21,22$ and $j = 14,15, \ldots, 22$) can be obtained from the nonzero elements of the first four rows by assigning x3 for x0 and superscript 2 for 1. The frequency equation is obtained by substituting $E = 0$ in equation (10.11) which reduces to a 20×20 determinant equation.

10.5 Numerical results and discussion

The frequency equation given in equation (1.15) is transcendental in nature with unknown frequency and wave number. By fixing the wave number, the solutions of the frequency equations are obtained. The material properties of PZT-5A used for the numerical calculation are given below:

$$C_{11} = 13.9 \times 10^{10}\,\mathrm{N\,m^{-2}};\; C_{12} = 7.78 \times 10^{10}\,\mathrm{N\,m^{-2}};$$

$$C_{13} = 7.43 \times 10^{10}\,\mathrm{N\,m^{-2}};\; C_{33} = 11.5 \times 10^{10}\,\mathrm{N\,m^{-2}};$$

$$C_{44} = 2.56 \times 10^{10}\,\mathrm{N\,m^{-2}};\; C_{66} = 3.06 \times 10^{10}\,\mathrm{N\,m^{-2}};$$

$$\beta_1 = 1.52 \times 10^{6}\,\mathrm{N\,K^{-1}m^{-2}};\; T_0 = 298\,\mathrm{K};$$

$$\beta_3 = 1.53 \times 10^{6}\,\mathrm{N\,K^{-1}m^{-2}},\; c_v = 420\,\mathrm{J\,kg^{-1}\,K^{-1}}$$

$$p_3 = -452 \times 10^{-6}\,\mathrm{C\,K^{-1}m^{-2}};\; e_{13} = -6.98\,\mathrm{C\,m^{-2}}$$

$$K_1 = K_3 = 1.5\,\mathrm{W\,m^{-1}K^{-1}};\; e_{33} = 13.8\,\mathrm{C\,m^{-2}}$$

$$e_{15} = 13.4\,\mathrm{C\,m^{-2}};\; \rho = 7750\,\mathrm{Kg\,m^{-2}};$$

$$\varepsilon_{11} = 60.0 \times 10^{-10}\,\mathrm{C^2\,N^{-1}\,m^{-2}};$$

$$\varepsilon_{33} = 5.47 \times 10^{-10}\,\mathrm{C^2\,N^{-1}\,m^{-2}}.$$

The non-dimensional frequency versus the wave number and thickness h are plotted in figures 10.2–10.5, which delineate the impacts of the underlying hydrostatic stress on the longitudinal vibrations of a hollow circular cylinder for the benefit of turning

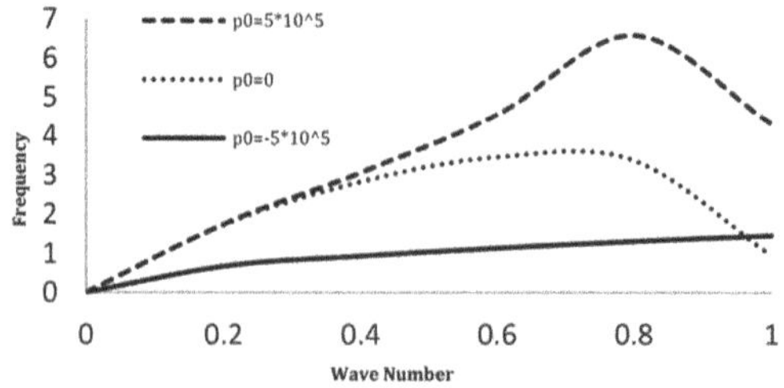

Figure 10.2. Variation of frequency versus wave number against hydrostatic stress $p0$ with $\Omega = 0$.

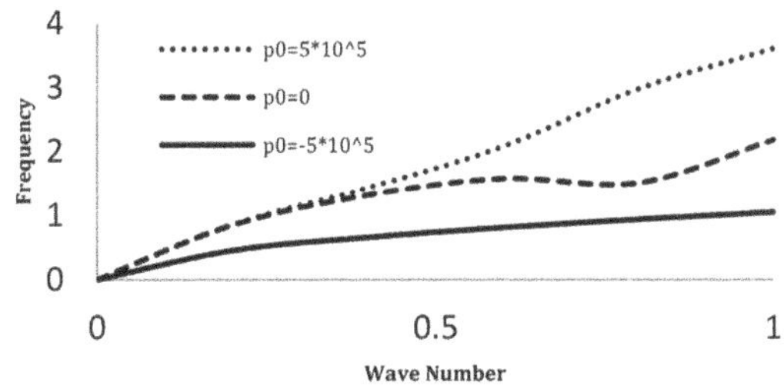

Figure 10.3. Variation of frequency versus wave number against hydrostatic stress $p0$ with $\Omega = 0.4$.

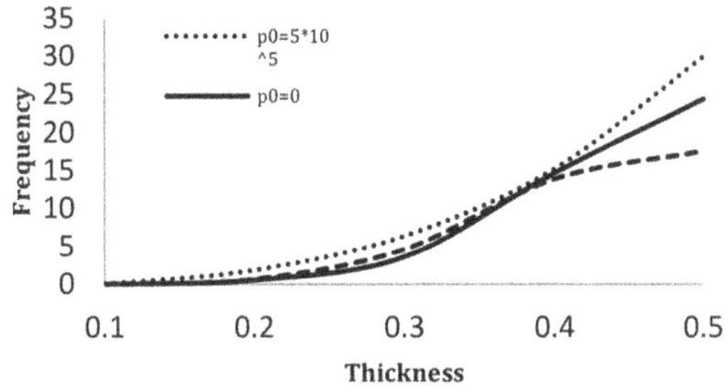

Figure 10.4. Variation of frequency versus thickness against hydrostatic stress $p0$ with $\Omega = 0$.

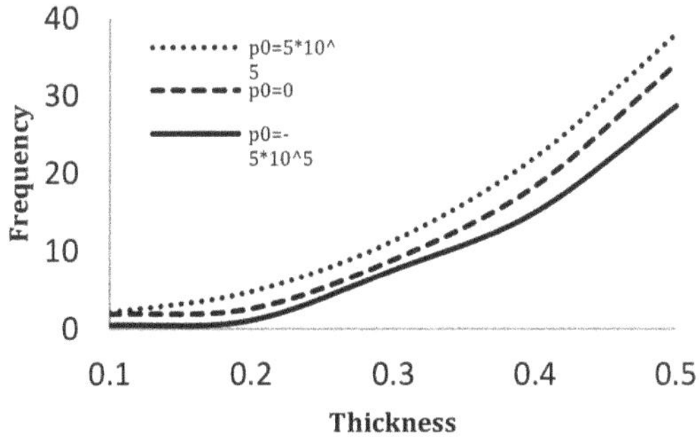

Figure 10.5. Variation of frequency versus thickness against hydrostatic stress $p0$ with $\Omega = 0.4$

parameter Ω. From figures 10.2 and 10.3 there are a type of comparative change gathered in non-dimensional frequency against the wave number which differ in rotational speed Ω, whereas the non-dimensional frequency expanding at the same time for the higher estimations of wave number to arrive as far as possible and again diminished in figure 10.1. From figures 10.4 and 10.5, there is a comparative change, which can be watched above all in non-dimensional frequency against the thickness which shifts in rotational speed Ω. The study represents the conduct of turning at first focussed on a hollow cylinder. Notwithstanding the idea of non-dimensional frequency against hydrostatic pressure ($p_0 = -5, 0, 5 \times 10^6$) is watched.

The non-dimensional frequency versus the wave number and thickness h is plotted in figures 10.6–10.9, which represent the impacts of the underlying hydrostatic stress on the longitudinal vibrations of a hollow circular cylinder for the

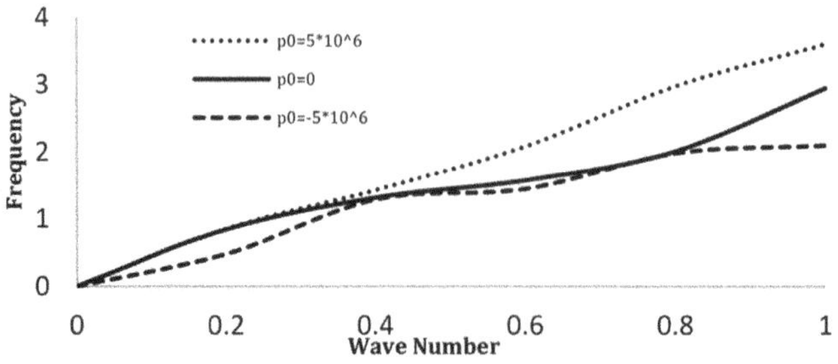

Figure 10.6. Variation of frequency versus wave number against hydrostatic stress $p0$ with $E = 0$

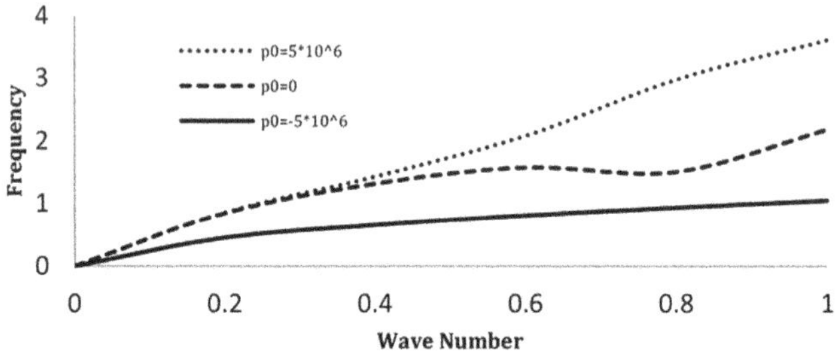

Figure 10.7. Variation of frequency versus wave number against hydrostatic stress $p0$ with $E = 0.5$.

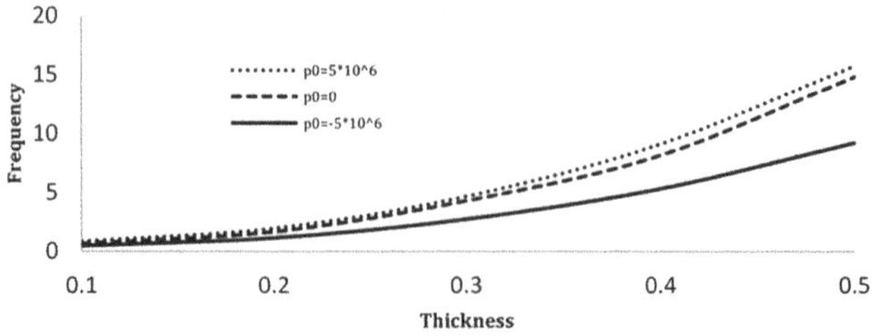

Figure 10.8. Variation of frequency versus thickness against hydrostatic stress $p0$ with $E = 0$.

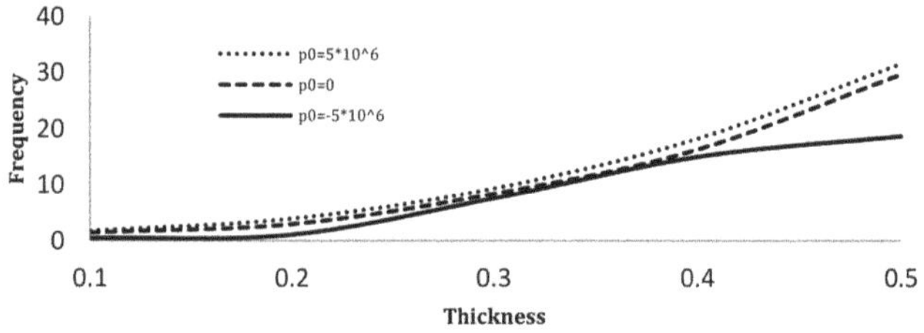

Figure 10.9. Variation of frequency versus thickness against hydrostatic stress $p0$ with $E = 0.5$.

estimation of electric parameter E. From figures 10.6 and 10.7 there is a slight change happening in non-dimensional frequency against the wave number which shifts in electric parameter E and the equivalent, which shows in hydrostatic pressure. Despite the fact that the non-dimensional frequency stays consistent by expanding in wave number, from figures 10.8 and 10.9 it is seen that the essential piece of non-dimensional frequency against the thickness shifts in electric parameter E which shows in hydrostatic pressure. From this it is seen that the conduct of electric parameter is focussed on a hollow cylinder. The idea of non-dimensional frequency against hydrostatic pressure (p_0=-5, 0, 5×10^6) consequently watched.

The non-dimensional frequency versus the thickness h against the with and without hydrostatic stress are plotted in 3D figures 10.10–10.13, which illustrate the effects of the initial hydrostatic stress on the longitudinal vibrations of LEMV and CFRP layers of the hollow circular cylinder. Figures 10.10 and 10.11 compare how the LEMV and CFRP layers vary without hydrostatic stress for increasing thickness of the cylinder. Figures 10.12 and 10.13 comparehowt the LEMV and CFRP layers vary with hydrostatic stress ($p_0 = 5 \times 10^6$) for increasing thickness of the cylinder. In both cases a linear nature was observed in LEMV layers against the influences of

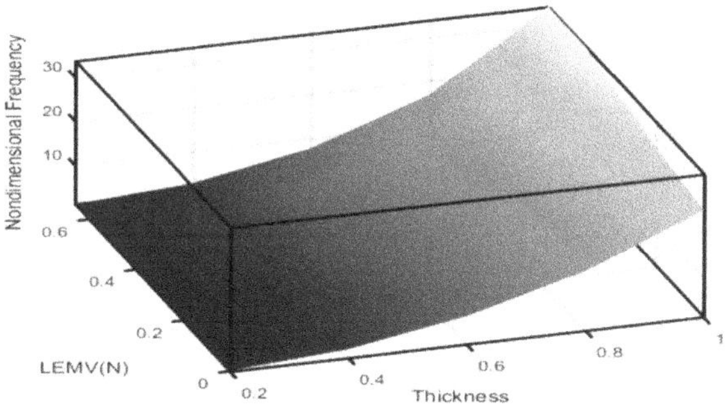

Figure 10.10. Variation of frequency versus thickness of a cylinder with LEMV layer and without hydtrostatic stress.

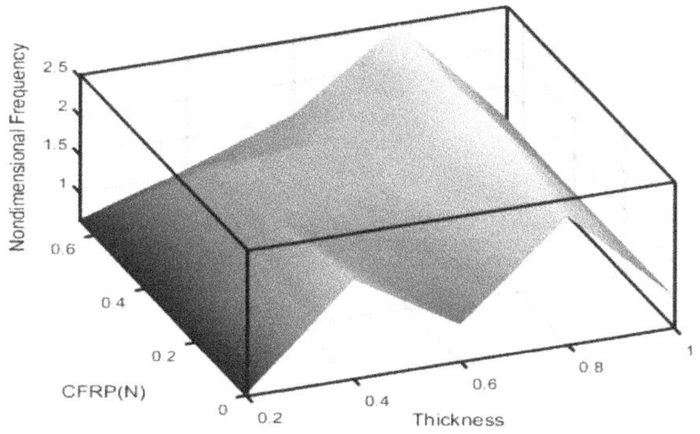

Figure 10.11. Variation of frequency versus thickness of a cylinder with CFRP layer and without hydrostatic stress.

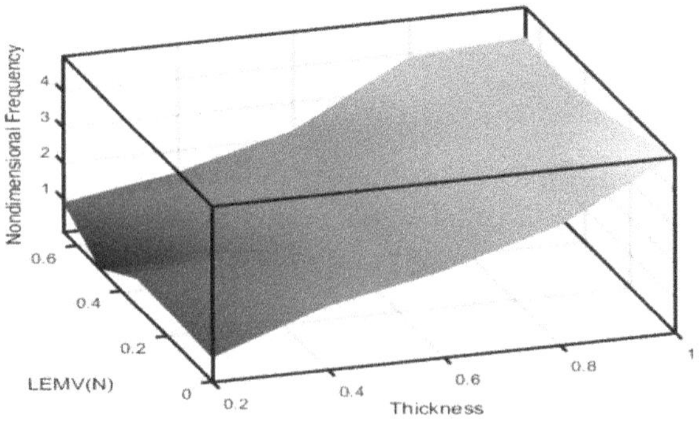

Figure 10.12. Variation of frequency versus thickness of a cylinder against hydrostatic stress in LEMV layer.

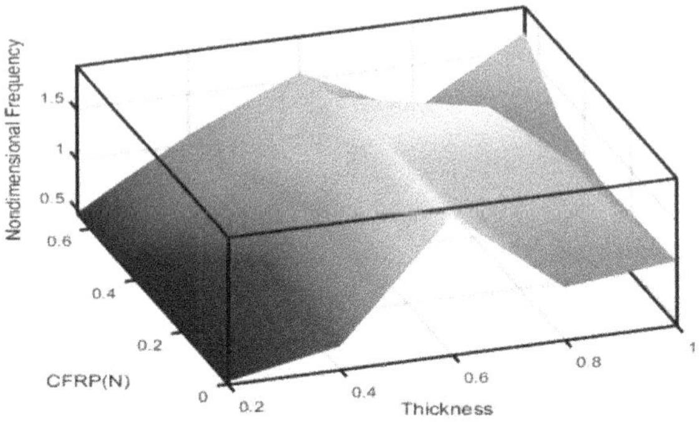

Figure 10.13. Variation of frequency versus thickness of a cylinder against hydrostatic stress in CFRP layer.

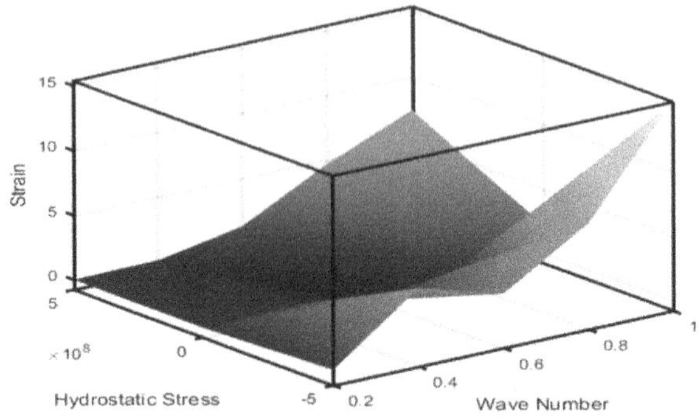

Figure 10.14. Variation of strain versus wave number against hydrostatic stress $p0$ with thermal parameter β.

with and without hydrostatic stress, but small deviations were noted in CFRP layers against the influences of with and without hydrostatic stress.

The three-dimensional figures 10.14–10.15 show the non-dimensional strain against the wave number and thickness h, which illustrate the effects of the initial

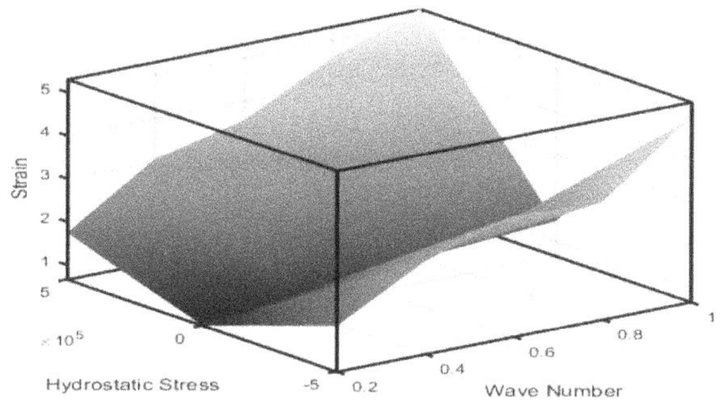

Figure 10.15. Variation of strain versus wave number against hydrostatic stress $p0$ without thermal parameter β.

hydrostatic stress on the longitudinal vibrations of a hollow circular cylinder for the value of thermal parameter β.

10.6 Axisymmetric vibration in a submerged piezoelectric cylindrical rod coated with thin film

The vibration of axisymmetric mode in a transversely isotropic submerged piezoelectric rod coated with thin film is studied in this section. The equations of motion are uncoupled by using potential functions in radial and axial directions. The surface area of the rod is coated by a perfectly conducting material and no slip boundary condition is employed along the solid–fluid interactions. The frequency equations are obtained for PZT-4 ceramic in longitudinal and flexural modes of vibration. The dispersion curves are drawn for computed stress, displacements and electric displacement. This type of study is important in construction of underwater rotating sensors. Also, the thin film coating and fluid can highly influence performance the wave medium.

Graff (1991) and Achenbach (1984) studied the wave propagation in elastic solids. Meeker and Meitzler (1964) reported detailed historical development of the problem. Tiersten (1969) developed the early notable contributions to the topic of the mechanics of piezoelectric solids. Parton and Kudryavtsev (1988) analysed the governing equations of piezoelectric materials. By using Fourier expansion collocation method, the wave propagation in infinite piezoelectric solid cylinders of arbitrary cross-section was constructed by Paul and Venkatesan (1987) and formulated by Nagaya (1981). The axially polarized piezoelectric cylinders with arbitrary boundary conditions were investigated by Ebenezer and Ramesh (2003). This result was extended by Botta and Cerri (2007) and compared with those for the effect of variable electric potential. Piezoelectric cylindrical transducers with radial polarization were developed by Kim and Lee (2007). The modelling of elastic waves in a fluid-loaded and immersed piezoelectric circular fibre was analysed by Selvamani (2017). The wave propagation in a fluid-loaded transversely isotropic cylinder was reported by Berliner and Solecki (1996). The propagation of Bleustein–Gulyaev wave in 6 mm piezoelectric materials loaded with viscous liquid was discussed by Guo and Sun (2008) using the theory of continuum mechanics and finite thickness that was reported by Qian *et al* (2010). Nagy (1995) studied the propagation of longitudinal guided waves in fluid-loaded transversely isotropic rod based on the superposition of partial waves. Guided waves in a transversely isotropic cylinder immersed in a fluid were analysed by Ahmad (2001). The effect of rotation in an axisymmetric vibration of a transversely isotropic solid bar has been studied by Selvamani and Ponnusamy (2012) immersed in an inviscid fluid. Selvamani and Ponnusamy (2015) discussed the wave propagation in a generalized piezothermoelastic rotating bar. The axisymmetric wave propagation in a cylinder coated with a piezoelectric layer was developed by Wang (2002). Sun and Cheng (1974). investigated the application for time delay devices. A theoretical model of the coated structure was investigated by Minagawa (1995) to predict attenuation characteristics for finding suitable modes for a guided wave inspection. Barshinger (2001) investigated the guided waves in pipes with viscoelastic coatings.

In the next section, the axisymmetric waves of a piezoelectric rod coated with thin film and submerged in inviscid fluid is studied using the constitutive equation of linear theory of elasticity and piezoelectric. The equations of motion are uncoupled by potential functions. The frequency equations are obtained for longitudinal and flexural modes of vibration for PZT-4 ceramic. The variation of dimensionless frequency, phase velocity and electric displacement are investigated and presented as dispersion curves.

10.7 Modelling of the problem

In the absence of body forces, the governing equations of motion can expressed as

$$\begin{aligned}\frac{\partial}{\partial r}\sigma_{rr}+\frac{\partial}{\partial z}\sigma_{rz}+\frac{\sigma_{rr}}{r}&=\rho\frac{\partial^2 u_r}{\partial t^2}\\ \frac{\partial}{\partial r}\sigma_{rz}+\frac{\partial}{\partial z}\sigma_{zz}+\frac{\sigma_{rz}}{r}&=\rho\frac{\partial^2 u_z}{\partial t^2}.\end{aligned} \tag{10.16}$$

The Gauss electric conduction equation without free charge is

$$\frac{1}{r}\frac{\partial}{\partial r}(rD_r)+\frac{\partial D_z}{\partial r}=0. \tag{10.17}$$

The coupled form of stress equations are given as

$$\begin{aligned}\sigma_{rr}&=c_{11}e_{rr}+c_{12}e_{\theta\theta}+c_{13}e_{zz}-e_{31}E_z,\\ \sigma_{zz}&=c_{13}e_{rr}+c_{13}e_{\theta\theta}+c_{33}e_{zz}-e_{33}E_z,\\ \sigma_{rz}&=2c_{44}e_{rz}-e_{15}E_r\end{aligned} \tag{10.18}$$

$$D_r=e_{15}e_{rz}+\varepsilon_{11}E_r,\ D_z=e_{31}(e_{rr}+e_{\theta\theta})+e_{33}e_{zz}+\varepsilon_{33}E_z \tag{10.19}$$

where σ_{rr}, $\sigma_{\theta\theta}$, σ_{zz}, $\sigma_{r\theta}$, $\sigma_{\theta z}$, σ_{rz} denotes stress and e_{rr}, $e_{\theta\theta}$, e_{zz}, $e_{r\theta}$, $e_{\theta z}$, e_{rz} stands for strain components, c_{11}, c_{12}, c_{13}, c_{33}, c_{44} and $c_{66}=(c_{11}-c_{12})/2$ are the five elastic constants, ρ is the mass density.

The strain e_{ij} is related to the displacements given by

$$e_{rr} = u_{r,\,r},\ \ e_{\theta\theta} = r^{-1}(u_r + u_{\theta,\,\theta}),\ \ e_{zz} = u_{z,\,z} \tag{10.20}$$

$$e_{r\theta} = u_{\theta,\,r} + r^{-1}(u_{r,\,\theta} - u_\theta),\ \ e_{z\theta} = (u_{\theta,\,z} + r^{-1}u_{z,\,\theta}),\ \ e_{rz} = u_{z,\,r} + u_{r,\,z}. \tag{10.21}$$

Inserting equations (10.3), (10.4) and (10.5) in equations (10.1) and (10.2), we get the following form

$$\begin{aligned} &c_{11}(u_{rr,\,r} + r^{-1}u_{r,\,r} - r^{-2}u_r) + c_{44}u_{r,\,zz} \\ &+ (c_{44} + c_{13})u_{z,\,rz} + (e_{31} + e_{15})V_{,rz} = \rho u_{r,\,tt} \end{aligned} \tag{10.22}$$

$$\begin{aligned} &c_{44}(u_{z,\,rr} + r^{-1}u_{z,\,r}) + r^{-1}(c_{44} + c_{13})(u_{r,\,z}) \\ &+ (c_{44} + c_{13})u_{r,\,rz} + c_{33}u_{z,\,zz} + e_{33}V_{,zz} \\ &+ e_{15}(V_{,rr} + r^{-1}V_{,r}) = \rho u_{z,\,tt} \end{aligned} \tag{10.23}$$

$$\begin{aligned} &e_{15}(u_{z,\,rr} + r^{-1}u_{z,\,r}) + (e_{31} + e_{15})(u_{r,\,zr} + r^{-1}u_{r,\,z}) \\ &+ e_{33}u_{z,\,zz} - \varepsilon_{33}V_{,zz} - \varepsilon_{11}(V_{,rr} + r^{-1}V_{,r}) = 0. \end{aligned} \tag{10.24}$$

10.8 Solutions of the field equation

The displacement variables are used to obtain the propagation of harmonic waves in a piezoelectric circular rod as Paul and Venkatesan (1987)

$$u_r(r,\,z,\,t) = \left(\phi_{,r}\right)e^{i(kz+\omega t)}\ V(r,\,z,\,t) = iVe^{i(kz+\omega t)}\ E_z(r,\,z,\,t) = E_{,z}e^{i(kz+\omega t)}$$

$$u_z(r,\,z,\,t) = \left(\frac{i}{a}\right)We^{i(kz+\omega t)}\ E_r(r,\,z,\,t) = -E_{,r}e^{i(kz+\omega t)} \tag{10.25}$$

where $i = \sqrt{-1}$, k is the wave number, ω is the angular frequency, $\phi(r)$, $W(r)$ are the displacement potentials and a is the geometrical parameter of the rod. The dimensionless quantities are defined as $x = r/a$, $\zeta = ka$, $\varpi^2 = \rho\omega^2a^2/c_{44}$, $\bar{c}_{11} = c_{11}/c_{44}$, $\bar{c}_{13} = c_{13}/c_{44}$, $\bar{c}_{33} = c_{33}/c_{44}$, $\bar{c}_{66} = c_{66}/c_{44}\bar{\varepsilon}_{11} = \varepsilon_{11}c_{44}/e_{33}^2$, $\bar{e}_{31} = e_{31}/e_{33}$, $\bar{e}_{15} = e_{15}/e_{33}$ and inserting equation (10.7) into equation (10.6), we get

$$(\bar{c}_{11}\,\nabla^2 + (\varpi^2 - \zeta^2))\phi - \zeta(1 + \bar{c}_{13})W - \zeta(\bar{e}_{31} + \bar{e}_{15})V = 0$$

$$\zeta(1 + \bar{c}_{13})\,\nabla^2\,\phi + (\nabla^2 + (\varpi^2 - \zeta^2\,\bar{c}_{33}))W + (\bar{e}_{15}\,\nabla^2 - \zeta^2)V = 0 \tag{10.26}$$

where

$$\nabla^2 = \frac{\partial^2}{\partial x^2} + x^{-1}\frac{\partial}{\partial x} + x^{-2}\frac{\partial^2}{\partial\theta^2}. \tag{10.27}$$

Equation (10.8) can be rewritten as

$$\begin{vmatrix} (\bar{c}_{11}\,\nabla^2 + (\varpi^2 - \zeta^2)) & -\zeta(1 + \bar{c}_{13}) & -\zeta(\bar{e}_{31} + \bar{e}_{15}) \\ \zeta(1 + \bar{c}_{13})\nabla^2 & (\nabla^2 + (\varpi^2 - \zeta^2\bar{c}_{33})) & (\bar{e}_{15}\,\nabla^2 - \zeta^2) \\ \zeta(\bar{e}_{31} + \bar{e}_{15})\nabla^2 & (\bar{e}_{15}\,\nabla^2 - \zeta^2) & (\zeta^2\bar{\varepsilon}_{33} - \bar{\varepsilon}_{11}\nabla^2) \end{vmatrix} (\phi,\, W,\, V) = 0. \tag{10.28}$$

Solving the above determinant, we get the following form

$$(P\,\nabla^6 + Q\,\nabla^4 + R\,\nabla^2 + S)(\phi,\, W,\, V) = 0 \tag{10.29}$$

where

$$P = c_{11}(\bar{e}_{15}^2 + \varepsilon_{11})\; Q = [(1 + \bar{c}_{11})\bar{\varepsilon}_{11} + \bar{e}_{15}^2]\varpi^2 + \begin{Bmatrix} 2(\bar{e}_{31} + \bar{e}_{15})\bar{c}_{13}\bar{e}_{15} - (1 + \bar{\varepsilon}_{11}\bar{c}_{33})\bar{c}_{11} \\ + \bar{c}_{13}^2\bar{\varepsilon}_{11} + 2\bar{c}_{13}\bar{\varepsilon}_{11} - 2\bar{e}_{15}\bar{c}_{11} + 2\bar{e}_{13}^2 \end{Bmatrix}\varsigma^2$$

$$R = \bar{\varepsilon}_{11}\varpi^4 - [(1 + \bar{c}_{13})\bar{\varepsilon}_{11} + (1 + \bar{c}_{11}) + (\bar{e}_{31} + \bar{e}_{15}) + 2\bar{e}_{15}]\varsigma^2\varpi^2 + \{\bar{c}_{11}(1 + \bar{c}_{33}\bar{\varepsilon}_{33}) \\ - [(\bar{e}_{31} + \bar{e}_{15})^2 + \bar{\varepsilon}_{11}] - 2\bar{e}_{31}(1 + \bar{c}_{13}) - \bar{c}_{13}\bar{\varepsilon}_{33}(\bar{c}_{33} + \bar{c}_{13}) + 2\bar{e}_{15}\}\varsigma^4$$

$$S = -\{(1 + \bar{c}_{33})\varsigma^6 - [2(1 + \bar{c}_{33})\bar{\varepsilon}_{33} + 1]\varsigma^4\varpi^2 + \bar{\varepsilon}_{33}\varsigma^2\varpi^4\}.$$

Solving equation (10.11), we get solutions for a circular rod as

$$\phi = \sum_{i=1}^{3} A_i J_n(\alpha_i a x)\cos n\theta \quad W = \sum_{i=1}^{3} a_i A_i J_n(\alpha_i a x)\cos n\theta$$
$$V = \sum_{i=1}^{3} b_i A_i J_n(\alpha_i a x)\cos n\theta. \tag{10.30}$$

Here $(\alpha_i a)^2 > 0$, $(i = 1, 2, 3)$ are the roots of the algebraic equation

$$A(\alpha a)^6 - B(\alpha a)^4 + C(\alpha a)^2 + D = 0. \tag{10.31}$$

The Bessel function J_n is used when the roots $(\alpha_i a)^2$, $(i = 1, 2, 3)$ are real or complex and the modified Bessel function I_n is used when the roots $(\alpha_i a)^2$, $(i = 1, 2, 3)$ are imaginary. If $(\alpha_4 a)^2 < 0$, the Bessel function J_n is replaced by the modified Bessel function I_n.

The constants a_i, b_i defined in equation (10.12) can be calculated from the equations

$$(1 + \bar{c}_{13})\varsigma a_i + (\bar{e}_{31} + \bar{e}_{15})\varsigma b_i = -(\bar{c}_{11}(\alpha_i a)^2 - \varpi^2 + \varsigma^2) \tag{10.32}$$

$$((\alpha_i a)^2 - \varpi^2 + \varsigma^2\bar{c}_{33})a_i + (\bar{e}_{15}(\alpha_i a)^2 + \varsigma^2)b_i = -(\bar{c}_{13} + 1)\varsigma(\alpha_i a)^2. \tag{10.33}$$

10.9 Boundary conditions and frequency equations

Here, the boundary conditions are considered for the coated surface as

$$\sigma_{rj} = -\delta_{j\,b}\, 2\, \mu'\, h' \left[\left(\frac{3\lambda' + 2\mu'}{\lambda' + 2\mu'} \right) U_{a,\,ab} + U_{b,\,aa} \right] + 2\, h'\, \rho'\, \ddot{U}_j\, V = 0 \qquad (10.34)$$

where λ' μ' ρ'&h' are Lamé's constants, density, thickness of the material coating, respectively, $\delta_{j\,b}$ is the Kronecker delta function with a, b taking the value of θ &z, and j taking r, θ &z. In order to get the axisymmetric waves a, b can take only z. Then the transformed boundary conditions are as follows

$$\sigma_{rr} = 2\, h'\, \rho'\, \ddot{U}\ \sigma_{rz} = -2\, h'\, \mu'\, G^2\, W_{,zz} + 2\, h'\, \rho'\, \ddot{W}\ V = 0 \text{ at } r = a \qquad (10.35)$$

where $G^2 = \frac{1 + C'_{12}}{C'_{11}}$

Inserting the solutions in equations (10.12) and (10.15) and in boundary condition equation (10.17), we get the following form

$$[B]\{X\} = \{0\} \qquad (10.36)$$

where $[B]$ is a 3×3 matrix of unknown wave amplitudes, and $\{X\}$ is an 3×1 column vector of the unknown amplitude coefficients B_1,B_2,B_3. When $|B| = 0$, the solution of equation (10.18) becomes nontrivial.

The components of $[M]$ are obtained as

$$B_{1i} = 2\bar{c}_{66}\{n(n-1) - \bar{c}_{11}(\alpha_i a)^2 - \varsigma(\bar{c}_{13}a_i + \bar{e}_{31}b_i)\}J_n(\alpha_i a) + 2\bar{c}_{66}(\alpha_i a)J_{n+1}(\alpha_i a),\ i = 1, 2$$

$$B_{13} = 2\bar{c}_{66}n\{(n-1)J_n(\alpha_4 a) - (\alpha_4 a)J_{n+1}(\alpha_4 a)\},$$
$$B_{14} = 2(\alpha_1 a)[(\rho' h'/a\rho(Ca)^2 - \overline{C_{66}})]J_n(\alpha_5 a)$$

$$B_{2i} = 2n\{(n-1)J_n(\alpha_i a) + (\alpha_i a)J_{n+1}(\alpha_i a)\},\ i = 1, 2,$$

$$B_{23} = \{[(\alpha_4 a)^2 - 2n(n-1)]J_n(\alpha_4 a) - 2(\alpha_4 a)J_{n+1}(\alpha_4 a)\}$$

$$B_{24} = 2(\alpha_1 a)[(\rho' h'/a\rho(Ca)^2 - \overline{C_{66}})]J_n(\alpha_5 a),$$
$$B_{3i} = ((\varsigma + a_i) + \bar{e}_{15}b_i)\{nJ_n(\alpha_i a) - (\alpha_i a)J_{n+1}(\alpha_i a)\},\ i = 1, 2$$

$$B_{33} = n\varsigma J_n(\alpha_4 a),\ B_{34} = 0.$$

10.10 Numerical results and investigations

The material properties are taken from Berlincourt *et al* (1964) for PZT-4 and gold.

$c_{11} = 13.9 \times 10^{10}\,\mathrm{N\,m^{-2}},\ c_{12} = 7.78 \times 10^{10}\,\mathrm{N\,m^{-2}},\ c_{13} = 7.43 \times 10^{10}\,\mathrm{N\,m^{-2}},$
$c_{33} = 11.5 \times 10^{10}\,\mathrm{N\,m^{-2}},\ c_{44} = 2.56 \times 10^{10}\,\mathrm{N\,m^{-2}},\ c_{66} = 3.06 \times 10^{10}\,\mathrm{N\,m^{-2}},$
$e_{31} = -5.2\,\mathrm{C\,m^{-2}},\ e_{33} = 15.1\,\mathrm{C\,m^{-2}},\ e_{15} = 12.7\,\mathrm{C\,m^{-2}}$
$\varepsilon_{11} = 6.46 \times 10^{-9}\,\mathrm{C^2\,N^{-1}\,m^{-2}},\ \varepsilon_{33} = 5.62 \times 10^{-9}\,\mathrm{C^2\,N^{-1}\,m^{-2}},\ \rho = 7500\,\mathrm{Kg\,m^{-2}}.$

In this problem, the non-dimensional frequencies of longitudinal and flexural modes are obtained by choosing $n = 0$ and $n = 1$. The notation used in the figures, namely FSM, and FASM denotes the flexural symmetric and antisymmetric mode. 1, 2 refer to the first and second mode.

The graphs are drawn between the dimensionless wave number and mechanical displacement for flexural modes with and without fluid medium in figures 10.16 and 10.17.

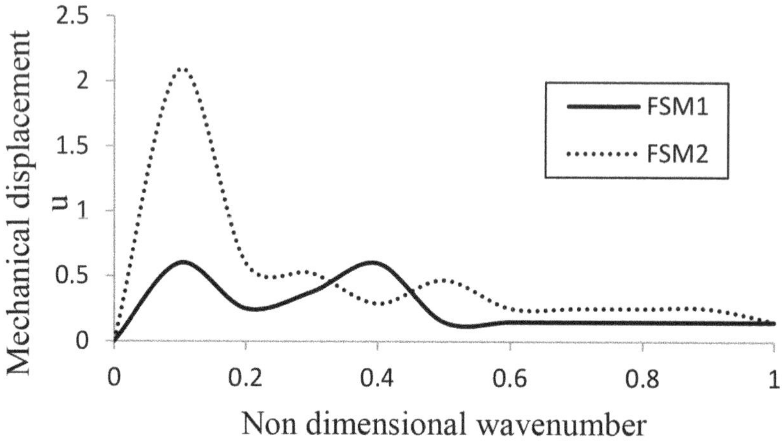

Figure 10.16. The effect of mechanical displacement along with wave number $|\varsigma|$ for flexural symmetric modes of piezoelectric cylindrical rod without coating.

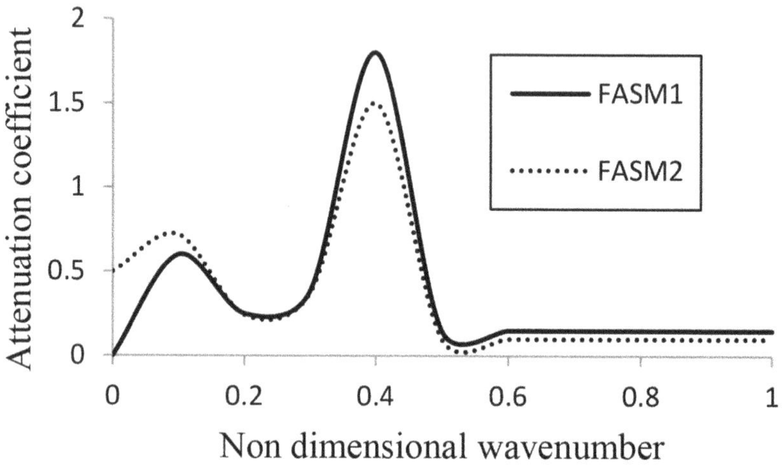

Figure 10.17. The effect of attenuation coefficient with non-dimensional wave number $|\varsigma|$ for flexural antisymmetric modes of piezoelectric cylindrical rod without coating.

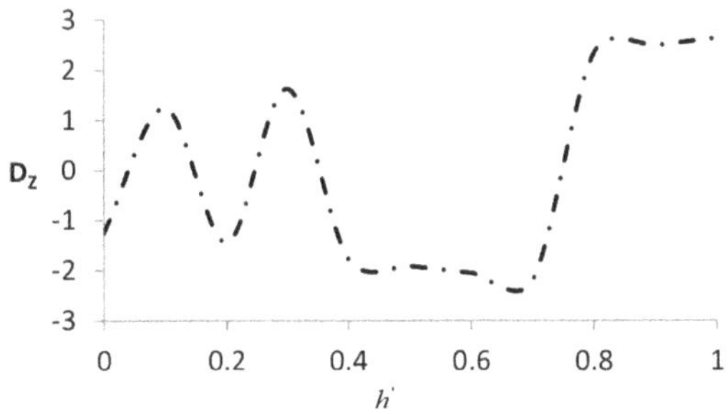

Figure 10.18. The influence of electric displacement with thickness of the coating material h' for piezoelectric rod with fluid.

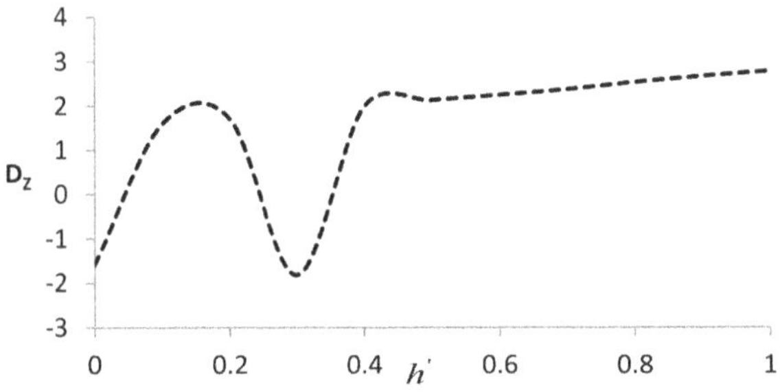

Figure 10.19. The influence of electric displacement with thickness of the coating material h' for piezoelectric rod without fluid.

Figure 10.16 shows the oscillation in the lower range of wave number that becomes linear propagation with respect to its higher wave number in different flexural modes of the rod with a fluid environment. But in figure 10.17, there is a small energy transfer between the modes in the lower range of wave number which might happen due to the coating of the rod and absence of fluid medium. The coating and fluid environment decrease the magnitude of the mechanical displacement, as in figures 10.16 and 10.17. The effect of electric displacement along with thickness of the coated layer is plotted in figures 10.18 and 10.19. From this, it is clear that the electric displacement is decreasing by increases of the rod thickness and again increasing and travelling in the wave propagation.

10.11 Conclusions

The three-dimensional theory is used to obtain the wave propagation of a submerged piezoelectric circular rod coated with thin film. The equations of motion are uncoupled by using three displacement potential functions. The frequency equations are obtained for longitudinal and flexural modes of vibration and the numerical results are studied for a PZT-4 material rod with gold coating. The graphs are obtained for mechanical displacement and electric displacements. From the graphical pattern, it is observed that the fluid and the coating of the piezoelectric rod highly influence the variations of the mechanical parameters in flexural symmetric and flexural antisymmetric modes.

References

Abbas I A and Othman M I A 2012 Generalized thermoelastic interaction in a fiber-reinforced anisotropic half-space under hydrostatic initial stress *J. Vib. Control* **18** 175–82

Abo-Dahab S M and Lotfy K H 2017 Two-temperature plane strain problem in a semiconducting medium under photo thermal theory *Waves Random Complex Medium* **27** 67–91

Abo-Dahab S, Lotfy K H and Gohaly A 2015 Rotation and magnetic field effect on surface waves propagation in an elastic layer lying overa generalized thermoelastic diffusive half-space with imperfect boundary *Math. Probl. Eng.* **2015** 671783

Abo-Dahab S M 2011 Reflection of P and SV waves from stress-free surface elastic half-space under influence of magnetic field and hydrostatic initial stress without energy dissipation *J. Vib. Control* **17** 2213–21

Abouelregal A E and Abo-Dahab S M 2012 Dual phase lag model of magneto-thermoelasticity infinite nonhomogeneous solid having a spherical cavity *J. Therm. Stresses* **35** 820–41

Acharya D P and Sengupta P R 1978 Magneto-thermo-elastic surface waves in initially stressed conducting media *Acta Geophys. Pol.* **26** 299–311

Achenbach J D 1984 *Wave Motion in Elastic Solids* (Amsterdam: North-Holland)

Ahmad F 2001 Guided waves in a transversely isotropic cylinder immersed in a fluid *J. Acoust Soc. Am.* **109** 886–90

Akbarov S D and Guz A N 2010 Axisymmetric longitudinal wave propagation in pre-stressed compound circular cylinders *Int. J. Eng. Sci.* **42** 769–91

Altenbach H and Eremeyev V A 2014 Vibrtion analysis of non-linear 6-parameter prestressed shells *Meccanica* **49** 1751–62

Barshinger J N 2001 Guided waves in pipes with viscoelastic coatings *PhD Dissertation* (The Pennsylvania State University)

Berlincourt D A, Curran D R and Jaffe H 1964 Piezoelectric and piezomagnetic materials and their function in transducers *Physical Acoustics: Principles and Methods* (New York: Academic Press) pp 169–267

Berliner J and Solecki R 1996 Wave propagation in a fluid-loaded transversely isotropic cylinder. Part I. Analytical formulation *Part II: numerical results J. Acoust. Soc. Am.* **99** 1841–53

Botta F and Cerri G 2007 Wave propagation in Reissner-Mindlin piezoelectric coupled cylinder with non-constant electric field through the thickness *Int. J. Solids Struct.* **44** 6201–19

Chadwick P and Windle D W 1984 Propagation of Rayleigh waves along isothermal and insulated boundaries *Proc. Royal Soc. Am.* **280** 47–71

De S N and Sengupta P R 1972 Magneto-elastic waves and disturbances in initially stressed conducting media *Pure Appl. Geophys.* **93** 41–54

Ebenezer D D and Ramesh R 2003 Analysis of axially polarized piezoelectric cylinders with arbitrary boundary conditions on flat surfaces *J. Acoust. Soc. Am.* **113** 1900–8

El-Sirafy I H, Abo-Dahab S M and Singh B 2012 Effects of voids and rotation on P wave in a thermoelastic half-space under Green–Naghdi theory *Math. Mech. Solids* **17** 243–53

Graff K F 1991 *Wave Motion in Elastic Solids* (New York: Dover)

Guo F L and Sun R 2008 Propagation of Bleustein–Gulyaev wave in 6mm piezoelectric materials loaded with viscous liquid *Int. J. Solids Struct.* **45** 3699–710

Hobinyand A D and Abbas I A 2015 Analytical solution of magneto-thermoelastic interaction in a fiber-reinforced anisotropic material *Eur. Phys. J.—Plus* **7** 131–424

Honarvar F, Enjilela E, Sinclair A N and Abbas Mimexami S 2007 Wave propagation in transversely isotropic cylinders *Int. J. Solids Mech.* **44** 5236–46

Iesan D 2008 A theory of prestressed thermoelastic Cosserat continua *J. Appl. Math. Mech.* **88** 306–19

Kim J O and Lee J G 2007 Dynamic characteristics of piezoelectric cylindrical transducers with radial polarization *J. Sound Vib.* **300** 241–9

Lord H W and Shulman Y 1967 A generalized dynamical theory of thermoelasticity *J. Mech. Phys. Solids.* **15** 299–309

Meeker T R and Meitzler A H 1964 Guided wave propagation in elongated cylinders and plates *Physical Acoustics.: Principles and Methods* (New York: Academic Press) pp 111–66

Minagawa S 1995 Propagation of harmonic waves in a layered elasto-piezoelectric composite *Mech. Mater.* **19** 165–70

Mirsky I 1965 Wave propagation in transversely isotropic circular cylinders, part I: theory, part II: numerical results *J. Acoust. Soc. Am.* **37** 1016–26

Nagaya K 1981 Dispersion of elastic waves in bars with polygonal cross-section *J. Acoust. Soc. Am.* **70** 763–70

Nagy B 1995 Longitudinal guided wave propagation in a transversely isotropic rod immersed in fluid *J. Acoust. Soc. Am.* **98** 454–7

Othman M I A and Song Y 2007 Reflection of plane waves from an elastic solid half-space under hydrostatic initial stress without energy dissipation *Int. J. Solids Struct.* **44** 5651–64

Othman M and Lotfy K H 2015 The influence of gravity on 2-D problem of two temperatures generalized thermoelastic medium with thermal relaxation *J. Comput. Theor. Nanosci.* **12** 2587–97

Othman M A, Abo-Dahab S M and Lotfy K H 2014 Gravitational effect and initial stress on generalized magneto-thermomicrostretch elastic solid for different theories *Appl. Math. Comput.* **230** 597–615

Paria G 1967 Magneto-elasticity and magneto-thermo-elasticity *Adv. Appl. Mech.* **10** 73–112

Parton V Z and Kudryavtsev B A 1988 *Electromagnetoelasticity* (New York: Gordon and Breach Science)

Paul H S and Venkatesan M 1987 Wave propagation in a piezoelectric ceramic cylinder of arbitrary cross section *J. Acoust. Soc. Am.* **82** 2013–20

Placidi L and Hutter K 2004 Thermodynamics of polycrystalline materials treated by the theory of mixtures with continuous diversity *Contin. Mech. Thermodyn.* **17** 409–51

Qian Z-H, Jin F, Li P and Hirose S 2010 Bleustein–Gulyaev waves in 6mm piezoelectric materials loaded with a viscous liquid layer of finite thickness *Int. J. Solids Struct.* **47** 3513–8

Selvamani R 2017 Modeling of elastic waves in a fluid-loaded and immersed piezoelectric circular fiber *Int. J. Appl. Comput. Math.* **3** 3263–77

Selvamani R and Ponnusamy P 2012 Effect of rotation in an axisymmetric vibration of a transversely isotropic solid bar immersed in an inviscid fluid *Mater. Phys. Mech.* **15** 97–106

Selvamani R and Ponnusamy P 2015 Wave propagation in a generalized piezothermoelastic rotating bar of circular cross-section *Multidiscip. Model. Mater. Struct.* **11** 216–37

Sengupta P R and Nath S 2011 Surface waves in fibre-reinforced anisotropic elastic media *Sadhana Proc.* **26** 363–70

Singh H and Sharma J N 1985 Generalized thermoelastic waves in transversely isotropic media *J. Acoust. Soc. Am.* **77** 1046–53

Sun C T and Cheng N C 1974 Piezoelectric waves on a layered cylinder *J. Appl. Phys.* **45** 4288–94
Tiersten H F 1969 *Linear Piezoelectric Plate Vibrations* (New York: Plenum Press)
Wang Q 2002 Axi-symmetric wave propagation in a cylinder coated with a piezoelectric layer *Int. J. Solids Struct.* **39** 3023–37
White J E and Tongtaow C 1981 Cylindrical waves in transversely isotropic media *J. Acoust. Soc. Am.* **70** 1147–55